Advanced Textbooks in Control and Signal Processing

Springer
London
Berlin
Heidelberg
New York
Barcelona
Hong Kong
Milan
Paris
Singapore
Tokyo

Series Editors

Other titles published in this series:

Genetic Algorithms
K.F. Man, K.S. Tang and S. Kwong

Model Predictive Control
E. F. Camacho and C. Bordons

Introduction to Optimal Estimation
E.W. Kamen and J. Su

Discrete-time Signal Processing
D. Williamson

Neural Networks for Modelling and Control of Dynamic Systems
M. Nørgaard, O. Ravn, N.K. Poulsen and L.K. Hansen

Modelling and Control of Robot Manipulators (2nd Edition)
L. Sciavicco and B. Siciliano

Soft Computing
L. Fortuna, G. Rizzotto, M. Lavorgna, G. Nunnari, M.G. Xibilia and R. Caponetto
Publication Due January 2001

Statistical Signal Processing
T. Chonavel and S. Vaton
Publication Due April 2001

L. H. Chiang, E. L. Russell and R. D. Braatz

Fault Detection and Diagnosis in Industrial Systems

With 81 Figures

Springer

Leo H. Chiang, MS
Richard D. Braatz, PhD
Department of Chemical Engineering, University of Illinois at Urbana-Champaign,
600 S. Mathews Avenue, Urbana, Illinois 61801-3792, USA

Evan L. Russell, PhD
ExxonMobil Upstream Reasearch Company, URC Building, Room C-312,
PO Box 2189, Houston, TX 77252, USA

ISBN 1-85233-327-8 Springer-Verlag London Berlin Heidelberg

British Library Cataloguing in Publication Data
Chiang, Leo H.
Fault detection and diagnosis in industrial systems. -
(Advanced textbooks in control and signal processing)
1. Fault location (Engineering) 2. Process control
I. Title II. Russell, Evan, 1972- III. Braatz, Richard D.,
1966-
670.4'2
ISBN 1852333278

Library of Congress Cataloging-in-Publication Data
Chiang, Leo H., 1975-
Fault detection and diagnosis in industrial systems / Leo H. Chiang, Evan L. Russell, and Richard D. Braatz.
p. cm. - (Advanced textbooks in control and signal processing)
Includes bibliographical references and index.
ISBN 1-85233-327-8 (alk. paper)
1. Chemical process control 2. Fault location (Engineering) I. Russell, Evan, 1972- II. Braatz, Richard D., 1966- III. Title. IV. Series.
TP155.75 .C465 2000
660'.2815—dc21

00-045601

Typesetting: Camera ready by authors
Printed and bound by Athenæum Press Ltd., Gateshead, Tyne & Wear
69/3830-543210 Printed on acid-free paper SPIN 10768812

Series Editors' Foreword

The topics of control engineering and signal processing continue to flourish and develop. In common with general scientific investigation, new ideas, concepts and interpretations emerge quite spontaneously and these are then discussed, used, discarded or subsumed into the prevailing subject paradigm. Sometimes these innovative concepts coalesce into a new sub-discipline within the broad subject tapestry of control and signal processing. This preliminary battle between old and new usually takes place at conferences, through the Internet and in the journals of the discipline. After a little more maturity has been acquired by the new concepts then archival publication as a scientific or engineering monograph may occur.

A new concept in control and signal processing is known to have arrived when sufficient material has developed for the topic to be taught as a specialised tutorial workshop or as a course to undergraduates, graduates or industrial engineers. The *Advanced Textbooks in Control and Signal Processing* Series is designed as a vehicle for the systematic presentation of course material for both popular and innovative topics in the discipline. It is hoped that prospective authors will welcome the opportunity to publish a structured presentation of either existing subject areas or some of the newer emerging control and signal processing technologies.

Fault detection and process monitoring is one of the new growth areas in process control. The reason for this development is not hard to find. New instrumentation and communications technologies have created a wealth of real-time data from processes in both new and existing manufacturing plant installations. Process operators are therefore keen to use this data to minimise plant downtime and optimise plant operations. The traditional routes to fault detection were model based and to use them the process has to be well understood. An alternative group of methods has emerged which do not require the use of an explicit model. This is the key basic construct for the data-driven paradigm. Model-free and non-parametric methods for fault detection, process optimisation and control design are currently at a particularly exciting stage of development.

This new advanced textbook by Chiang, Russell and Braatz primarily tackles the data-driven routes to Fault Detection and Diagnosis. It is an outgrowth of a prior *Advances in Industrial Control* monograph; Russell, Chiang and Braatz. *Data-driven Techniques for Fault Detection and Diagnosis in Chemical Processes*, 2000, ISBN 1-85233-258-1. The new textbook expands the material of the monograph and gives a fuller presentation of some of the alternative model-based methods, the analytical methods, and of the knowledge-based techniques.

This allows the reader to compare and contrast the different approaches to the problem of fault detection and diagnosis. Thus the text is suitable for advanced courses for process, chemical and control engineers.

M.J. Grimble and M.A. Johnson
Industrial Control Centre
Glasgow, Scotland, U.K.
October, 2000

Preface

Modern manufacturing facilities are large scale, highly complex, and operate with a large number of variables under closed-loop control. Early and accurate fault detection and diagnosis for these plants can minimize downtime, increase the safety of plant operations, and reduce manufacturing costs. Plants are becoming more heavily instrumented, resulting in more data becoming available for use in detecting and diagnosing faults. Univariate control charts (e.g., Shewhart charts) have a limited ability to detect and diagnose faults in such multivariable processes. This has led to a surge of academic and industrial effort concentrated on developing more effective process monitoring methods. A large number of these methods are being regularly applied to real industrial systems, which makes these techniques suitable for coverage in undergraduate and graduate courses.

This textbook presents the theoretical background and practical techniques for process monitoring. The intended audience is engineering students and practicing engineers. The book is appropriate for a first-year graduate or advanced undergraduate course in process monitoring. Numerous simple examples and a simulator for a large-scale industrial plant are used to illustrate the methods. As the most effective method for learning the techniques is by applying them, the Tennessee Eastman plant simulator has been made available at `http://brahms.scs.uiuc.edu`. Readers are encouraged to collect process data from the simulator, and then apply a range of process monitoring techniques to detect, isolate, and diagnose various faults. The process monitoring techniques can be implemented using commercial software packages such as the MATLAB PLS Toolbox and ADAPTx.

What were the goals in writing this textbook? Although much effort has been devoted to process monitoring by both academics and industrially employed engineers, books on the subject are still rather limited in coverage. These books usually focus entirely on one type of approach such as statistical quality control (Montgomery (1991), Park and Vining (2000)) or analytical methods (Chen and Patton (1999), Gertler (1998), Patton, Frank, and Clark (1989)). Some books treat both statistical and analytical methods (Himmelblau (1978), Basseville and Nikiforov (1993)), but do not cover knowledge-based methods. Wang (1999) covers both statistical and knowledge-based methods, but does not cover analytical methods. Many process monitoring methods of practical importance, such as those based on canonical variate analysis and Fisher discriminant analysis, are described in hardly any books on process monitoring (an exception is Russell, Chiang, and Braatz (2000)).

While many of these books do an excellent job covering their intended topics, it was our opinion that there was a need for a single textbook that covers data-driven, analytical, and knowledge-based process monitoring methods. Part of the motivation for this is that many engineering curricula do not have sufficient space for courses on each of these topics in isolation. But all of these methods are becoming increasingly important in practice, and should be studied by engineering students who plan to work in industry. These include mechanical, electrical, industrial, chemical, nuclear, manufacturing, control, aerospace, quality, and reliability engineers, as well as applied statisticians.

The proportion of coverage given to each topic is based on our own experience (all three authors have applied process monitoring methods to industrial systems with hundreds of measured variables), as well as on the industrial experience of other engineers as described in person or in publications. The first chapter gives an overview of process monitoring procedures and methods. Chapter 2 provides background in multivariate statistics, including univariate control charts and a discussion of data requirements. Chapter 3 discusses pattern classification, including discriminant analysis and feature extraction, which are fundamental to fault diagnosis techniques.

Chapters 4-7 cover data-driven process monitoring methods. **Principal component analysis** (PCA) and **partial least squares** (PLS) are multivariate statistical methods that generalize the univariate control charts that have been applied for decades. **Fisher discriminant analysis** (FDA) is a fault diagnosis method based on the pattern classification literature. **Canonical variate analysis** (CVA) is a subspace identification method that has been used in process monitoring in a similar manner to PCA and PLS. These four methods represent the state of the art in data-driven process monitoring methods, which are the methods most heavily used in many chemical and manufacturing industries. One reason for the popularity of data-driven methods is that they do not require first-principles models, the development of which is usually costly or time-consuming. For this reason, these methods are also the predominant methods that have been applied to large-scale systems. In Chapters 8-10 the methods are compared through application to a large-scale system, the Tennessee Eastman plant simulator. This gives the readers an understanding of the strengths and weaknesses of various approaches, as well as some realistic homework problems.

Chapter 11 describes analytical methods, including parameter estimation, state estimation, and parity relations. While not as pervasive as data-driven methods in many industries, in some cases a first-principles model is available, and analytical methods are suited to using these models for process monitoring. Also, in most engineering curricula it is the analytical approach that is most closely related to topics covered in other control courses. Chapter 12 describes knowledge-based methods, including causal analysis, expert systems, and pattern recognition. Knowledge-based methods are especially suited to systems in which there are inadequate data to apply a data-driven method, but qualitative or semi-qualitative models can be derived from causal modeling of the system, expert knowledge, or fault-symptom examples. Each of the data-driven, analytical, and knowledge-based approaches have strengths and limitations. Incorporating several techniques for process monitoring can be beneficial in many applications. Chapter 12 also discusses various combinations of process monitoring techniques.

The authors thank International Paper, DuPont, and the National Center for Supercomputing Applications for funding over the past three years this textbook was being written.

L.H.C., E.L.R., R.D.B
Urbana, Illinois

Contents

Part I

Introduction

1. Introduction

In the process and manufacturing industries, there has been a large push to produce higher quality products, to reduce product rejection rates, and to satisfy increasingly stringent safety and environmental regulations. Process operations that were at one time considered acceptable are no longer adequate. To meet the higher standards, modern industrial processes contain a large number of variables operating under closed-loop control. The standard process controllers (PID controllers, model predictive controllers, *etc.*) are designed to maintain satisfactory operations by compensating for the effects of disturbances and changes occurring in the process. While these controllers can compensate for many types of disturbances, there are changes in the process which the controllers cannot handle adequately. These changes are called **faults**. More precisely, a fault is defined as an unpermitted deviation of at least one characteristic property or variable of the system [140].

The types of faults occurring in industrial systems include process parameter changes, disturbance parameter changes, actuator problems, and sensor problems [162]. Catalyst poisoning and heat exchanger fouling are examples of process parameter changes. A disturbance parameter change can be an extreme change in the concentration of a process feed stream or in the ambient temperature. An example of an actuator problem is a sticking valve, and a sensor producing biased measurements is an example of a sensor problem. To ensure that the process operations satisfy the performance specifications, the faults in the process need to be detected, diagnosed, and removed. These tasks are associated with **process monitoring**. **Statistical process control** (SPC) addresses the same issues as process monitoring, but to avoid confusion with standard process control, the methods mentioned in this text will be referred to as **process monitoring methods**.

The goal of process monitoring is to ensure the success of the planned operations by recognizing anomalies of the behavior. The information not only keeps the plant operator and maintenance personnel better informed of the status of the process, but also assists them to make appropriate remedial actions to remove the abnormal behavior from the process. As a result of proper process monitoring, downtime is minimized, safety of plant operations is improved, and manufacturing costs are reduced. As industrial systems have become more highly integrated and complex, the faults occurring in modern processes present monitoring challenges that are not readily addressed using univariate control charts (e.g., Shewhart charts, see Section 2.3). The weaknesses of univariate control charts for detecting faults in multivariate processes have led to a surge of research literature concentrated on developing better methods for process monitoring. This growth of research activity can also be explained by the fact that industrial systems are becoming more heavily instrumented, resulting in larger quantities of data available for use in process monitoring, and that modern computers are becoming more powerful. The availability of data collected during various operating and fault conditions is essential to process monitoring. The storage capacity and computational speed of modern computers enable process monitoring algorithms to be computed when applied to large quantities of data.

1.1 Process Monitoring Procedures

The four procedures associated with process monitoring are: **fault detection, fault identification, fault diagnosis**, and **process recovery**. There appears to be no standard terminology for these procedures as the terminology varies across disciplines; the terminology given by Raich and Cinar [272] is adopted here. **Fault detection** is determining whether a fault has occurred. Early detection may provide invaluable warning on emerging problems, with appropriate actions taken to avoid serious process upsets. **Fault identification** is identifying the observation variables most relevant to diagnosing the fault. The purpose of this procedure is to focus the plant operator's and engineer's attention on the subsystems most pertinent to the diagnosis of the fault, so that the effect of the fault can be eliminated in a more efficient manner. **Fault diagnosis** is determining which fault occurred, in other words, determining the cause of the observed out-of-control status. Isermann [138] more specifically defines fault diagnosis as determining the type, location, magnitude, and time of the fault. The fault diagnosis procedure is essential to the counteraction or elimination of the fault. **Process recovery,** also called **intervention,** is removing the effect of the fault, and it is the procedure needed to close the **process monitoring loop** (see Figure 1.1). Whenever a fault is detected, the fault identification, fault diagnosis, and process recovery procedures are employed in the respective sequence; otherwise, only the fault detection procedure is repeated.

While all four procedures may be implemented in a process monitoring scheme, this is not always necessary. For example, a fault may be diagnosed (fault diagnosis) without identifying the variables immediately affected by the fault (fault identification). Additionally, it is not necessary to automate all four procedures. For instance, an automated fault identification procedure may be used to assist the plant operators and engineers to diagnose the fault (fault diagnosis) and recover normal operation. Often the goal of process monitoring is to incorporate the plant operators and engineers into the process monitoring loop efficiently rather than to automate the monitoring scheme entirely.

After a fault occurs, the in-control operations can often be recovered by reconfiguring the process, repairing the process, or retuning the controllers. Once a fault has been properly diagnosed, the optimal approach to counteract the fault may not be obvious. A feasible approach may be to retune the standard process controllers. Several methods have been developed to evaluate controller performance [66, 111, 162, 274, 295, 312], and these can be used to determine which controllers in the process need to be retuned to restore satisfactory performance. In the case of a sensor problem, a sensor reconstruction technique can be applied to the process to restore in-control operations [77]. Even though process recovery is an important and necessary component of the process monitoring loop, process recovery is not the focus of this book.

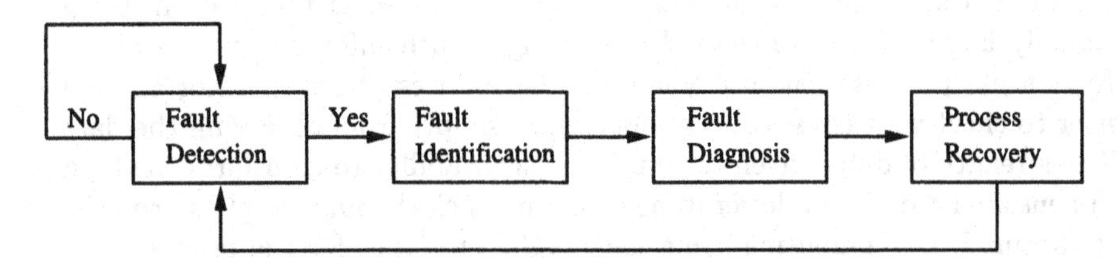

Fig. 1.1. A schema of the process monitoring loop

1.2 Process Monitoring Measures

A typical process monitoring scheme contains one or more measures, based on developments from statistical theory, pattern classification theory, information theory, and/or systems theory. These measures in some way represent the state or behavior of the process. The idea is to convert on-line data collected from the process into a few meaningful measures, and thereby assist the operators in determining the status of the operations and if necessary in diagnosing the faults. For fault detection, limits may be placed on some of the measures, and a fault is detected whenever one of the evaluated measures is outside the limits. In this way, the measures are able to define the in-control process behavior and accordingly the out-of-control status. By developing measures that accurately characterize the behavior of each observation variable, the measure value of one variable can be compared against the measure values for other variables to determine the variable most affected by the fault. Faults can also be diagnosed by developing and comparing measures that accurately represent the different faults of the process.

The goal of process monitoring is to develop measures that are maximally **sensitive** and **robust** to all possible faults. Faults are manifested in several ways; however, and it is highly unlikely that all faults occurring in a process can be effectively detected and diagnosed with only a few measures. Since each measure characterizes a fault in a different manner, one measure will be more sensitive to certain faults and less sensitive to other faults relative to other measures. This motivates using multiple process monitoring measures, with the proficiency of each measure determined for the particular process and the possible faults at hand.

Process monitoring measures can be classified as being associated with one or more of three approaches; namely, **data-driven**, **analytical**, and **knowledge-based**. The data-driven measures are derived directly from process data. Modern industrial systems, whether an entire industrial plant or a single paper machine, are large-scale systems. With the heavy instrumentation typical of modern processes, large-scale systems produce an exceptionally large amount of data. Even though much information is available from these processes, it is beyond the capabilities of an operator or engineer to effectively assess process operations simply from observing the data. The strength of data-driven techniques is their ability to transform the high-dimensional data into a lower dimension, in which the important information is captured. By computing some meaningful statistics for the process operators and engineers, a process monitoring scheme for a large-scale system can be improved significantly. The main drawback of data-driven measures is that their proficiency is highly dependent on the quantity and quality of the process data.

Unlike the data-driven approach, the analytical approach uses mathematical models often constructed from first principles. The analytical approach is applicable to information-rich systems, where satisfactory models and enough sensors are available. Most analytical measures are based on parameter estimation, observer-based design, and/or parity relations. Most applications of the analytical approach have been to systems with a relatively small number of inputs, outputs, and states. It is difficult to apply the analytical approach to **large-scale systems** (*i.e.*, systems containing a large number of inputs, outputs, and/or states) because it requires detailed models in order to be effective [73, 141, 360]. Detailed models for large-scale systems are expensive to obtain given all the crosscouplings associated with a multivariable system [137]. The main advantage of the analytical approach is the ability to incorporate physical understanding of the process into the process monitoring scheme. In other words, when detailed analytical models are available, the analytical measures can significantly outperform the data-driven measures.

The knowledge-based approach uses qualitative models to develop process monitoring measures. The knowledge-based approach is especially well suited for systems in which detailed mathematical models are not available. Most knowledge-based measures are based on causal analysis, expert systems, and/or pattern recognition. Like the analytical approach, most applications of the knowledge-based approach have been to systems with a relatively small number of inputs, outputs, and states. Constructing the fault models for a large-scale system can require a large amount of effort [8, 360]. Software packages are being developed to enable the knowledge-based approach to be more easily applied to complex systems.

1.3 Process Monitoring Methods

The proficiency of the data-driven, analytical, and knowledge-based approaches depends on the quality and type of available models, and on the quantity and quality of data available. These aspects along with the advantages and disadvantages of various methods are discussed in this textbook.

Traditional monitoring methods consisted of **limit sensing** and **discrepancy detection**. Limit sensing raises an alarm when observations cross predefined thresholds, and has been applied traditionally because it is easy to implement and understand. Limit sensing, however, lacks sensitivity to some process upsets because it ignores interactions between the process variables for the various sensors [73, 138]. Discrepancy detection raises an alarm by comparing simulated to actual observed values. Discrepancy detection highly depends on model accuracy, and model inaccuracies are unavoidable in practice. Since it is difficult to distinguish genuine faults from errors in the model, discrepancy detection can lack robustness [73]. As discussed in Section 1.2, robust discrepancy detection statistics have been studied, however, effective statistics are difficult to obtain, especially for large-scale systems.

Limit sensing determines thresholds for each observation variable without using any information from the other variables, and in this way is identical to the univariate statistical techniques discussed in Section 2.3. These methods ignore the correlations among the observation variables (**spacial correlations**) and the correlations among measurements of the same variable taken at different times (**serial correlations**). (Note that spacial correlations also refer to correlations between different measurements taken at essentially the same physical location.) Process data are spacially correlated because there is often a large number of sensor readings taken throughout the process and the variability of the process variables is restricted to a lower dimension (for example, due to phase equilibria or conservation laws, such as the material and energy balances) [76]. Also, process data are serially correlated because the sampling intervals are relatively small and the standard process controllers are unable to remove all the systematic trends due to inertial components, such as tanks, reactors, and recycle streams. Because limit sensing does not take into account the spacial correlations, it lacks sensitivity to many faults occurring in industrial systems [142, 143], and because limit sensing also ignores the serial correlations, it can lack robustness [112].

The need to handle spacial correlations has led to the development and employment of process monitoring statistics based on multivariate statistical techniques. **Principal component analysis** (PCA) is the most widely used data-driven technique for monitoring industrial systems. PCA is a dimensionality reduction technique for process monitoring which has been heavily studied and applied to industrial systems over the past decade. PCA is an optimal dimensionality reduction technique in terms of capturing the variance of the data, and it accounts for correlations among variables [142, 143]. The lower-dimensional representations of the data produced by PCA can improve the proficiency of detecting and diagnosing faults using multivariate statistics. The structure abstracted by PCA can be useful in identifying either the variables responsible for the fault and/or the variables most affected by the fault. In cases where most of the information in the data can be captured in only two or three dimensions, which can be true for some processes [207], the dominant process variability can be visualized with a single plot (for example, see Figure 4.3). Irrespective of how many dimensions are required in the lower-dimensional space, other plots (e.g., T^2 and Q charts) can be used which look similar to univariate control charts but are based on multivariate statistics. These control charts can help the operators and engineers to interpret significant trends in the process data [177].

Fisher discriminant analysis (FDA) is a dimensionality reduction technique developed and studied within the **pattern classification** community [74]. FDA determines the portion of the observation space that is most effective in discriminating amongst several data classes. Discriminant analysis is applied to this portion of the observation space for fault diagnosis. The dimensionality reduction technique is applied to the data in all the classes simultaneously. Thus, all fault class information is utilized when the discriminant function is evaluated for each class and better fault diagnosis performance is expected. The theoretical developments for FDA suggest that it should be more effective than PCA for diagnosing faults.

Partial least squares (PLS) are data decomposition methods for *maximizing covariance* between predictor (independent) block and predicted (dependent) block for each component. PLS attempts to find loading and score vectors that are correlated with the predicted block X while describing a large amount of the variation in the predictor block Y [343]. A popular application of PLS is to select X to contain sensor data and Y to contain only product quality data [207]. Similar to PCA, such inferential models (also known as soft sensors) can be used for detecting, identifying, and diagnosing faults [207, 259, 260]. Another application of PLS primarily focusing on fault diagnosis is to define Y as class membership [46]. This PLS method is known as discriminant partial least squares.

The process monitoring statistics based on PCA, PLS, and FDA can be extended to include serial correlations by augmenting the data collected at a particular time instant to the data collected during several of the previous consecutive sampling instances. An alternative method to address serial correlations is to average the measurements over many data points (this method has the similar philosophy of CUSUM and EWMA charts, see Section 2.3 for a brief discussion). Another simple approach is to use a larger sampling interval. However, these approaches do not utilize the useful developments made in system identification theory for quantifying serial correlation. A class of system identification methods that produces state variables directly from the data are called **subspace algorithms**. The subspace algorithm based on **canonical variate analysis** (CVA) is particularly attractive because of its close relationship to PCA, FDA, and PLS. These relationships motivate the deviation of CVA-based statistics for fault detection, identification, and diagnosis that take serial correlations into account.

The measures for PCA, FDA, PLS, and CVA can be calculated based entirely on the data. When a detailed first-principles or other mathematical model is available, the analytical approach can provide more effective process monitoring than data-driven techniques. Based on the measured input and output, the analytical methods generate features using detailed mathematical models. Commonly used features include residuals, parameter estimates, and state estimates. Faults are detected or diagnosed by comparing, either directly and after some transformation, the observed features with the features associated with normal operating conditions.

Analytical methods that use residuals as features are commonly referred to as **analytical redundancy** methods. The residuals are the outcomes of consistency checks between the plant observations and a mathematical model. In the preferred situation, the residuals or transformations of the residuals will be relatively large when faults are present, and small in the presence of disturbances, noise, and/or modeling errors. This allows the definition of thresholds to detect the presence of faults [87, 101, 221].

The three main ways to generate residuals are **parameter estimation**, **observers**, and **parity relations** [94].

1. **Parameter estimation**. For parameter estimation, the residuals are the difference between the nominal model parameters and the estimated model parameters. Deviations in the model parameters serve as the basis for detecting and isolating faults [20, 135, 136, 163].
2. **Observers**. The observer-based method reconstructs the output of the system from the measurements or a subset of the measurements with the aid of observers. The difference between the measured outputs and the estimated outputs is used as the vector of residuals [86, 54, 68].
3. **Parity relations**. This method checks the consistency of the mathematical equations of the system with the measurements. The parity relations

are subjected to a linear dynamic transformation, with the transformed residuals used for detecting and isolating faults [63, 101, 226, 227].

The analytical approach requires accurate quantitative mathematical model in order to be effective. For large-scale systems, such information may not be available or it may be too costly and time-consuming to obtain. An alternative method for process monitoring is to use knowledge-based methods such as causal analysis, expert systems, and pattern recognition. These techniques are based on qualitative models, which can be obtained through causal modeling of the system, expert knowledge, a detailed description of the system, or fault-symptom examples. Causal analysis techniques are based on the causal modeling of fault-symptom relationships. Qualitative and semi-quantitative relationships in these causal models can be obtained without using first principles. Causal analysis techniques including **signed directed graphs** and the **symptom tree** are primarily used for diagnosing faults.

Expert systems are used to imitate the reasonings of human expert when diagnosing faults. The experience from a domain expert can be formulated in terms of rules, which can be combined with the knowledge from first principles or a structural description of the system for diagnosing faults. Expert systems are able to capture human diagnostic associations that are not readily translated into mathematical or causal models.

Pattern recognition techniques use association between data patterns and fault classes without an explicit modeling of internal process states or structure. Examples include **artificial neural networks** and **self-organizing maps**. These techniques are related to the data-driven techniques (PCA, PLS, FDA, and CVA) in terms of modeling the relationship between data patterns and fault classes. The data-driven techniques are dimensionality reduction techniques based on rigorous multivariate statistics. On the other hand, neural networks and self-organizing maps are black box methods that learn the pattern based entirely from the training sessions.

All measures based on data-driven, analytical, and knowledge-based approaches have their advantages and disadvantages, so that no single approach is best for all applications. Usually the best process monitoring scheme employs multiple statistics or methods for fault detection, identification, and diagnosis [73]. Efforts have been made to incorporate several techniques for process monitoring. This can be beneficial in many applications.

1.4 Book Organization

This book is an introduction to techniques for detecting, identifying, and diagnosing faults in industrial systems. This includes descriptions of all three of the main approaches to process monitoring: data-driven, analytical, and knowledge-based. All of these approaches are becoming increasingly important in practice, and it is necessary for engineering students and industrially-employed engineers to understand the strengths and weaknesses of all the approaches and to understand how to apply them. Many examples are used to compare the effectiveness and illustrate how to apply various process monitoring methods. These include a chemical plant, a gravity tank problem where a number of leaks can occur, and a water recirculation system with a centrifugal pump driven by a DC motor.

The book is organized into five parts. Part I (this chapter) is an introduction to process monitoring approaches. Part II provides the background necessary to understand the process monitoring methods described later in the book. Chapter 2 provides an introduction to multivariate statistics, and Chapter 3 covers pattern classification. Part III describes the data-driven process monitoring methods: PCA, FDA, PLS, and CVA. The methods as described in the literature are extended in cases where the process monitoring statistics were incomplete or inadequate. Part IV describes the application of the process monitoring methods to the Tennessee Eastman process. The Tennessee Eastman process is described in Chapter 8, while Chapter 9 states how the methods are applied to the Tennessee Eastman process. The results of the methods applied to the simulated data are discussed in Chapter 10. Part V describes the analytical and knowledge-based approaches. Chapter 11 describes analytical methods based on parameter estimation, observer-based design, and parity relations. Chapter 12 describes knowledge-based methods based on causal analysis, expert systems, and pattern recognition. This is followed by a discussion of combinations of multiple process monitoring techniques. Application examples in Part V include a gravity tank problem where a number of leaks can occur, and a water recirculation system with a centrifugal pump driven by a DC motor.

Part II

Background

2. Multivariate Statistics

2.1 Introduction

The effectiveness of the data-driven measures depends on the characterization of the process data variations. There are two types of variations for process data: **common cause** and **special cause** [245]. The common cause variations are those due entirely to random noise (e.g., associated with sensor readings), whereas special cause variations account for all the data variations not attributed to common cause. Standard process control strategies may be able to remove most of the special cause variations, but these strategies are unable to remove the common cause variations, which are inherent to process data. Since variations in the process data are inevitable, statistical theory plays a large role in most process monitoring schemes.

The application of statistical theory to monitor processes relies on the assumption that the characteristics of the data variations are relatively unchanged unless a fault occurs in the system. By the definition of a fault as an abnormal process condition (see Chapter 1), this is a reasonable assumption. It implies that the properties of the data variations, such as the mean and variance, are repeatable for the same operating conditions, although the actual values of the data may not be very predictable. The repeatability of the statistical properties allows thresholds for certain measures, effectively defining the out-of-control status, to be determined automatically. This is an important step to automating a process monitoring scheme.

The purpose of this chapter is to illustrate how to use statistical methods for monitoring processes, in particular methods using the multivariate T^2 statistic. This chapter begins in Section 2.2 by describing the data pretreatment procedure, which is typically performed before determining the statistical parameters (mean, covariance, *etc.*) for the data. The traditional approach to statistical process monitoring using univariate statistics is discussed in Section 2.3. Then in Section 2.4, the T^2 statistic is described along with its advantages over univariate statistics for process monitoring. It is shown in Section 2.5 how to apply the T^2 statistic with statistically-derived thresholds, in order to automate the fault detection procedure and to remove outliers from the training data. In Section 2.6, the applicability of the T^2 statistic is determined in terms of the amount of data available to calculate the statistical parameters.

2.2 Data Pretreatment

To extract the information in the data relevant to process monitoring effectively, it is often necessary to pretreat the data in the training set. The **training set** contains off-line data available for analysis prior to the online implementation of the process monitoring scheme and is used to develop the measures representing the in-control operations and the different faults. The pretreatment procedures consist of three tasks: **removing variables**, **autoscaling**, and **removing outliers**.

The data in the training set may contain variables that have no information relevant to monitoring the process, and these variables should be removed before further analysis. For instance, it may be known *a priori* that certain variables exhibit extremely large measurement errors, such as those due to improper sensor calibrations, or some of the variables may be physically separate from the portion of the process that is being monitored. In these instances, the proficiency of the process monitoring method can be improved by removing the inappropriate variables.

Process data often need to be scaled to avoid particular variables dominating the process monitoring method, especially those methods based on dimensionality reduction techniques, such as PCA and FDA. For example, when performing an unscaled dimensionality reduction procedure on temperature measurements varying between 300K and 320K and concentration measurements varying between 0.4 and 0.5, the temperature measurements would dominate even though the temperature measurements may be no more important than the concentration measurements for monitoring the process.

Autoscaling standardizes the process variables in a way that ensures each variable is given equal weight before the application of the process monitoring method. It consists of two steps. The first step is to subtract each variable by its sample mean because the objective is to capture the variation of the data from the mean. The second step is to divide each variable of the mean-centered data by its standard deviation. This step scales each variable to unit variance, ensuring that the process variables with high variances do not dominate. When autoscaling is applied to new process data, the mean to be subtracted and the standard deviation to be divided are taken from the training set.

Outliers are isolated measurement values that are erroneous. These values may significantly influence the estimation of statistical parameters and other parameters related to a given measure. Removing the outliers from the training set can significantly improve the estimation of the parameters and should be an essential step when pretreating the data [255]. Obvious outliers can be removed by plotting and visually inspecting the data for outlying points. More rigorous methods based on statistical thresholds can be employed for removing outliers, and a method for doing this using the T^2 statistic is discussed in Section 2.5. For simplicity of presentation only, it is assumed in the remainder of this book that the data has been pretreated, unless otherwise stated.

2.3 Univariate Statistical Monitoring

A univariate statistical approach to limit sensing can be used to determine the thresholds for each **observation variable** (a process variable observed through a sensor reading), where these thresholds define the boundary for in-control operations and a violation of these limits with on-line data would indicate a fault. This approach is typically employed using a **Shewhart chart** [10, 70, 230] (see Figure 2.1) and has been referred to as **limit sensing** [73] and **limit value checking** [138]. The values of the upper and lower control limits on the Shewhart chart are critical to minimizing the rate of **false alarms** and the rate of **missed detections.** A **false alarm** is an indication of a fault, when in actuality a fault has not occurred; a **missed detection** is no indication of a fault, though a fault has occurred. For fault detection, there is an inherent tradeoff between minimizing the false alarm and missed detection rates. Tight threshold limits for an observation variable result in a high false alarm and low missed detection rate, while limits which are too spread apart result in a low false alarm and a high missed detection rate.

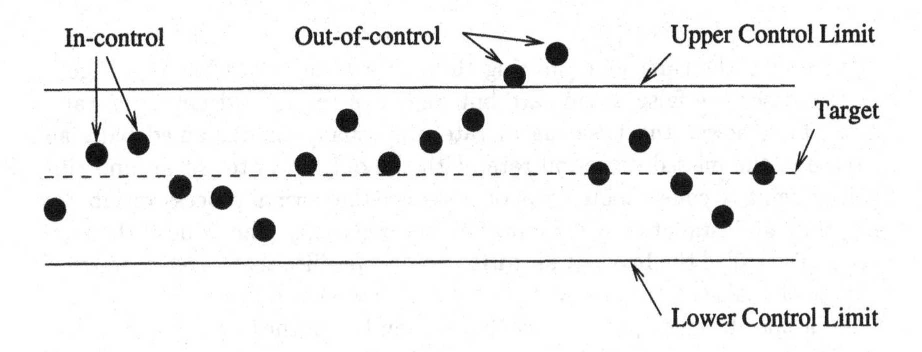

Fig. 2.1. An illustration of the Shewhart chart. The black dots are observations.

Given certain threshold values, statistical hypothesis theory can be applied to predict the false alarm and missed detection rates based on the statistics of the data in the training sets. Consider the case where there can potentially be a single fault i (the more general case of multiple fault classes will be treated thoroughly in the next chapter). Let ω represents the *event* of an in-control operation and ω_i represents the *event* of a specific fault, i. Consider a single observation x with the null hypothesis (assign x as ω) and the alternative hypothesis (assign x as ω_i), the false alarm rate is equal to the type I error, and the missed detection rate for fault i is equal to the type II error [230]. This is illustrated graphically in Figure 2.2.

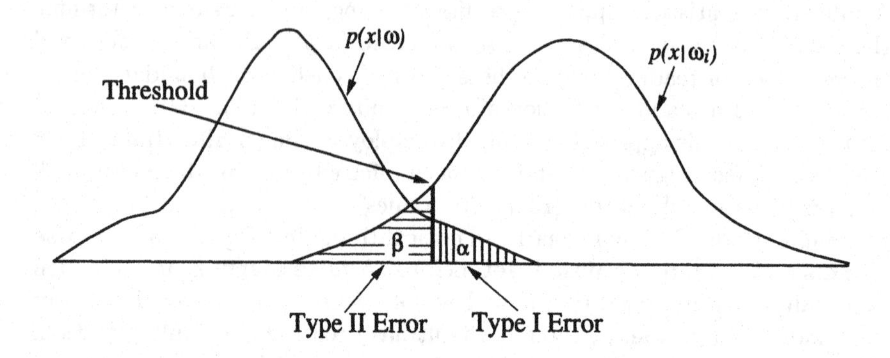

Fig. 2.2. The type I and type II error regions for the null hypothesis (assign x as ω) and the alternative hypothesis (assign x as ω_i). The probability density function for x conditioned on ω is $p(x|\omega)$; the probability density function for x conditioned on ω_i is $p(x|\omega_i)$. The probability of a type I error is α and the probability of a type II error is β. Using Bayesian decision theory [74], these notions can be generalized to include *a priori* probabilities of ω and ω_i.

Increasing the threshold (shifting the vertical line to the right in Figure 2.2) decreases the false alarm rate but increases the missed detection rate. Attempts to lower the false alarm rate are usually accompanied with an increase in the missed detection rate, with the only ways to get around this tradeoff being to collect more data, or to reduce the normal process variability (e.g., through installation of sensors of higher precision). The value of the type I error, also called the **level of significance** α, specifies the degree of tradeoff between the false alarm rate and the missed detection rate.

As a specific example, assume for the null hypothesis that any deviations of the process variable x from a desired value μ are due to inherent measurement and process variability described by a normal distribution with standard deviation σ:

$$p(x) = \frac{1}{\sigma\sqrt{2\pi}} \exp\left[-\frac{(x-\mu)^2}{2\sigma^2}\right]. \tag{2.1}$$

The alternative hypothesis is that $x \neq \mu$. Assuming that the null hypothesis is true, the probabilities that x is in certain intervals are

$$\Pr\{x < (\mu - c_{\alpha/2}\sigma)\} = \alpha/2 \tag{2.2}$$

$$\Pr\{x > (\mu + c_{\alpha/2}\sigma)\} = \alpha/2 \tag{2.3}$$

$$\Pr\{(\mu - c_{\alpha/2}\sigma) \leq x \leq (\mu + c_{\alpha/2}\sigma)\} = 1 - \alpha \tag{2.4}$$

where $c_{\alpha/2}$ is the standard normal deviate corresponding to the $(1 - \alpha/2)$ percentile. The standard normal deviate is calculated using the cumulative standard normal distribution [120]; the standard normal deviates corresponding to some common α values are listed in Table 2.1.

Table 2.1. Some typical standard normal deviate values

$\alpha/2$	$c_{\alpha/2}$
0.00135	3.00
0.0025	2.81
0.005	2.58
0.01	2.33
0.025	1.96

The lower and upper thresholds for the process variable x are $\mu - c_{\alpha/2}\sigma$ and $\mu + c_{\alpha/2}\sigma$, respectively. Figure 2.3 illustrates the application of Shewhart chart to monitor the Mooney viscosity of an industrial elastomer [245]. The desired value μ is 50.0; a standard deviation value of $\sigma = 0.5$ is known to characterize the intrinsic variability associated with the sampling procedure. Since all the data points fall inside the upper and lower control limit lines corresponding to $c_{\alpha/2} = 3.0$, the process is said to be "in control".

As long as the sample mean and standard deviation of the training set accurately represent the true statistics of the process, the thresholds using (2.2) and (2.3) should result in a false alarm rate equal to α when applied to on-line data. If 20,000 data points were collected during "in control" operation defined by $c_{\alpha/2} = 3.0$, 27 data points would be expected to fall above the upper control limit, while 27 data points would be expected to fall below the lower control limit. Some typical α values for fault detection are 0.005, 0.01, and 0.05. It has been suggested that even if x does not follow a normal distribution, the limits derived from (2.2) and (2.3) are effective as long as the data in the training set are an accurate representation of the variations during normal operations [171].

Process monitoring schemes based on Shewhart charts may not provide adequate false alarm and missed detection rates. These rates can be improved by employing measures that incorporate observations from multiple consecutive time instances, such as the **cumulative sum** (CUSUM) and **exponentially-weighted moving average** (EWMA) **charts** [80, 230, 245].

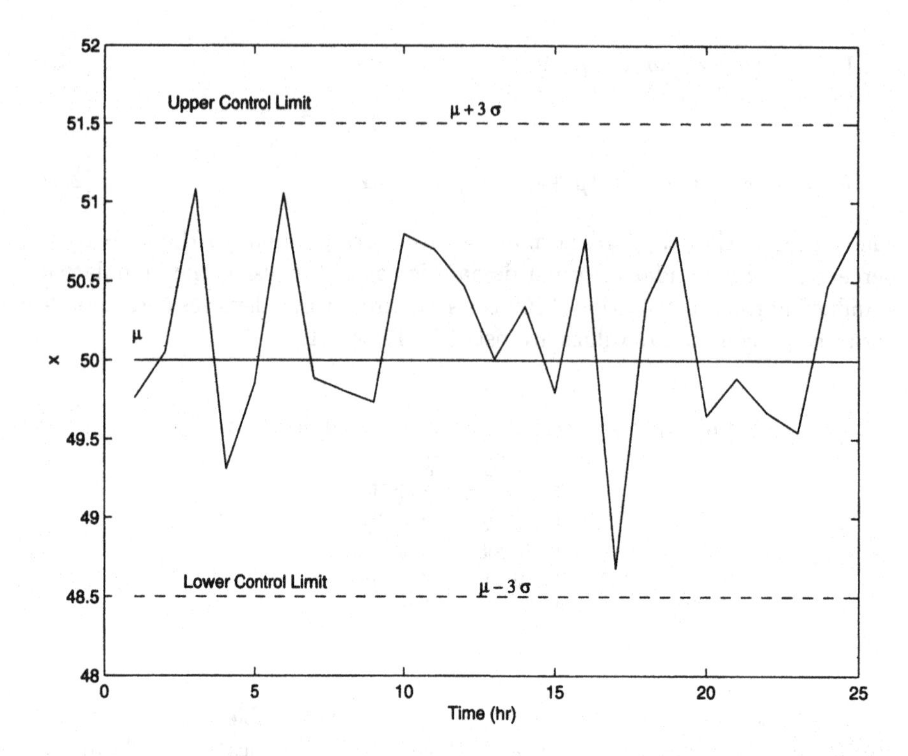

Fig. 2.3. Shewhart chart for the Mooney viscosity data taken from [245]

For a given false alarm rate, these methods can increase the sensitivity to faults over the measures using the Shewhart charts and accordingly decrease the missed detection rate, but at the expense of increasing the **detection delay**, which is the amount of time expended between the start of the fault and time to detection. This suggests that the CUSUM and EWMA charts are better suited for faults producing small persistent process shifts, and the Shewhart charts are better for detecting faults producing sudden large process shifts.

The univariate statistical charts (Shewhart, CUSUM, and EWMA) determine the thresholds for each observation variable individually without considering the information contained in the other variables. As discussed in Section 1.3, because these methods ignore the correlation between variables, they do not accurately characterize the behavior of most modern industrial processes. The next section describes the multivariate T^2 statistic, which takes into account the correlations between the variables.

2.4 T^2 Statistic

Let the data in the training set, consisting of m observation variables and n observations for each variable, be stacked into a matrix $X \in \mathcal{R}^{n \times m}$, given by

$$X = \begin{bmatrix} x_{11} & x_{12} & \cdots & x_{1m} \\ x_{21} & x_{22} & \cdots & x_{2m} \\ \vdots & \vdots & \cdots & \vdots \\ x_{n1} & x_{n2} & \cdots & x_{nm} \end{bmatrix}, \tag{2.5}$$

then the sample covariance matrix of the training set is equal to

$$S = \frac{1}{n-1} X^T X. \tag{2.6}$$

An eigenvalue decomposition of the matrix S,

$$S = V \Lambda V^T, \tag{2.7}$$

reveals the correlation structure for the covariance matrix, where Λ is diagonal and V is orthogonal ($V^T V = I$, where I is the identity matrix) [104]. The projection $\mathbf{y} = V^T \mathbf{x}$ of an observation vector $\mathbf{x} \in \mathcal{R}^m$ decouples the observation space into a set of uncorrelated variables corresponding to the elements of $\mathbf{y}$. The variance of the ith element of $\mathbf{y}$ is equal to the ith eigenvalue in the matrix Λ. Assuming S is invertible and with the definition

$$\mathbf{z} = \Lambda^{-1/2} V^T \mathbf{x}, \tag{2.8}$$

the Hotelling's T^2 statistic is given by [143]

$$T^2 = \mathbf{z}^T \mathbf{z}. \tag{2.9}$$

The matrix V rotates the major axes for the covariance matrix of $\mathbf{x}$ so that they directly correspond to the elements of $\mathbf{y}$, and Λ scales the elements of $\mathbf{y}$ to produce a set of variables with unit variance corresponding to the elements of $\mathbf{z}$. The conversion of the covariance matrix is demonstrated graphically in Figure 2.4 for a two-dimensional observation space ($m = 2$).

The T^2 statistic is a scaled squared 2-norm of an observation vector $\mathbf{x}$ from its mean. The scaling on $\mathbf{x}$ is in the direction of the eigenvectors and is inversely proportional to the standard deviation along the eigenvectors. This allows a scalar threshold to characterize the variability of the data in the entire m-dimensional observation space. Given a level of significance, appropriate threshold values for the T^2 statistic can be determined automatically by applying the probability distributions discussed in the next section.

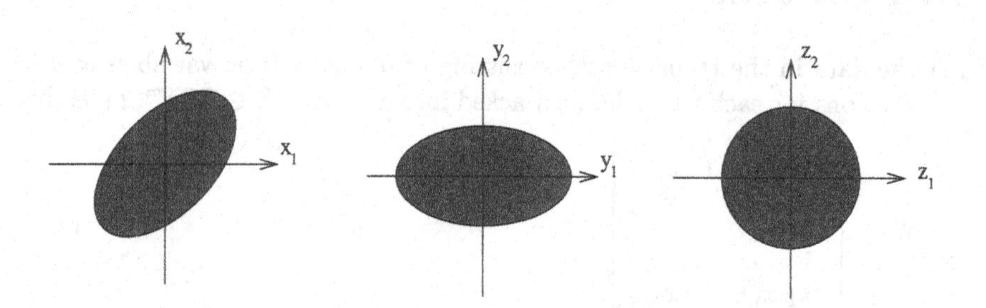

Fig. 2.4. A graphical illustration of the covariance conversion for the T^2 statistic

2.5 Thresholds for the T^2 Statistic

Appropriate thresholds for the T^2 statistic based on the level of significance, α, can be determined by assuming the observations are randomly sampled from a multivariate normal distribution. If it is assumed additionally that the sample mean vector and covariance matrix for normal operations are equal to the actual mean vector and covariance matrix, respectively, then the T^2 statistic follows a χ^2 distribution with m degrees of freedom [209],

$$T_\alpha^2 = \chi_\alpha^2(m). \tag{2.10}$$

The set $T^2 \leq T_\alpha^2$ is an elliptical confidence region in the observation space, as illustrated in Figure 2.5 for two process variables $m = 2$. Applying (2.10) to process data produces a confidence region defining the in-control status whereas an observation vector projected outside this region indicates that a fault has occurred. Given a level of significance α, Figure 2.5 illustrates the conservatism eliminated by employing the T^2 statistic versus the univariate statistical approach outlined in Section 2.3. As the degree of correlation between the process variables increases, the elliptical confidence region becomes more elongated and the amount of conservatism eliminated by using the T^2 statistic increases.

When the actual covariance matrix for the in-control status is not known but instead estimated from the sample covariance matrix (2.6), faults can be detected for observations taken outside the training set using the threshold given by

$$T_\alpha^2 = \frac{m(n-1)(n+1)}{n(n-m)} F_\alpha(m, n-m) \tag{2.11}$$

where $F_\alpha(m, n-m)$ is the upper $100\alpha\%$ critical point of the F-distribution with m and $n-m$ degrees of freedom [209]. For a given level of significance, the upper in-control limit in (2.11) is larger (more conservative) than the

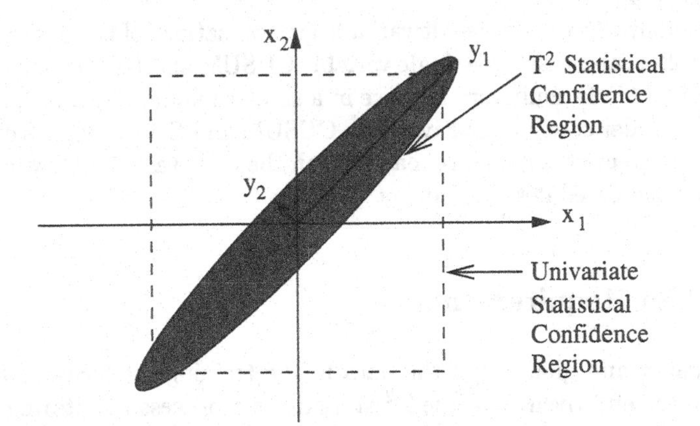

Fig. 2.5. A comparison of the in-control status regions using the T^2 statistic (2.9) and the univariate statistics (2.2) and (2.3) for two process variables [272, 307]

limit in (2.10), and the two limits approach each other as the amount of data increases ($n \to \infty$) [308].

When the sample covariance matrix (2.6) is used, the outliers in the training set can be detected using the threshold given by

$$T_\alpha^2 = \frac{(n-1)^2(m/(n-m-1))F_\alpha(m, n-m-1)}{n(1+(m/(n-m-1))F_\alpha(m, n-m-1)}. \tag{2.12}$$

For a given level of significance, the upper in-control limit in (2.12) is smaller (less conservative) than the limit in (2.10), and the two limits approach each other as the amount of data increases ($n \to \infty$) [308]. Equation (2.12) is also appropriate for detecting faults during process startup, when the covariance matrix is determined recursively on-line because no data are available *a priori* to determine the in-control limit.

The upper control limits in (2.10), (2.11), and (2.12) assume that the observation at one time instant is statistically independent to the observations at other time instances. This can be a bad assumption for short sampling intervals. However, if there are enough data in the training set to capture the normal process variations, the T^2 statistic can be an effective tool for process monitoring even if there are mild deviations from the normality or statistical independence assumptions [30, 171].

There are several extensions that are usually not studied in the process control literature, but for which there are rigorous statistical formulations. In particular, lower control limits can be derived for T^2 [308] which can detect shifts in the covariance matrix (although the upper control limit is usually used to detect shifts in mean, it can also detect changes in the covariance matrix) [114].

The above T^2 tests are multivariable generalizations of the Shewhart chart used in the scalar case. The single variable CUSUM and EWMA charts can be generalized to the multivariable case in a similar manner [171, 203, 292, 338]. As in the scalar case, the multivariable CUSUM and EWMA charts can detect small persistent changes more readily than the multivariable Shewhart chart, but with increased detection delay.

2.6 Data Requirements

The quality and quantity of the data in the training set have a large influence on the effectiveness of the T^2 statistic as a process monitoring tool. An important question concerning the training set is, "How much data is needed to statistically populate the covariance matrix for m observation variables?" This question is answered here by determining the amount of data needed to produce a threshold value sufficiently close to the threshold obtained by assuming infinite data in the training set.

For a given level of significance α, a threshold based on infinite observations in the training set, or equivalently an exactly known covariance matrix, can be computed using (2.10), and the threshold for n observations in the training set is calculated using (2.11). The relative error produced by these two threshold values,

$$\epsilon = \frac{\frac{m(n-1)(n+1)}{n(n-m)}F_\alpha(m, n-m) - \chi^2_\alpha(m)}{\chi^2_\alpha(m)}, \tag{2.13}$$

indicates the sufficiency of the data amount n, where a large ϵ implies that more data should be collected. Table 2.2 shows the data requirements using (2.13) for various numbers of observation variables, where $\epsilon = 0.10$ and $\alpha = 0.5$; this implies that the medians of the T^2 statistic using (2.10) and (2.11) differ by less than 10%. The table indicates that the required number of observations is approximately 10 times the dimensionality of the observation space. The data requirements given in Table 2.2 do not take into account sensitivities that occur when some diagonal elements of Λ in (2.8) are small. In such cases the accuracy of the estimated values of the corresponding diagonal elements of the inverse of Λ will be poor, which will give erratic values for T^2 in (2.9). This motivates the use of the dimensionality reduction techniques described in Part III of this book.

Table 2.2. The amount of data n required for various number of observation variables m where $\epsilon = 0.10$ and $\alpha = 0.5$

Number of Observation Variables m	Data Requirement n
1	19
2	30
3	41
4	52
5	63
10	118
25	284
50	559
100	1110
200	2210

2.7 Homework Problems

1. Read the original article by Hotelling on the T^2 statistic [126]. How much of the results of this chapter were anticipated by Hotelling? Suggest reasons why these ideas took so long to work their way into industrial process applications.
2. Write a short report on the lower control limits for the T^2 statistic discussed by [308]. For what types of processes and faults will such limits be useful? Give a specific process example (list process, sensors, actuators, *etc.*). Suggest reasons why most of the process control and statistics literature ignores the lower control limit. Justify your statements.
3. Write a short report on the single variable CUSUM and EWMA control charts, including the mathematical expressions for the upper control limits in terms of a distribution function and assumptions on the noise statistics. You are welcome to use any books or journal articles on statistical quality control.
4. Extend the report in Problem 3 to the case of multivariate systems.
5. Consider the photographic process with the covariance matrix given in Table 1 of Jackson and Mulholdkar [145]. Reproduce as much as possible the results reported in the subsequent tables. Discuss the relative merits of the multivariate T^2 compared to scalar Shewhart charts for that process.

3. Pattern Classification

3.1 Introduction

Today's processes are heavily instrumented, with a large amount of data collected on-line and stored in computer databases. Much of the data are usually collected during out-of-control operations. When the data collected during the out-of-control operations have been previously diagnosed, the data can be categorized into separate *classes* where each class pertains to a particular fault. When the data have not been previously diagnosed, cluster analysis may aid the diagnoses of the operations during which the data were collected [299], and the data can be categorized into separate classes accordingly. If hyperplanes can separate the data in the classes as shown in Figure 3.1, these **separating planes** can define the boundaries for each of the fault regions. Once a fault is detected using on-line data observations, the fault can be diagnosed by determining the fault region in which the observations are located. Assuming the detected fault is represented in the database, the fault can be properly diagnosed in this manner.

This assignment of data to one of several categories or classes is the problem addressed by **pattern classification** theory [74]. The typical pattern classification system assigns an observation vector to one of several classes via three steps: **feature extraction**, **discriminant analysis**, and **maximum selection** (see Figure 3.2). The objective of the feature extraction step is to increase the robustness of the pattern classification system by reducing the dimensionality of the observation vector in a way that retains most of the information discriminating amongst the different classes. This step is especially important when there is a limited amount of quality data available. Using the information in the reduced-dimensional space, the discriminant calculator computes for each class the value of the **discriminant function**, a function quantifying the relationship between the observation vector and a class. By selecting the class with the maximum discriminant function value, the discriminant functions indirectly serve as the separating planes shown in Figure 3.1; however, in general the decision boundaries will not be linear.

The objective of this chapter is to provide an overview of the statistical approach to pattern classification. The focus of this chapter is on parametric approaches to pattern classification. Assuming the statistical distributions of

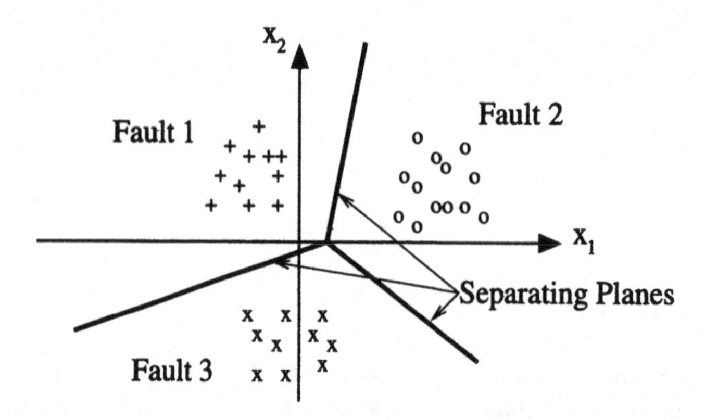

Fig. 3.1. A graphical illustration of the separating plane approach to pattern classification

the classes are known, an optimal pattern classification system can be developed using a parametric approach, while nonparametric approaches, such as the nearest neighbor rule [55], are suboptimal [74]. Pattern classification theory has been a key factor in developing fault diagnosis methods [270, 272], and the background in this chapter is important to understanding the fault diagnosis methods discussed in Part III. This chapter proceeds in Section 3.2 by presenting the optimal discriminant analysis technique for normally distributed classes. Section 3.3 discusses the feature extraction step.

3.2 Discriminant Analysis

The pattern classification system assigns an observation to class i with the maximum discriminant function value

$$g_i(\mathbf{x}) > g_j(\mathbf{x}) \qquad \forall j \neq i \tag{3.1}$$

where $g_j(\mathbf{x})$ is the discriminant function for class j given a data vector $\mathbf{x} \in \mathcal{R}^m$. The statistics of the data in each class can provide analytical measures to determine the optimal discriminant functions in terms of minimizing the *error rate*, the average probability of error. With ω_i being the event of class i (for example, a fault condition), the error rate can be minimized by using the discriminant function [74]

$$g_i(\mathbf{x}) = P(\omega_i|\mathbf{x}) \tag{3.2}$$

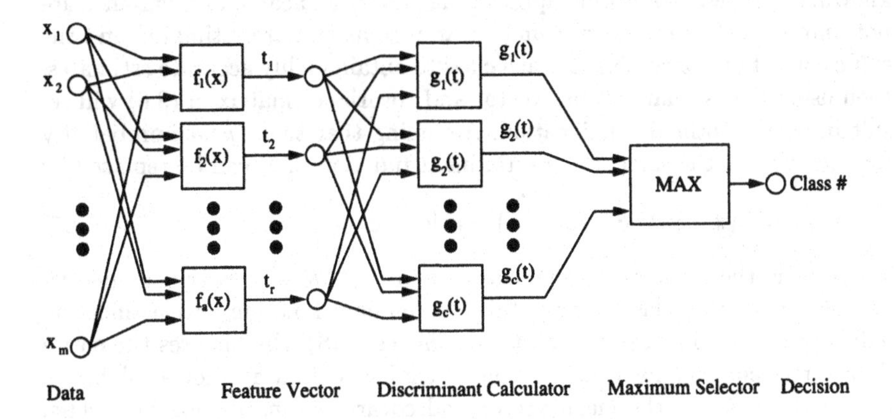

Fig. 3.2. A schema of a typical pattern classification system, where $f_i(x)$ are the feature extraction functions and $g_i(t)$ are the discriminant analysis functions

where $P(\omega_i|\mathbf{x})$ is the *a posteriori* probability of $\mathbf{x}$ belonging to class i. This is equivalent to choosing the separating curves to be the points at which the *a posteriori* probabilities are equal.

Using Bayes' rule,

$$P(\omega_i|\mathbf{x}) = \frac{p(\mathbf{x}|\omega_i)P(\omega_i)}{p(\mathbf{x})} \tag{3.3}$$

where $P(\omega_i)$ is the *a priori* probability for class ω_i, $p(\mathbf{x})$ is the probability density function for $\mathbf{x}$, and $p(\mathbf{x}|\omega_i)$ is the probability density function for $\mathbf{x}$ conditioned on ω_i. It can be shown that identical classification occurs when (3.2) is replaced by [74]

$$g_i(\mathbf{x}) = \ln p(\mathbf{x}|\omega_i) + \ln P(\omega_i). \tag{3.4}$$

If the data for each class is normally distributed, $p(\mathbf{x}|\omega_i)$ is given by

$$p(\mathbf{x}|\omega_i) = \frac{1}{(2\pi)^{m/2}\left[\det(\Sigma_i)\right]^{1/2}} \exp\left[-\frac{1}{2}(\mathbf{x}-\mu_i)^T\Sigma_i^{-1}(\mathbf{x}-\mu_i)\right] \tag{3.5}$$

where m is the number of measurement variables, and μ_i and Σ_i are the mean vector and covariance matrix for class i, respectively [74]. Substituting (3.5) into (3.4) gives

$$g_i(\mathbf{x}) = -\frac{1}{2}(\mathbf{x}-\mu_i)^T\Sigma_i^{-1}(\mathbf{x}-\mu_i) - \frac{m}{2}\ln 2\pi - \frac{1}{2}\ln[\det(\Sigma_i)] + \ln P(\omega_i) \tag{3.6}$$

This equation assumes that the mean vector and covariance matrix are known. In process monitoring applications, the true mean and covariance are not known. If the mean vector and covariance matrix are estimated and the sufficient data are available for each class to obtain highly accurate estimates, then using the estimated mean vector and covariance matrix in (3.6) will result in nearly optimal classification. Assuming that the *a priori* probability for each class is the same, the discriminant function (3.6) can be replaced by

$$g_i(\mathbf{x}) = -(\mathbf{x} - \bar{\mathbf{x}}_i)^T S_i^{-1}(\mathbf{x} - \bar{\mathbf{x}}_i) - \ln[\det(S_i)] \tag{3.7}$$

where $\bar{\mathbf{x}}_i$ is the mean vector for class i and $S_i \in \mathcal{R}^{m \times m}$ is the sample covariance matrix for class i. Using this discriminant function for classification will be referred to as **multivariate statistics** (MS) when it uses the entire data dimensionality for classification. If sufficient data are not available to accurately estimate the mean vector and covariance matrix for each class, then (3.6) will result in suboptimal classifications. In this case dimensionality reduction can be used to improve classification, as described in the next section.

Assuming that the *a priori* probability for each class is the same and the total amount of variability in each class is the same, an identical classification occurs when (3.6) is replaced by

$$g_i(\mathbf{x}) = -T_i^2 = -(\mathbf{x} - \mu_i)^T S_i^{-1}(\mathbf{x} - \mu_i) \tag{3.8}$$

where T_i^2 is the T^2 statistic for class i (see last chapter). By using the threshold T_α^2 in (2.11), the values for each $g_i(\mathbf{x})$ in (3.8) can be converted to levels of significance which implicitly account for the uncertainties in the mean vector and covariance matrix for each class.

3.3 Feature Extraction

The objective of the pattern classification system is to minimize the *misclassification rate*, the number of incorrect classifications divided by the total number of classifications, whenever it is applied to *testing data*, data independent of the training set. The dimensionality reduction of the feature extraction step can play a key role in minimizing the misclassification rate for observations outside the training set, especially when the dimensionality of the observation space m is large and the number of observations in each class n is small. If the statistical parameters such as the mean and covariance of the classes are known exactly, from an information point of view the entire observation space should be maintained for the discriminant analysis step. In reality, inaccuracies in the statistical parameters of the classes exist. Consequently, the amount of information obtained in some directions of the observation space, specifically those that do not add much information in discriminating the data in the training set, may not outweigh the inaccuracies in the statistical parameters, and the elimination of these directions in the feature extraction step can decrease the misclassification rate when applied to data independent of the training set.

The dimensionality reduction of the feature extraction step can also be motivated using system identification theory [199]. In system identification, it is shown that the accuracy of a model can be improved by decreasing the number of independent model parameters. This is due to the fact that the mean-squared error of the parameter estimates is reduced by decreasing the number of independent model parameters. By decreasing the number of independent parameters, the variance contribution of the parameter estimates on the mean-squared error is decreased more than the bias contribution is increased. These same arguments can be applied to the feature extraction step. For normally distributed classes, the covariance matrix has $m(m+1)/2$ independent parameters. Reducing the data dimensionality reduces the number of independent parameters in the covariance matrix. This increases the bias of the estimate of the covariance matrix, but decreases the variance. When the decrease in the variance contribution to the parameter error outweighs the increase in the bias contribution, the dimensionality reduction results in better covariance estimates and possibly lower misclassification rates when applied to data outside the training set.

Once the dimensionality reduction has been performed, classification is performed by applying discriminant analysis to the reduced-dimensional space. Applications of discriminant analysis to various reduced-dimensional spaces will be described in Part III. In particular, Chapter 5 describes a procedure for optimally reducing the dimensionality in terms of pattern classification.

3.4 Homework Problems

1. Derive Equation 3.4.
2. Derive Equation 3.6.
3. Derive Equation 3.8.
4. Explain in detail how to use (3.8) to compute levels of significance for each class i.
5. Consider the case where all the class covariance matrices in (3.5) are equal, $\Sigma_i = \Sigma$. Show that the discriminant function (3.6) can be replaced by a discriminant function which is linear in $\mathbf{x}$ without changing the classification. With this linear discriminant function, show that the equations $g_i(\mathbf{x}) = g_j(\mathbf{x})$ define separating planes as shown in Figure 3.1. Derive the equations for the separating curves when the class covariance matrices are not equal. What are the shapes of these separating curves?

Part III

Data-driven Methods

4. Principal Component Analysis

4.1 Introduction

By projecting the data into a lower-dimensional space that accurately characterizes the state of the process, dimensionality reduction techniques can greatly simplify and improve process monitoring procedures. **Principal component analysis** (PCA) is such a dimensionality reduction technique. It produces a lower-dimensional representation in a way that preserves the correlation structure between the process variables, and is optimal in terms of capturing the variability in the data.

The application of PCA as a dimensionality reduction tool for monitoring industrial processes has been studied by several academic and industrial engineers [177, 260]. Applications of PCA to plant data have been conducted at DuPont and other companies, with much of the results published in conference proceedings and journal articles [169, 260, 259, 343]. Several academics have performed similar studies based on data collected from computer simulations of processes [75, 117, 157, 183, 204, 207, 269, 270, 272, 307]. For some applications, most of the variability in the data can be captured in two or three dimensions [207], and the process variability can be visualized with a single plot. This one-plot visualization and the structure abstracted from the multidimensional data assist the operators and engineers in interpreting the significant trends of the process data [177].

For the cases when most of the data variations cannot be captured in two or three dimensions, methods have been developed to automate the process monitoring procedures [209, 260, 272]. The application of PCA in these methods is motivated by one or more of three factors. First, PCA can produce lower-dimensional representations of the data which better generalize to data independent of the training set than that using the entire dimensionality of the observation space, and therefore, improve the proficiency of detecting and diagnosing faults. Second, the structure abstracted by PCA can be useful in identifying either the variables responsible for the fault and/or the variables most affected by the fault. Third, PCA can separate the observation space into a subspace capturing the systematic trends of the process and a subspace containing essentially the random noise. Since it is widely accepted that certain faults primarily affect one of the two subspaces [77, 345, 346], applying one measure developed for one subspace and another measure developed for the other subspace can increase the sensitivity of the process monitoring scheme to faults in general. The three aforementioned attributes of PCA are further discussed later in this chapter.

The purpose of this chapter is to describe the PCA methods for process monitoring. It begins in Section 4.2 by defining PCA and in Section 4.3 by discussing the different methods which can be used to automatically determine the order of the PCA representation. Sections 4.4, 4.5, and 4.6 discuss the PCA developments for fault detection, identification, and diagnosis, respectively. In Section 4.7 is a discussion of **dynamic PCA** (DPCA), which takes into account serial correlations in the process data. Section 4.8 discusses other PCA-based process monitoring methods.

4.2 Principal Component Analysis

PCA is a linear dimensionality reduction technique, optimal in terms of capturing the variability of the data. It determines a set of orthogonal vectors, called **loading vectors**, ordered by the amount of variance explained in the loading vector directions. Given a training set of n observations and m process variables stacked into a matrix X as in (2.5), the loading vectors are calculated by solving the stationary points of the optimization problem

$$\max_{\mathbf{v} \neq 0} \frac{\mathbf{v}^T X^T X \mathbf{v}}{\mathbf{v}^T \mathbf{v}} \tag{4.1}$$

where $\mathbf{v} \in \mathcal{R}^m$. The stationary points of (4.1) can be computed via the singular value decomposition (SVD)

$$\frac{1}{\sqrt{n-1}} X = U \Sigma V^T \tag{4.2}$$

where $U \in \mathcal{R}^{n \times n}$ and $V \in \mathcal{R}^{m \times m}$ are unitary matrices, and the matrix $\Sigma \in \mathcal{R}^{n \times m}$ contains the non-negative real **singular values** of decreasing magnitude along its main diagonal ($\sigma_1 \geq \sigma_2 \geq \cdots \geq \sigma_{\min(m,n)} \geq 0$), and zero offdiagonal elements. The loading vectors are the orthonormal column vectors in the matrix V, and the variance of the training set projected along the ith column of V is equal to σ_i^2. Solving (4.2) is equivalent to solving an eigenvalue decomposition of the sample covariance matrix S,

$$S = \frac{1}{n-1} X^T X = V \Lambda V^T \tag{4.3}$$

where the diagonal matrix $\Lambda = \Sigma^T \Sigma \in \mathcal{R}^{m \times m}$ contains the non-negative real **eigenvalues** of decreasing magnitude ($\lambda_1 \geq \lambda_2 \geq \cdots \geq \lambda_m \geq 0$) and the ith eigenvalue equals the square of the ith singular value (*i.e.*, $\lambda_i = \sigma_i^2$).

In order to optimally capture the variations of the data while minimizing the effect of random noise corrupting the PCA representation, the loading vectors corresponding to the a largest singular values are typically retained. The motivation for reducing the dimensionality of the PCA representation is analogous to the arguments given in Section 3.3 for pattern classification. Selecting the columns of the loading matrix $P \in \mathcal{R}^{m \times a}$ to correspond to the loading vectors associated with the first a singular values, the projections of the observations in X into the lower-dimensional space are contained in the **score matrix**,

$$T = XP, \tag{4.4}$$

and the projection of T back into the m-dimensional observation space,

$$\hat{X} = TP^T. \tag{4.5}$$

The difference between X and $\hat{X}$ is the residual matrix E:

$$E = X - \hat{X}. \tag{4.6}$$

The **residual matrix** captures the variations in the observation space spanned by the loading vectors associated with the $m - a$ smallest singular values. The subspaces spanned by $\hat{X}$ and E are called the **score space** and **residual space**, respectively. The subspace contained in the matrix E has a small signal-to-noise ratio, and the removal of this space from X can produce a more accurate representation of the process, $\hat{X}$.

Defining $\mathbf{t_i}$ to be i^{th} column of T in the training set, the following properties can be shown (see Homework Problem 5) [259]

1. $\text{Var}(\mathbf{t_1}) \geq \text{Var}(\mathbf{t_2}) \geq \cdots \geq \text{Var}(\mathbf{t_a})$.
2. $\text{Mean}(\mathbf{t_i}) = 0;\ \forall i$.
3. $\mathbf{t_i}^T \mathbf{t_k} = 0;\ \forall i \neq k$.
4. There exists no other orthogonal expansion of a components that captures more variations of the data.

A new observation (column) vector in the testing set, $\mathbf{x} \in \mathcal{R}^m$, can be projected into the lower-dimensional score space $t_i = \mathbf{x}^T \mathbf{p_i}$ where $\mathbf{p_i}$ is the ith loading vector (see Figure 4.1). The transformed variable t_i is also called the ith **principal component** of $\mathbf{x}$ [147]. To distinguish between the transformed variables and the transformed observation, the transformed variables will be called **principal components** and the individual transformed observations will be called **scores**. The statistical properties listed above allow each of the scores to be monitored separately using the univariate statistical procedures discussed in Section 2.3. With the vectors projected into the lower dimensional space using PCA, only a variables needed to be monitored, as compared with m variables without the use of PCA. When enough data are collected in the testing set, the score vectors $\mathbf{t_1}, \mathbf{t_2}, \ldots, \mathbf{t_a}$ can be formed. If these score vectors do not satisfy the four properties listed above, the testing set is most likely collected during different operating conditions than for the training set. This abstraction of structure from the multidimensional data is a key component of the score contribution method for fault identification discussed in Section 4.5.

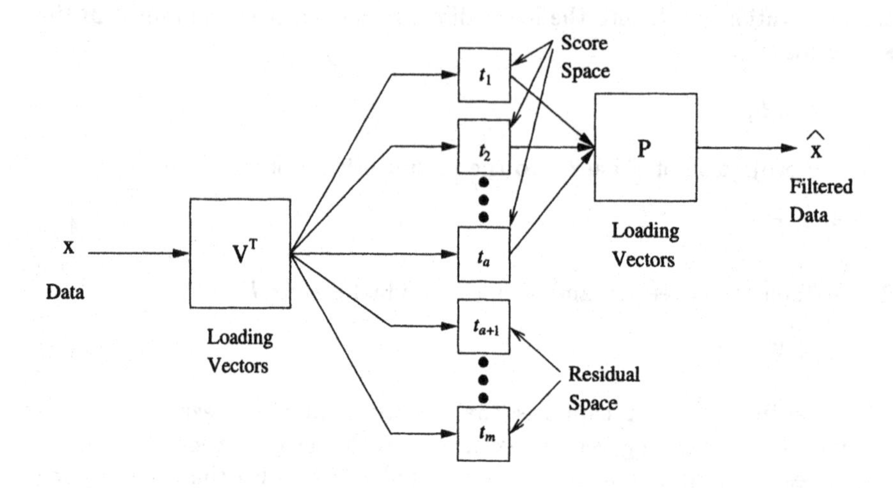

Fig. 4.1. The projection of the observation vector **x** into the score and residual spaces, and the computation of the filtered observation $\hat{\mathbf{x}}$

The application of PCA will be illustrated using Fisher's classic data set [45, 82]. The data set consists of three classes, with each class containing $m = 4$ measurements and $n = 50$ observations (see Table 4.1 and Figure 4.2).

Class 3 data were used to construct X as in (2.5). After autoscaling X and solving (4.3), we have

$$\Lambda = \begin{bmatrix} 1.92 & 0 & 0 & 0 \\ 0 & 0.96 & 0 & 0 \\ 0 & 0 & 0.88 & 0 \\ 0 & 0 & 0 & 0.24 \end{bmatrix}, \tag{4.7}$$

and

$$V = \begin{bmatrix} 0.64 & -0.29 & 0.052 & -0.71 \\ 0.64 & -0.23 & 0.25 & 0.69 \\ 0.34 & 0.33 & -0.88 & 0.11 \\ 0.25 & 0.87 & 0.41 & -0.09 \end{bmatrix}. \tag{4.8}$$

The total variance for X projected along V is equal to the trace of Λ, which is 4.0. The ith value in the diagonal of Λ indicates the amount of variance

Table 4.1. Statistics of Fisher's data [45, 82]

Class 1: Iris Virginica	Mean	Std. Deviation	Range
Sepal length	6.59	0.64	4.9–7.9
Sepal width	2.98	0.32	2.2–3.8
Petal length	5.55	0.55	4.5–6.9
Petal width	2.03	0.27	1.4–2.5
Class 2: Iris Versicolor	**Mean**	**Std. Deviation**	**Range**
Sepal length	5.94	0.52	4.9–7.0
Sepal width	2.77	0.31	2.0–3.4
Petal length	4.29	0.47	3.0–5.1
Petal width	1.33	0.20	1.0–1.8
Class 3: Iris Setosa	**Mean**	**Std. Deviation**	**Range**
Sepal length	5.01	0.35	4.3–5.8
Sepal width	3.43	0.38	2.3–4.4
Petal length	1.46	0.17	1.0–1.9
Petal width	0.30	0.40	0.1–3.0

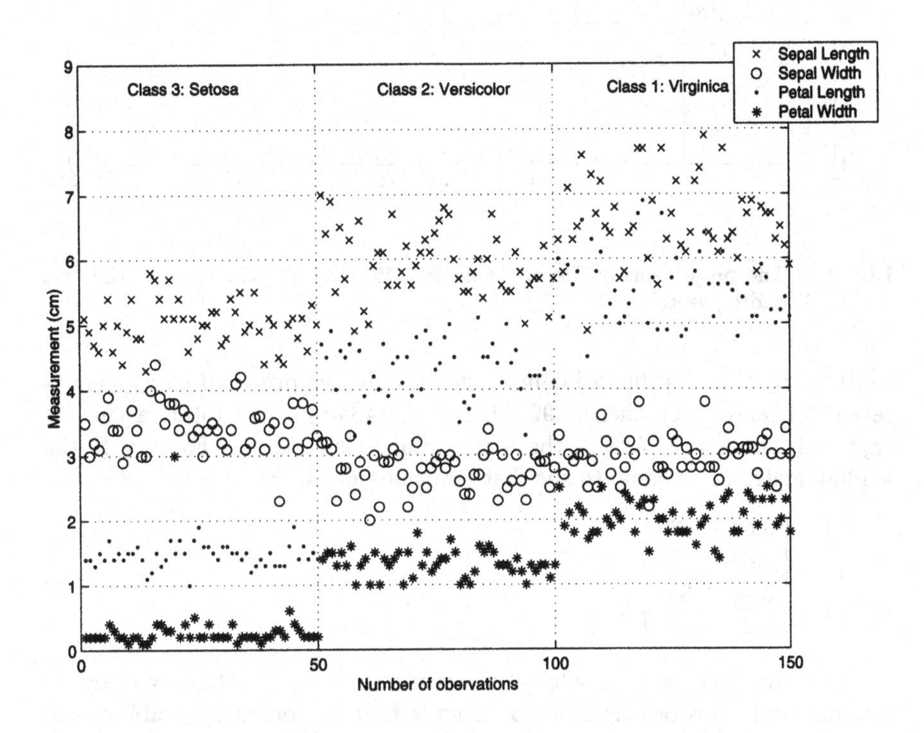

Fig. 4.2. Plot of Fisher's data [82, 45]

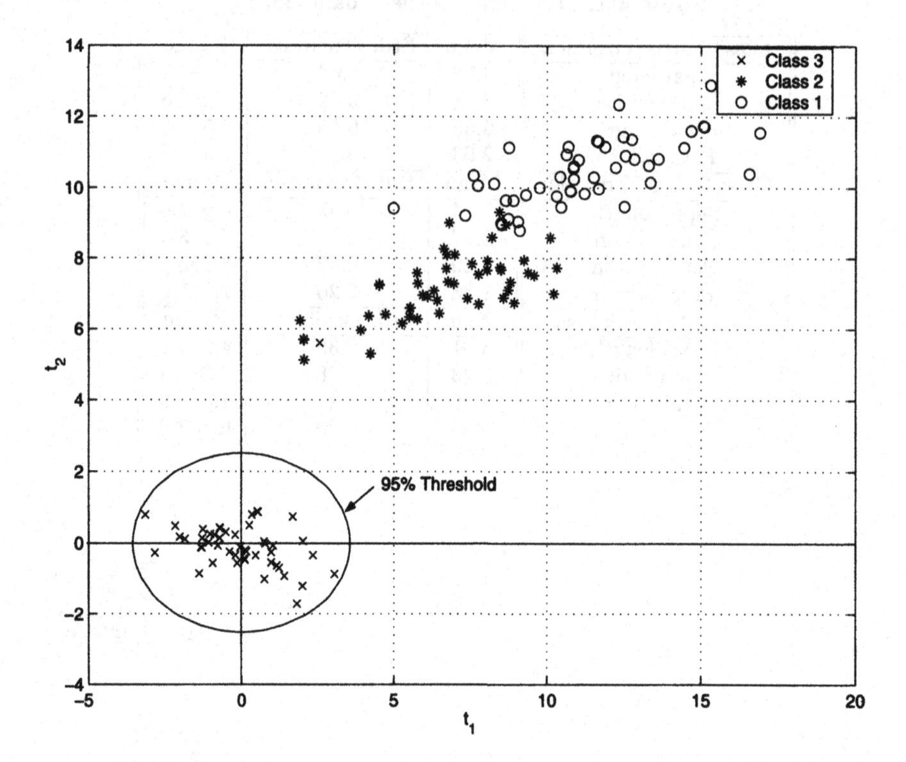

Fig. 4.3. The projections of Fisher's data [82, 45] for three classes onto the first two PCA loading vectors

captured by the ith principal component. If only one principal component is retained (*i.e.*, $a = 1$), then $(1.92/4.0)100\% = 48.0\%$ of the total variance is captured. For $a = 2$, 72% of the total variance is captured. For $a = 2$, the loading matrix P is equal to the first two columns of V:

$$P = \begin{bmatrix} 0.64 & -0.29 \\ 0.64 & -0.23 \\ 0.34 & 0.33 \\ 0.25 & 0.87 \end{bmatrix}. \tag{4.9}$$

The score matrix T is calculated according to (4.4). The advantage of retaining only two principal components is that the process variability can be visualized by plotting t_2 versus t_1 (see Figure 4.3).

It is easy to verify that $\mathrm{Var}(\mathbf{t_1}) \geq \mathrm{Var}(\mathbf{t_2})$ by observing that the variation along the horizontal axis is much greater than that of the vertical axis for the Class 3 data in Figure 4.3. The ellipsoid and the data for Class 3 are centered at the origin, which indicates that $\mathrm{Mean}(\mathbf{t_1}) = \mathrm{Mean}(\mathbf{t_2}) = 0$. It is straightforward to verify that $\mathbf{t_1}$ and $\mathbf{t_2}$ are orthogonal to each other.

A threshold defines an elliptical confidence region for data belonging to Class 3 (the calculation of the threshold will be described in Section 4.4). In this example, statistics predict that there is a 95% probability that a Class 3 data point should fall inside the ellipsoid. It is clearly shown in Figure 4.3 that PCA is able to separate Class 3 data from Classes 1 and 2, except for the apparent outlier located at $(t_1, t_2) = (2.5, 5.6)$.

4.3 Reduction Order

It is commonly accepted and with certain assumptions theoretically justified [345] that the portion of the PCA space corresponding to the larger singular values describes most of the *systematic* or *state variations* occurring in the process, and the portion of the PCA space corresponding to the smaller singular values describes the *random noise*. By appropriately determining the number of loading vectors, a, to maintain in the PCA model, the systematic variations can be decoupled from the random variations, and the two types of variations can be monitored separately, as discussed in Section 4.4. Several techniques exist for determining the value of the reduction order a [117, 144, 267, 315], but there appears to be no dominant technique. The methods for determining a described here are:

1. the percent variance test,
2. the scree test,
3. parallel analysis, and
4. the PRESS statistic.

The **percent variance** method determines a by calculating the smallest number of loading vectors needed to explain a specific minimum percentage of the total variance. (Recall that the variance associated with the ith loading vector is equal to the square of the singular value, σ_i^2.) Because this minimum percentage is chosen arbitrarily, it may be too low or too high for a particular application.

The **scree test** assumes that the variance, σ_i^2, corresponding to the random noise forms a linear profile. The dimension of the score space a is determined by locating the value of σ_i^2 where the profile is no longer linear. The identification of this break can be ambiguous, and thus, this method is difficult to automate. It is especially ambiguous when several breaks from linearity occur in the profile.

Parallel analysis determines the dimensionality by comparing the variance profile to that obtained by assuming independent observation variables. The reduction order is determined as the point at which the two profiles cross. This approach ensures that the significant correlations are captured in the score space, and it is particularly attractive since it is intuitive and easy to automate. Ku, Storer, and Georgakis [183] recommend the parallel analysis method, because in their experience, it performs the best overall.

The dimension of the score space can also be determined using a **cross-validation** procedure with the **prediction residual sum of squares** (PRESS) statistic [347],

$$PRESS(i) = \frac{1}{mn}||X - \hat{X}||_F^2 \tag{4.10}$$

where i is the number of loading vectors retained to calculate $\hat{X}$ and $||\cdot||_F$ is the Frobenius norm (the square root of the sum of squares of all the elements). For the implementation of this technique, the training set is divided into groups. The PRESS statistic for one group is computed based on various dimensions of the score space, i, using all the other groups. This is repeated for each group, and the value i associated with the minimum average PRESS statistic determines the dimension of the score space.

4.4 Fault Detection

As discussed in Section 2.4, the T^2 statistic can be used to detect faults for multivariate process data. Given an observation vector $\mathbf{x}$ and assuming that $\Lambda = \Sigma^T\Sigma$ is invertible, the T^2 statistic in (2.9) can be calculated directly from the PCA representation (4.2)

$$T^2 = \mathbf{x}^T V(\Sigma^T\Sigma)^{-1} V^T\mathbf{x}. \tag{4.11}$$

This follows from the fact that the V matrix in (2.7) can be computed to be identical to the V matrix in (4.2), and the σ_i^2 are equal to the diagonal elements of Λ. When the number of observation variables is large and the amount of data available is relatively small, the T^2 statistic (4.11) tends to be an inaccurate representation of the in-control process behavior, especially in the loading vector directions corresponding to the smaller singular values. Inaccuracies in these smaller singular values have a huge effect on the calculated T^2 statistic because the square of the singular values is inverted in (4.11). Additionally, the smaller singular values are prone to errors because these values contain small signal-to-noise ratios and the associated loading vector directions often suffer from a lack of excitation. Therefore, in this case the loading vectors associated only with the larger singular values should be retained in calculating the T^2 statistic.

By including in the matrix P the loading vectors associated only with the a largest singular values, the T^2 statistic for the lower-dimensional space can be computed [143]

$$T^2 = \mathbf{x}^T P\Sigma_a^{-2}P^T\mathbf{x}. \tag{4.12}$$

where Σ_a contains the first a rows and columns of Σ. The T^2 statistic (4.12) measures the variations in the score space only. If the actual mean and covariance are known, the T^2 statistic threshold derived from (2.10) is

$$T_\alpha^2 = \chi_\alpha^2(a). \tag{4.13}$$

When the actual covariance matrix is estimated from the sample covariance matrix, the T^2 statistic threshold derived from (2.11) is

$$T_\alpha^2 = \frac{a(n-1)(n+1)}{n(n-a)} F_\alpha(a, n-a). \tag{4.14}$$

To detect outliers in the training set, the threshold derived from (2.12) is

$$T_\alpha^2 = \frac{(n-1)^2 (a/(n-a-1)) F_\alpha(a, n-a-1)}{n(1 + (a/(n-a-1)) F_\alpha(a, n-a-1))}. \tag{4.15}$$

Because the T^2 statistic in (4.12) is not affected by the inaccuracies in the smaller singular values of the covariance matrix, it is able to better represent the normal process behavior and provides a more robust fault detection measure when compared to the T^2 statistic in (4.11). Using the arguments in Section 4.3, the T^2 statistic (4.12) can be interpreted as measuring the systematic variations of the process, and a violation of the threshold would indicate that the systematic variations are out of control.

For the example in the last section, we have $n = 50$ and $a = 2$. According to an F–distribution table [120], $F_{0.05}(2, 48) = 3.19$. The threshold T_α^2 is equal to 6.64 according to (4.14). The elliptical confidence region, as shown in Figure 4.3, is given by

$$T^2 = \mathbf{x}^T P \Sigma_a^{-2} P^T \mathbf{x} \le 6.64, \tag{4.16}$$

with

$$\Sigma_a^2 = \begin{bmatrix} 1.92 & 0 \\ 0 & 0.96 \end{bmatrix}. \tag{4.17}$$

The equation

$$\mathbf{t} = P^T \mathbf{x} \tag{4.18}$$

converts this region into the ellipse in Figure 4.3. Inserting (4.18) into (4.16) gives

$$\mathbf{t}^T \Sigma_a^{-2} \mathbf{t} \le 6.64 \tag{4.19}$$

or

$$\frac{t_1^2}{1.92} + \frac{t_2^2}{0.96} \le 6.64 \tag{4.20}$$

where $\mathbf{t} = [t_1 \; t_2]^T$.

Data from Classes 1 and 2 are used to illustrate that the PCA model is able to detect data that do not come from Class 3. The data sets for Classes

1 and 2 are first autoscaled according to the mean and standard deviation of Class 3. Equation 4.4 is used to calculate the score matrices for Classes 1 and 2. As shown in Figure 4.3, the mean of each score vector for Classes 1 and 2 is not equal to zero. Indeed, all the data points for Classes 1 and 2 fall outside the elliptical confidence region, indicating data from Classes 1 and 2 are indeed different from the Class 3 data.

The T^2 statistic in (4.11) is overly sensitive to inaccuracies in the PCA space corresponding to the smaller singular values because it directly measures the variation along *each* of the loading vectors. In other words, it *directly* measures the scores corresponding to the smaller singular values. The portion of the observation space corresponding to the $m - a$ smallest singular values can be monitored more robustly by using the Q statistic [145, 144, 150, 176, 348]

$$Q = \mathbf{r}^T\mathbf{r}, \qquad \mathbf{r} = (I - PP^T)\mathbf{x}, \tag{4.21}$$

where $\mathbf{r}$ is the residual vector, a projection of the observation $\mathbf{x}$ into the residual space. Since the Q statistic does not directly measure the variations along each loading vector but measures the total sum of variations in the residual space, the Q statistic does not suffer from an over-sensitivity to inaccuracies in the smaller singular values [145]. The Q statistic, also known as the **squared prediction error** (SPE), is a squared 2-norm measuring the deviation of the observations to the lower-dimensional PCA representation.

The distribution for the Q statistic has been approximated by Jackson and Mudholkar [145]

$$Q_\alpha = \theta_1 \left[\frac{h_0 c_\alpha \sqrt{2\theta_2}}{\theta_1} + 1 + \frac{\theta_2 h_0 (h_0 - 1)}{\theta_1^2} \right]^{1/h_0} \tag{4.22}$$

where $\theta_i = \sum_{j=a+1}^{n} \sigma_j^{2i}$, $h_0 = 1 - \frac{2\theta_1\theta_3}{3\theta_2^2}$, and c_α is the normal deviate corresponding to the $(1 - \alpha)$ percentile. Given a level of significance, α, the threshold for the Q statistic can be computed using (4.22) and be used to detect faults.

Within the context of Section 4.3, the Q statistic measures the random variations of the process, for example, that associated with measurement noise. The threshold (4.22) can be applied to define the normal variations for the random noise, and a violation of the threshold would indicate that the random noise has significantly changed. The T^2 and Q statistics along with their appropriate thresholds detect different types of faults, and the advantages of both statistics can be utilized by employing the two measures together. When these two statistics are utilized along with their respective thresholds, it produces a cylindrical in-control region, as illustrated for $a = 2$ in Figure 4.4. The figure indicates that the 'x' data were collected during in-control operations, the 'o' data represent T^2 statistic violation, and the '+' data represent Q statistic violation.

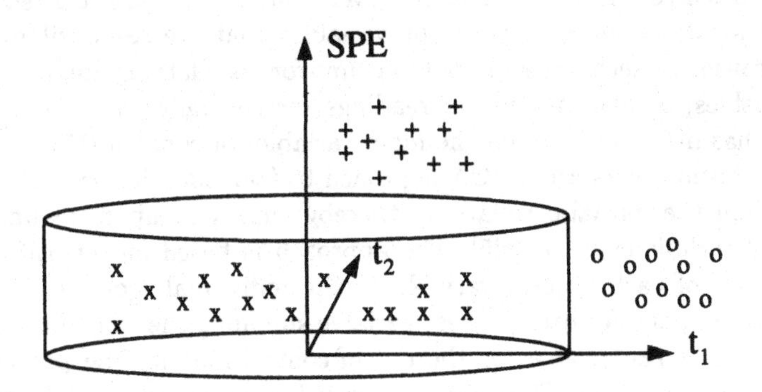

Fig. 4.4. A graphical illustration for fault detection using the Q and T^2 statistics

4.5 Fault Identification

Once a fault has been detected, the next step is to determine the cause of the out-of-control status. The task of diagnosing the fault can be rather challenging when the number of process variables is large, and the process is highly integrated. Also, many of the measured variables may deviate from their set-points for only a short time period when a fault occurs, due to control loops bringing the variables back to their set-points (even though the fault is persisting in the system). This type of systems behavior can disguise the fault, making it difficult for an automated fault diagnosis algorithm to correctly isolate the correct fault acting on the system.

The objective of fault identification is to determine which observation variables are most relevant to diagnosing the fault, thereby focusing the plant operators and engineers on the subsystem(s) most likely where the fault occurred. This assistance provided by the fault identification scheme in locating the fault can effectively incorporate the operators and engineers in the process monitoring scheme and significantly reduce the time to recover in-control operations.

Traditionally, univariate statistical techniques were employed for fault identification. Given an observation vector $\mathbf{x}$, the normalized errors for each variable x_j were calculated as

$$e_j = (x_j - \mu_j)/s_j \tag{4.23}$$

where μ_j is the mean and s_j is the standard deviation of the jth variable. These normalized errors were plotted on the same graph, and thresholds based on the level of significance were used to detect the out-of-control variables, as discussed in Section 2.3. However, univariate statistical techniques for fault identification can leave out variables that are responsible for the fault because the techniques do not account for correlations among the process variables, or can give alarm readings for so many variables that the engineer has little guidance on the main variables of concern [171].

Contribution plots are a PCA approach to fault identification that takes into account the spacial correlations, thereby improving upon the univariate statistical techniques [171, 225]. The approach is based on quantifying the contribution of each process variable to the individual scores of the PCA representation, and for each process variable summing the contributions *only* of those scores responsible for the out-of-control status. The procedure is applied in response to a T^2 violation, and it is summarized as follows:

1. Check the normalized scores $(t_i/\sigma_i)^2$ for the observation $\mathbf{x}$ and determine the $r \leq a$ scores responsible for the out-of-control status. For instance, those scores with $(t_i/\sigma_i)^2 > \frac{1}{a}(T^2_\alpha)$. (Recall that t_i is the score of the observation projected onto the ith loading vector, and σ_i is the corresponding singular value.)
2. Calculate the *contribution* of each variable x_j to the out-of-control scores t_i

$$cont_{i,j} = \frac{t_i}{\sigma_i^2} p_{i,j}(x_j - \mu_j) \tag{4.24}$$

   where $p_{i,j}$ is the $(i, j)^{th}$ element of the loading matrix P.
3. When $cont_{i,j}$ is negative, set it equal to zero.
4. Calculate the *total contribution* of the jth process variable, x_j,

$$CONT_j = \sum_{i=1}^{r}(cont_{i,j}). \tag{4.25}$$

5. Plot $CONT_j$ for all m process variables, x_j, on a single graph.

The variables responsible for the fault can be prioritized or ordered by the total contribution values $CONT_j$, and the plant operators and engineers can immediately focus on those variables with high $CONT_j$ values and use their process knowledge to determine the cause of the out-of-control status. While the overall variable contribution approach can be applied to the portion of the observation space corresponding to the $m - a$ smallest singular values, it is not practical because the total contribution values $CONT_j$ would be overly sensitive to the smaller singular values.

Wise *et al.* [346] developed a PCA approach to fault identification which is based on quantifying the total variation of each of the process variables in the residual space. Assuming that the $m - a$ smallest singular values are all equal, the variance for each variable x_j inside the residual space can be estimated as [346]

$$\hat{s}_j^2 = \sum_{i=a+1}^{p} p_{i,j}\sigma_i^2. \tag{4.26}$$

Given q new observations, the variance of the jth variable outside the PCA model space can be tested where

$$s_j^2/\hat{s}_j^2 > F_\alpha(q-a-1, n-a-1) \tag{4.27}$$

would indicate an out-of-control variable, where s_j^2 and $\hat{s}_j^2$ are the variance estimates of the jth variable for the new and training set observations, respectively, and $F_\alpha(q-a-1, n-a-1)$ is the $(1-\alpha)$ percentile limit using the F distribution [120]. Equation 4.27 is testing the null hypothesis, with the null hypothesis being $s_j = \hat{s}_j$ and the one-sided alternative hypothesis being $s_j > \hat{s}_j$. The one-sided alternative hypothesis is accepted (*i.e.*, the null hypothesis is rejected) if (4.27) holds [120]. In most of the times, the variable that is responsible for a fault has a larger variance than it has in the training set (*i.e.*, $s_j > \hat{s}_j$). However, this is not always true. For example, a broken sensor may give constant reading, indicating that $s_j < \hat{s}_j$. This motivates the use of two-sided hypothesis testing, with the null hypothesis being $s_j = \hat{s}_j$ and the two-sided alternative hypothesis being $s_j \neq \hat{s}_j$. We conclude $\hat{s}_j \neq s_j$ if [120]

$$s_j^2/\hat{s}_j^2 > F_{\alpha/2}(q-a-1, n-a-1) \tag{4.28}$$

or

$$\hat{s}_j^2/s_j^2 > F_{\alpha/2}(n-a-1, q-a-1). \tag{4.29}$$

In addition, a large shift in the mean inside the residual space occurs if [346, 120]

$$\frac{\mu_j - \hat{\mu}_j}{\hat{s}_j\sqrt{\frac{1}{q-a} + \frac{1}{n-a}}} > t_{\alpha/2}(q+n-2a-2) \tag{4.30}$$

or

$$\frac{\mu_j - \hat{\mu}_j}{\hat{s}_j\sqrt{\frac{1}{q-a} + \frac{1}{n-a}}} < -t_{\alpha/2}(q+n-2a-2), \tag{4.31}$$

where μ_j and $\hat{\mu}_j$ are the means of x_j for the new and training set observations, respectively, and $t_{\alpha/2}(q+n-2a-2)$ is the $(1-\alpha/2)$ percentile limit using the t distribution. Equations 4.30 and 4.31 are testing the null hypothesis, with the null hypothesis being $\mu_j = \hat{\mu}_j$ and the alternative hypothesis being $\mu_j \neq \hat{\mu}_j$. The alternative hypothesis is accepted if (4.30) or (4.31) holds [120].

The variables responsible for the out-of-control status, detected by the Q statistic, can be identified using (4.27), (4.30), and (4.31). In addition, the variables can be prioritized using the expression values (4.27), (4.30), and (4.31) where the variable with the largest expression value is given priority. In [346], sensor failures are detected and identified using (4.27), (4.30), and (4.31). Other PCA-based methods developed specifically for detecting sensor failures are discussed elsewhere [77, 239].

The fault identification approaches using (4.27), (4.30), and (4.31) require a group of $q >> 1$ observations. As discussed in Section 2.3, measures based on several consecutive observations are able to increase the robustness and sensitivity over measures based on only a single observation, but result in a slower response time for larger process shifts. A fault identification measure based on an observation vector at a single time instant is the normalized error

$$RES_j = r_j/\hat{s}_j \tag{4.32}$$

where r_j is the jth variable of the residual vector. The values of (4.32) can be used to prioritize the variables where the variable with the highest normalized error is given priority. The measure (4.32), when compared to (4.27), (4.30), and (4.31), is able to indicate the current status of the process immediately after a large process shift more accurately.

4.6 Fault Diagnosis

The previous section discussed fault identification methods, which identify the variables associated with the faulty subsystem. Although these methods assist in diagnosing the faults, it may take a substantial amount of time and process expertise on behalf of the plant operators and engineers before the fault is properly diagnosed. Much of this time and expertise can be eliminated by employing an automated fault diagnosis scheme. One approach is to construct separate PCA models for each process unit [117]. A fault associated with a particular process unit is assumed to occur if the PCA model for that unit indicates that the process is out of control. While this approach can narrow down the cause of abnormal process operations, it will not unequivocally diagnose the cause. This distinguishes these *fault isolation* techniques (which are based on non-supervised classification) from the fault diagnosis techniques (which are based on supervised classification) described below.

Several researchers have proposed techniques to use principal component analysis for fault diagnosis. The simplest approach is to construct a single PCA model and define regions in the lower-dimensional space which classifies whether a particular fault has occurred [346]. This approach is unlikely to be effective when a significant number of faults can occur [360].

The way in which a pattern classification system can be applied to diagnose faults automatically was described in Chapter 3 how a pattern. The feature extraction step was shown to be important especially when the dimensionality of the data is large and the quantity of quality data is relatively small (see Section 3.3). A PCA approach which can handle a larger number of faults than using a single PCA model is to develop a separate PCA model based on data collected during each specific fault situation, and then apply the Q [181], T^2 [269], or other statistics [269, 270, 272, 360] to each PCA model to predict which fault or faults most likely occurred. This approach is a combination of principal component analysis and discriminant analysis [270]. Various discriminant functions for diagnosing faults are discussed in the following.

One way to use PCA for fault diagnosis is to derive *one* model based on the data from *all* fault classes. Stacking the data for all fault classes into matrix X, the loading matrix P can be calculated based on (4.2) or (4.3). The maximum likelihood classification for an observation $\mathbf{x}$ is fault class i with the maximum **score discriminant**, which is derived from (3.6) to be

$$\begin{aligned} g_i(\mathbf{x}) = &-\tfrac{1}{2}(\mathbf{x}-\bar{\mathbf{x}}_i)^T P\left(P^T S_i P\right)^{-1} P^T(\mathbf{x}-\bar{\mathbf{x}}_i) + \ln(p_i) \\ &-\tfrac{1}{2}\ln\left[\det\left(P^T S_i P\right)\right] \end{aligned} \tag{4.33}$$

where $\bar{\mathbf{x}}_i$ is the mean vector for class i,

$$\bar{\mathbf{x}}_i = \frac{1}{n_i}\sum_{\mathbf{x}_j \in \mathcal{X}_i} \mathbf{x}_j, \tag{4.34}$$

n_i is the number of data points in fault class i, $\mathcal{X}_i$ is the set of vectors $\mathbf{x}_j$ which belong to the fault class i, and $S_i \in \mathcal{R}^{m\times m}$ is the sample covariance matrix for fault class i, as defined in (2.6).

If P is selected to include all of the dimensions of the data (*i.e.*, $P = V \in \mathcal{R}^{m\times m}$) and the overall likelihood for all fault classes is the same, Equation 4.33 reduces to the discriminant function for multivariate statistics (MS) as defined in (3.7). MS selects the most probable fault class based on maximizing the discriminant function (3.7). MS also serves as a benchmark for the other statistics, as the dimensionality should only be reduced if it decreases the misclassification rate for a testing set.

The **score discriminant, residual discriminant**, and **combined discriminant** are three discriminant analysis techniques used with *multiple* PCA models [269]. Assuming the PCA models retain the important variations in discriminating between the faults, an observation $\mathbf{x}$ is classified as being in the fault class i with the maximum score discriminant

$$g_i(\mathbf{x}) = -\frac{1}{2}\mathbf{x}^T P_i \Sigma_{a,i}^{-2} P_i^T \mathbf{x} - \frac{1}{2}\ln[\det(\Sigma_{a,i}^2)] + \ln(p_i) \tag{4.35}$$

where P_i is the loading matrix for fault class i, $\Sigma_{a,i}$ is the diagonal matrix Σ_a as shown in (4.12) for fault class i ($\Sigma_{a,i}^2$ is the covariance matrix of $P_i\mathbf{x}$), and p_i is the overall likelihood of fault class i [150, 272]. Note that (4.35) assumes that the observation vector $\mathbf{x}$ has been autoscaled according to the mean and standard deviation of the training set for fault class i. Equation 4.35 is based on the discriminant function (3.6).

The matrices P_i, $\Sigma_{a,i}$, and p_i in (4.35) depend solely on fault class i, that is, the discriminant function for *each* fault class is derived *individually*. A weakness of this approach is that useful information for other classes is not utilized when each model is derived. In general, the reduction order a for each fault class is different. This indicates that the discriminant function (4.35) for each fault class i is evaluated based on different dimensions of the projected data $P_i^T\mathbf{x}$. This inconsistency can result in relatively high misclassification rates.

In contrast to (4.35), the projection matrix P in (4.33) not only utilizes information from all fault classes, but also projects the data onto the same dimensions for each class. Because of these properties, the discriminant function (4.33) can significantly outperform (4.35) for diagnosing faults. To distinguish the *one-model* PCA with the *multi-model* PCA, we will refer to the one-model PCA as **PCA1** and the multi-model PCA as **PCAm** throughout the book.

Assuming that the overall likelihood for all fault classes is the same and the sample covariance matrix of $P_i\mathbf{x}$ for all classes is the same, the use of the score discriminant (4.35) reduces to use of the T_i^2 statistic, where

$$T_i^2 = \mathbf{x}^T P_i \Sigma_{a,i}^{-2} P_i^T \mathbf{x} \tag{4.36}$$

(similarly as shown in Section 3.2). In this case, the score discriminant will select the fault class as that which corresponds to the minimum T_i^2 statistic.

Assuming that the important variations in discriminating between the faults are contained in the residual space for each fault class, it is most likely that an observation is represented by the fault class i with the minimum **residual discriminant**

$$Q_i/(Q_\alpha)_i \tag{4.37}$$

where the subscript i indicates fault class i. If the important variations in discriminating between the faults are contained both within the score and residual space, then an observation is most likely to be represented by the fault class i with the minimum **combined discriminant**

$$c_i[T_i^2/(T_\alpha^2)_i] + (1 - c_i)[Q_i/(Q_\alpha)_i] \tag{4.38}$$

where c_i is a weighting factor between 0 and 1 for fault class i. Assuming an out-of-control observation does not represent a new fault, each of these discriminant analysis techniques (4.35), (4.37), and (4.38) can be used to diagnose the fault.

When a fault is diagnosed as fault i, it is *not* likely to represent a new fault when

$$[T_i^2/(T_\alpha^2)_i] << 1 \tag{4.39}$$

and

$$[Q_i/(Q_\alpha)_i] << 1. \tag{4.40}$$

These conditions indicate that the observation is a good match to fault model i. If either of these conditions is not satisfied (for example, $[T_i^2/(T_\alpha^2)_i]$ or $[Q_i/(Q_\alpha)_i]$ is greater than 1), then the observation is not accurately represented by fault class i and it is likely that the observation represents a new fault.

Before the application of a pattern classification system to a fault diagnosis scheme, it is useful to assess the likelihood of successful diagnosis. In [270, 272], Raich and Cinar describe a quantitative measure of similarity between the covariance structures of two classes. The measure, referred to as the **similarity index**, for Classes 1 and 2 is calculated as

$$f = \frac{1}{m}\sum_{j=1}^{m} \tilde{\sigma}_j \tag{4.41}$$

where $\tilde{\sigma}_j$ is the jth singular value of $V_1^T V_2$ and the matrices V_1 and V_2 contain all m loading vectors for Classes 1 and 2, respectively. The value of f ranges between 0 and 1, where a value near 0 indicates a *lack of similarity* and a value equal to 1 indicates an *exact similarity* [179]. While a high similarity does not guarantee misdiagnosis, a low similarity does generally indicate a low probability of misdiagnosis. The similarity index can be applied to PCA models by replacing V_1 and V_2 with the loading matrix P_1 for Class 1 and the loading matrix P_2 for Class 2, respectively.

In [270, 272], a measure of class similarity using the overlap of the mean for one class into the score space of another class is developed from [212]. Define $\mu_1 \in \mathcal{R}^m$ and $\mu_2 \in \mathcal{R}^m$ to be the means of Classes 1 and 2, respectively, $P \in \mathcal{R}^{m \times a}$ as the projection matrix containing the a loading vectors for Class 2, ρ as the fraction of the explained variance in the data used to build the second PCA model, and $\bar{\Sigma} \in \mathcal{R}^{a \times a}$ as the covariance in a model directions for the second PCA model. The test statistic, referred to as the **mean overlap**, for Classes 1 and 2 is

$$m = \frac{\mathbf{r}^T\mathbf{r}}{(1-\rho)\mathbf{t}^T\bar{\Sigma}^{-1}\mathbf{t}} \tag{4.42}$$

where $\mathbf{t} = P^T(\mu_1 - \mu_2)$ is the approximation of μ_1 by the second model and $\mathbf{r} = P\mathbf{t} - \mu_1$ is the residual error in μ_1 unexplained by the second model. The threshold for (4.42) can be determined from the following distribution

$$m_\alpha = F_\alpha(m-a, n-a) \tag{4.43}$$

where n is the number of model observations for Class 2. In simulations, Raich and Cinar found that the mean overlap was not as successful as the similarity index for indicating pairwise misdiagnosis [270, 272].

Multiple faults occurring within the same time window are likely to happen for many industrial processes. The statistics for *detecting* a single fault are directly applicable for detecting multiple faults because the threshold in (4.14) depends only on the data from the normal operating conditions (Fault 0). The task of *diagnosing* multiple faults is rather challenging and the proficiencies of the fault diagnosis statistics depend on the nature of the combination of the faults. A straightforward approach for diagnosing multiple faults is to introduce new models for each combination of interest; this approach could describe combinations of faults that produce models that are not simply the consensus of component models [270, 272]. The disadvantage of this approach is that the number of combinations grows exponentially with the number of faults. For a detailed discussion of diagnosing multiple faults, refer to the journal articles [270, 272].

4.7 Dynamic PCA

The PCA monitoring methods discussed previously assume implicitly that the observations at one time instant are statistically independent to observations at past time instances. For typical industrial processes, this assumption is valid only for long sampling times, *i.e.*, 2 to 12 hours. This suggests that a method taking into account the serial correlations in the data is needed in order to implement a process monitoring method with fast sampling times. A simple method to check whether correlations are present in the data is through the use of an autocorrelation chart of the principal components [272, 336]. If significant autocorrelation is shown in the autocorrelation chart, the following approaches can be used. One approach to address this issue is to incorporate EWMA/CUSUM charts with PCA (see Section 4.8). Another approach is to average the measurements over a number of data points. Alternatively, PCA can be used to take into account the serial correlations by augmenting each observation vector with the previous h observations and stacking the data matrix in the following manner,

$$X(h) = \begin{bmatrix} \mathbf{x}_t^T & \mathbf{x}_{t-1}^T & \cdots & \mathbf{x}_{t-h}^T \\ \mathbf{x}_{t-1}^T & \mathbf{x}_{t-2}^T & \cdots & \mathbf{x}_{t-h-1}^T \\ \vdots & \vdots & \ddots & \vdots \\ \mathbf{x}_{t+h-n}^T & \mathbf{x}_{t+h-n-1}^T & \cdots & \mathbf{x}_{t-n}^T \end{bmatrix} \tag{4.44}$$

where $\mathbf{x}_t^T$ is the m-dimensional observation vector in the training set at time interval t. By performing PCA on the data matrix in (4.44), a multivariate **autoregressive** (AR), or ARX model if the process inputs are included, is extracted directly from the data [183, 343]. To see this, consider a simple example of a single input single output (SISO) process, which is described by the ARX(h) model

$$y_t = \alpha_1 y_{t-1} + \cdots + \alpha_h y_{t-h} + \beta_0 u_t + \beta_1 u_{t-1} + \cdots + \beta_h u_{t-h} + e_t \tag{4.45}$$

where y_t and u_t are the output and input at time t, respectively, $\alpha_1, \ldots, \alpha_h, \beta_1, \ldots, \beta_h$ are constant coefficients, and e_t is a white noise process with zero mean [336, 343]. Mathematically, the ARX(h) model states that the output at time t is linearly related to the past h inputs and outputs. With $\mathbf{x}_t^T = [y_t \;\; u_t]$, the matrix $X(h)$ in (4.44) becomes:

$$X(h) = \begin{bmatrix} y_t & u_t & y_{t-1} & u_{t-1} & \cdots & y_{t-h} & u_{t-h} \\ y_{t-1} & u_{t-1} & y_{t-2} & u_{t-2} & \cdots & y_{t-h-1} & u_{t-h-1} \\ \vdots & \vdots & \vdots & \vdots & \ddots & \vdots & \vdots \\ y_{t+h-n} & u_{t+h-n} & y_{t+h-n-1} & u_{t+h-n-1} & \cdots & y_{t-n} & u_{t-n} \end{bmatrix}. \tag{4.46}$$

The ARX(h) model indicates that the first column of $X(h)$ is linearly related to the remaining columns. In the noise-free case the matrix formed in (4.46) would be rank deficient (*i.e.*, not full rank). When PCA is applied to $X(h)$ using (4.3), the eigenvector corresponding to the zero eigenvalue would reveal the ARX(h) correlation structure [183]. In the case where noise is present, the matrix will be nearly rank deficient. The eigenvector corresponding to a nearly zero eigenvalue will be an approximation of the ARX(h) correlation structure [183, 240].

Note that the Q statistic is then the squared prediction error of the ARX model. If enough lags h are included in the data matrix, the Q statistic is statistically independent from one time instant to the next, and the threshold (4.22) is theoretically justified. This method of applying PCA to (4.44) is referred to as **dynamic PCA** (DPCA). When multi-model PCAm is used with (4.44) for diagnosing faults, it will be referred to as **DPCAm**. Note that a statistically justified method can be used for selecting the number of lags h to include in the data for our studies (see Section 7.5). The method for automatically determining h described in [183] is not used here. Experience indicates that $h = 1$ or 2 is usually appropriate when DPCA is used for process monitoring. The fault detection and diagnosis measures for static PCA generalize directly to DPCA. For fault identification, the measures for each observation variable can be calculated by summing the values of the measures corresponding to the previous h lags.

It has been stated that in practice the presence of serial correlations in the data does not compromise the effectiveness for the static PCA method when there are enough data to represent all the normal variations of the process [171]. Irrespective of this claim, including lags in the data matrix as in (4.44) can result in the PCA representation correlating more information.

Therefore, as long as there are enough data to justify the added dimensionality of including h lags, DPCA is expected to perform better than PCA for detecting faults from serially correlated data, and this has been confirmed by testing PCA and DPCA on the Tennessee Eastman problem [183].

4.8 Other PCA-based Methods

The EWMA and CUSUM charts have been generalized to the multivariate case [56, 202, 207, 258, 351, 116], and these generalizations can be applied to the PCA-based T^2 statistic in (4.12). Applying these methods can result in increased sensitivity and robustness of the process monitoring scheme, as discussed in Section 2.3. EWMA and CUSUM charts use data from consecutive observations. If a large number of observations is required, an increase in the detection delay can be expected.

The process monitoring measures discussed so far are for continuous processes. Process monitoring measures for batch processes have been developed with the most heavily studied being **multiway PCA**, [243, 343, 39]. Multiway PCA is a three-dimensional extension of the PCA approach. The three dimensions of the array represent the observation variables, the time instances, and the batches, respectively, whereas PCA methods for continuous processes contain only two dimensions, the observation variables and the time instances. Details and applications of multiway PCA are provided in the references [243, 343, 39].

PCA is a linear dimensionality reduction technique, which ignores the nonlinearities that may exist in the process data. Industrial processes are inherently nonlinear; therefore, in some cases nonlinear methods for process monitoring may result in better performance compared to the linear methods. Kramer [172] has generalized PCA to the nonlinear case by using autoassociative neural networks (this is called **nonlinear principal component analysis**). Dong and McAvoy [71] have developed a nonlinear PCA approach based on principal curves and neural networks that produce independent principal components. It has been shown that for certain data nonlinearities these nonlinear PCA neural networks are able to capture more variance in a smaller dimension compared to the linear PCA approach. A comparison of three neural network approaches to process monitoring has been made [76]. Neural networks can also be applied in a pattern classification system to capture the nonlinearities in the data. A text on using neural networks as a pattern classifier is **Neural Networks for Pattern Recognition** by Bishop [29]. Although neural networks potentially can capture more information in a smaller-dimensional space than the linear dimensionality reduction techniques, an accurate neural network typically requires much more data and computational time to train, especially for large-scale systems.

4.9 Homework Problems

1. Read an article on the use of multiway PCA (e.g., [39, 93, 243, 343]) and write a report describing in detail how the technique is implemented and applied. Describe how the computations are performed and how the statistics are computed. Formulate both fault detection and diagnosis versions of the algorithm. For what types of processes are these algorithms suited? Provide some hypothetical examples.
2. Describe in detail how to blend PCA with CUSUM and EWMA, including the equations for the thresholds.
3. Read an article on the use of PCA for diagnosing sensor faults (e.g., [77, 239]) and write a report describing in detail how the technique is implemented and applied. Compare and contrast the techniques described in the paper with the techniques described in this book.
4. Read an article on the application of nonlinear PCA (e.g., [172, 71]) and write a report describing in detail how the technique is implemented and applied. Describe how the computations are performed and how the statistics are computed. For what types of processes are these algorithms suited? Provide some hypothetical examples.
5. Prove the properties 1-4 given below Equation 4.6.
6. Section 5 of [145] describes several alternatives to the Q statistic for quantifying deviations outside of those quantified by the T^2 statistic. Describe these statistics in detail, including their thresholds, advantages, and disadvantages. [Note: one of the statistics is closely related to the T_r^2 statistic in Chapter 7.]
7. Apply PCA to the original Class 3 data set reported by Fisher [82], and construct Figure 4.3 including the confidence ellipsoid. Now reapply PCA and reconstruct the figure for the case where the outlier at $(t_1, t_2) = (2.5, 5.6)$ is removed from the Class 3 data set. Compare the confidence ellipsoids obtained in the two cases. Comment on the relative importance of removing the outlier from the Class 3 data set before applying PCA.
8. Read the article [100] which describes the use of structured residuals and PCA to isolate and diagnose faults, and write a report describing in detail how the technique is implemented and applied. Compare and contrast the approach with the techniques described in this book.

5. Fisher Discriminant Analysis

5.1 Introduction

In the pattern classification approach to fault diagnosis outlined in Chapter 3, it was described how the dimensionality reduction of the feature extraction step can be a key factor in reducing the misclassification rate when a pattern classification system is applied to new data (data independent of the training set). The dimensionality reduction is especially important when the dimensionality of the observation space is large while the numbers of observations in the classes are relatively small. A PCA approach to dimensionality reduction was discussed in the previous chapter. Although PCA contains certain optimality properties in terms of fault detection, it is not as well-suited for fault diagnosis because it does not take into account the information between the classes when determining the lower-dimensional representation. **Fisher discriminant analysis** (FDA), a dimensionality reduction technique that has been extensively studied in the pattern classification literature, takes into account the information between the classes and has advantages over PCA for fault diagnosis [46, 277].

This chapter begins in Section 5.2 by defining FDA and presenting some of its optimality properties for pattern classification. An information criterion for FDA is developed in Section 5.3 for automatically determining the order of dimensionality reduction. In Section 5.4, it is described how FDA can be used for fault detection and diagnosis. PCA and FDA are compared in Section 5.5 both theoretically and in application to some data sets. Section 5.6 describes **dynamic FDA** (DFDA), an approach based on FDA that takes into account serial (temporal) correlations in the data.

5.2 Fisher Discriminant Analysis

For fault diagnosis, data collected from the plant during specific faults are categorized into classes, where each class contains data representing a particular fault. FDA is a linear dimensionality reduction technique, optimal in terms of maximizing the separation amongst these classes [74]. It determines a set of linear transformation vectors, ordered in terms of maximizing the scatter between the classes while minimizing the scatter within each class.

Define n as the number of observations, m as the number of measurement variables, p as the number of classes, and n_j as the number of observations in the jth class. Represent the vector of measurement variables for the ith observation as $\mathbf{x_i}$. If the training data for all classes have already been stacked into the matrix $X \in \mathcal{R}^{n \times m}$ as in (2.5), then the transpose of the ith row of X is the column vector $\mathbf{x_i}$.

To understand Fisher discriminant analysis, first we need to define various matrices that quantify the total scatter, the scatter within classes, and the scatter between classes. The **total-scatter matrix** is [74, 129]

$$S_t = \sum_{i=1}^{n} (\mathbf{x_i} - \bar{\mathbf{x}})(\mathbf{x_i} - \bar{\mathbf{x}})^T \tag{5.1}$$

where $\bar{\mathbf{x}}$ is the **total mean vector**

$$\bar{\mathbf{x}} = \frac{1}{n} \sum_{i=1}^{n} \mathbf{x_i}. \tag{5.2}$$

With $\mathcal{X}_j$ defined as the set of vectors $\mathbf{x_i}$ which belong to the class j, the **within-scatter matrix** for class j is

$$S_j = \sum_{\mathbf{x_i} \in \mathcal{X}_j} (\mathbf{x_i} - \bar{\mathbf{x}}_\mathbf{j})(\mathbf{x_i} - \bar{\mathbf{x}}_\mathbf{j})^T \tag{5.3}$$

where $\bar{\mathbf{x}}_\mathbf{j}$ is the mean vector for class j:

$$\bar{\mathbf{x}}_\mathbf{j} = \frac{1}{n_j} \sum_{\mathbf{x_i} \in \mathcal{X}_j} \mathbf{x_i}. \tag{5.4}$$

The **within-class-scatter matrix** is

$$S_w = \sum_{j=1}^{p} S_j \tag{5.5}$$

and the **between-class-scatter matrix** is

$$S_b = \sum_{j=1}^{p} n_j (\bar{\mathbf{x}}_\mathbf{j} - \bar{\mathbf{x}})(\bar{\mathbf{x}}_\mathbf{j} - \bar{\mathbf{x}})^T. \tag{5.6}$$

The total-scatter matrix is equal to the sum of the between-scatter matrix and the within-scatter matrix [74],

$$S_t = S_b + S_w. \tag{5.7}$$

The objective of the first FDA vector is to maximize the scatter between classes while minimizing the scatter within classes:

$$\max_{\mathbf{v} \neq 0} \frac{\mathbf{v}^T S_b \mathbf{v}}{\mathbf{v}^T S_w \mathbf{v}} \tag{5.8}$$

assuming invertible S_w where $\mathbf{v} \in \mathcal{R}^m$. The second FDA vector is computed so as to maximize the scatter between classes while minimizing the scatter within classes among all axes perpendicular to the first FDA vector, and so on for the remaining FDA vectors. It can be shown that the linear transformation vectors for FDA can be calculated by computing the stationary points of the optimization problem (5.8) [74, 129]. The FDA vectors are equal to the eigenvectors $\mathbf{w_k}$ of the generalized eigenvalue problem

$$S_b \mathbf{w_k} = \lambda_k S_w \mathbf{w_k} \tag{5.9}$$

where the eigenvalues λ_k indicate the degree of overall separability among the classes by projecting the data onto $\mathbf{w_k}$. Any software package that does matrix manipulations, such as MATLAB [109, 110] or IMSL [132], has subroutines for computing the generalized eigenvalues and eigenvectors. Because the direction and not the magnitude of $\mathbf{w_k}$ is important, the Euclidean norm (square root of the sum of squares of each element) of $\mathbf{w_k}$ can be chosen to be equal to 1 ($||\mathbf{w_k}|| = 1$).

The FDA vectors can be computed from the generalized eigenvalue problem as long as S_w is invertible. This will almost always be true provided that the number of observations n is significantly larger than the number of measurements m (the case in practice). Since S_w is expected to be invertible for applications of FDA to fault diagnosis, methods to calculate the FDA vectors for the case of non-invertible S_w are only cited here [45, 123, 305].

The first FDA vector is the eigenvector associated with the largest eigenvalue, the second FDA vector is the eigenvector associated with the second largest eigenvalue, and so on. A large eigenvalue λ_k indicates that when the data in the classes are projected onto the associated eigenvector $\mathbf{w_k}$ there is overall a large separation of the class means relative to the class variances, and consequently, a large degree of separation among the classes along the direction $\mathbf{w_k}$. Since the rank of S_b is less than p, there will be at most $p-1$ eigenvalues which are not equal to zero, and FDA provides useful ordering of the eigenvectors only in these directions.

It is useful to write the goal of FDA more explicitly in terms of a linear transformation. Define the matrix $W_p \in \mathcal{R}^{m \times (p-1)}$ with the $p-1$ FDA vectors as columns. Then the linear transformation of the data from m-dimensional space to $(p-1)$-dimensional space is described by

$$\mathbf{z_i} = W_p^T \mathbf{x_i} \tag{5.10}$$

where $\mathbf{z_i} \in \mathcal{R}^{(p-1)}$. FDA computes the matrix W_p such that data $\mathbf{x_1}, \ldots, \mathbf{x_n}$ for the p classes are optimally separated when projected into the $p-1$ dimensional space. In the case where p is equal to 2, this is equivalent to projecting the data onto a line in the direction of the vector $\mathbf{w}$, for which the projected data are the best separated.

5.3 Reduction Order

No reduction of dimensionality would be needed if the covariance matrix and mean vector were known exactly (see Section 3.3). Errors in the sample covariance matrix (2.6) occur in practice, however, and the dimensionality reduction provided by FDA may be necessary to reduce the misclassification rate when the pattern classification system is applied to new data (data independent of the training set). A popular method for selecting the reduction order for dimensionality reduction methods is to use **cross-validation** [98, 343]. This approach separates the data into multiple sets: the training set, and the testing (or validation) set. The dimensionality reduction procedure is applied to the data in the training set, and then its performance is evaluated by applying the reduced-dimension model to the data in the testing set for each reduction order. The reduction order is selected to optimize the performance based on the testing set. For example, if the goal is fault diagnosis, the order of the reduced model would be specified by minimizing the misclassification rate of the testing set.

Cross-validation is not always practical in fault diagnosis applications because there may not be enough data to separate into two sets. In this situation, it is desirable to determine the order of the dimensionality reduction using all the data in the training set. Variations on cross-validation that split the data into larger numbers of sets (such as "leave-one-out" cross-validation [344]) are computationally expensive.

As discussed in Section 3.3, the error of a model can be minimized by choosing the number of independent parameters so that it optimally trades off the bias and variance contributions on the mean-squared error. In an effort to minimize the mean-squared error, criteria in the form

$$(\text{prediction error term}) + (\text{model complexity term}) \tag{5.11}$$

have been minimized to determine the appropriate model order [199]. The **Akaike's information criterion** (AIC), popularly applied in system identification for optimally selecting the model order (for an example, see Section 7.6), can be derived in the form (5.11) [199]. In (5.11), the **prediction error term** is a function of the estimated model parameters and the data in the training set, and the **model complexity term** is a function of the number of independent parameters and the amount of data in the training set. In system identification, the prediction error term is usually chosen as the average squared prediction error for the model, but in general, the choice of the complexity term is subjective [199].

A strength of the AIC is that it relies only on information in one set of data (the training data), unlike cross-validation which requires either additional data or a partitioning of the original data set. A criterion in the form (5.11) can be developed for automatically selecting the order for FDA using the information only in the training set [46, 277]. The order can be determined by computing the dimensionality a that minimizes the information criterion

$$f_m(a) + \frac{a}{\tilde{n}} \tag{5.12}$$

where $f_m(a)$ is the misclassification rate (the proportion of misclassifications, which is between 0 and 1) for the training set by projecting the data onto the first a FDA vectors, and $\tilde{n}$ is the average number of observations per class. The misclassification rate of the training set, $f_m(a)$, indicates the amount of information contained in the first a FDA vectors beneficial for pattern classification. While the misclassification rate of the training set typically decreases as a increases, for new data (data independent of the training set), the misclassification rate initially decreases and then increases above a certain order due to overfitting the data. The model complexity term $a/\tilde{n}$ is added in (5.12) to penalize the increase of dimensionality.

The scaling of the reduction order a by the average number of observations per class, $\tilde{n}$, has some intuitive implications. To illustrate this, consider the case where the number of observations in each class is the same, $n_j = \tilde{n}$. It can be shown using some simple algebra that the inclusion of the $a/\tilde{n}$ term in (5.12) ensures that the order selection procedure produces a value for a less than or equal to $\tilde{n}$. In words, this constraint prevents the lower-dimensional model from having a higher dimensionality than justified by the number of observations in each class.

The model complexity term $a/\tilde{n}$ can also be interpreted in terms of the total number of misclassifications per class. Defining $m(a)$ as the total number of misclassifications in the training set for order a and assuming that $n_j = \tilde{n}$, the information criterion (5.12) can be written as

$$\frac{m(a)}{p\tilde{n}} + \frac{a}{\tilde{n}} \tag{5.13}$$

where $n = p\tilde{n}$ is the total number of observations. Let us consider the case where it is to be determined whether a reduction order of $a + 1$ should be preferred over a reduction order of a. Using the information criterion (5.13) and recalling that a smaller value for the information criterion is preferred, a reduction order of $a + 1$ is preferred if

$$\frac{m(a+1)}{p\tilde{n}} + \frac{a+1}{\tilde{n}} < \frac{m(a)}{p\tilde{n}} + \frac{a}{\tilde{n}}. \tag{5.14}$$

This is equivalent to

$$\frac{m(a)}{p} - \frac{m(a+1)}{p} > 1. \tag{5.15}$$

The complexity term does not allow the reduction order to be increased merely by decreasing the number of misclassifications, but only if the decrease in the total number of misclassifications *per class* is greater than 1.

The above analyses indicate that the scaling of a in the model complexity term $a/\tilde{n}$ in the information criterion (5.12) is reasonable. This is confirmed by application in Chapter 10 (for example, see Figure 10.21, where the information criterion correctly captures the shape and slope of the misclassification rate curves for the testing sets).

5.4 Fault Detection and Diagnosis

When FDA is applied for pattern classification, the dimensionality reduction technique is applied to the data in *all* the classes simultaneously. More precisely, denote $W_a \in \mathcal{R}^{m \times a}$ as the matrix containing the eigenvectors $\mathbf{w_1}, \mathbf{w_2}, \ldots, \mathbf{w_a}$ computed from (5.9). The discriminant function can be derived from (3.6) to be [97]

$$g_j(\mathbf{x}) = -\frac{1}{2}(\mathbf{x} - \bar{\mathbf{x}}_\mathbf{j})^T W_a \left(\frac{1}{n_j - 1} W_a^T S_j W_a\right)^{-1} W_a^T (\mathbf{x} - \bar{\mathbf{x}}_\mathbf{j}) + \ln(p_i)$$
$$-\frac{1}{2} \ln \left[\det \left(\frac{1}{n_j - 1} W_a^T S_j W_a\right)\right] \tag{5.16}$$

where S_j, $\bar{\mathbf{x}}_\mathbf{j}$, and n_j are defined in (5.3) and (5.4). In contrast to PCA1 (see Section 4.6), FDA uses the class information to compute the reduced-dimensional space, so that the discriminant function (5.16) exploits that class information to a far greater degree than can be done by PCA. In contrast to PCAm, FDA utilizes *all* p fault class information when evaluating the discriminant function or each class.

FDA can also be applied to *detect* faults by defining an additional class of data, that collected during normal operating conditions, to the fault classes. The proficiency of fault detection using (5.16) depends on the similarity between the data from the normal operating conditions and the data from the fault classes in the training sets. When there exists a transformation W such that the data from the normal operating conditions can be reasonably separated from the other fault classes, using FDA for fault detection will produce small missed detection rates for the known fault classes. Equation 5.16 does not take into account unknown faults associated with data outside of the lower-dimensional space defined by the FDA vectors, so (5.16) may not detect these kinds of faults. It is best to use (5.16) with a residual-based FDA statistic (see Homework Problem 2), which together can detect both faults associated with data inside the space defined by the FDA vectors, and faults associated with data outside of this space. This joint use of two FDA statistics is similar to the joint use of the PCA Q or T^2 statistics, as discussed in Chapter 4. The advantage of using the FDA statistics instead of the PCA statistics is that the fault classification information can be taken into account to improve the ability to detect faults. The disadvantage is that the FDA statistics require that fault classification information to define its lower-dimensional space (defined by W).

As mentioned in Section 5.2, only the first $p - 1$ eigenvectors in FDA maximize the scatter between the classes while minimizing the scatter within each class. The rest of the $m-p+1$ eigenvectors corresponding to the zero eigenvalues are not ordered by the FDA objective (5.8). The ranking of these generalized eigenvectors is determined by the particular software package implementing the eigenvalue decomposition algorithm, which does not order the eigenvectors in a manner necessarily useful for classification. However, more than $p-1$ dimensions in a lower-dimensional space may be useful for classification, and a procedure to select vectors beyond the first $p-1$ FDA vectors can be useful. Here two methods are described which use PCA to compute additional vectors for classification.

One method is to use FDA for the space defined by the first $p-1$ eigenvectors, and to use the PCA1 vectors for the rest of the $m-p+1$ vectors, ordered from the PCA vectors associated with the highest variability to the vectors associated with the lower variability. If the reduction order $a \leq p-1$, Equation 5.16 is used directly. If $a \geq p$, the alternative discriminant function is used:

$$\begin{aligned} g_j(\mathbf{x}) = & -\tfrac{1}{2}(\mathbf{x}-\bar{\mathbf{x}}_\mathbf{j})^T W_{mix,a}\left(\frac{1}{n_j-1} W_{mix,a}^T S_j W_{mix,a}\right)^{-1} W_{mix,a}^T(\mathbf{x}-\bar{\mathbf{x}}_\mathbf{j}) \\ & -\tfrac{1}{2}\ln\left[\det\left(\frac{1}{n_j-1} W_{mix,a}^T S_j W_{mix,a}\right)\right] + \ln(p_i) \end{aligned} \tag{5.17}$$

where $W_{mix,a} = [W_{p-1}\ P_{a-p+1}]$, and P_{a-p+1} is the first $a-p+1$ columns of the PCA1 loading matrix P (defined in Section 4.6). When this method is used for diagnosing faults, it will be referred to as the **FDA/PCA1 method.** Recall from Section 4.2 that the variances associated with the loading vectors in PCA are ranked in descending order. Given that the vectors from PCA1 can be useful in a classification procedure (see Section 4.6), incorporating the first $a-p+1$ PCA1 loading vectors into the FDA/PCA1 method may provide additional information for discriminating amongst classes.

Another method to define an additional $m-p+1$ vectors is to apply PCA1 to the residual space of FDA, defined by

$$R = X(I - W_{p-1}W_{p-1}^T). \tag{5.18}$$

As before, if the reduction order $a \leq p-1$, Equation 5.16 is used directly. If $a \geq p$, then the alternative discriminant function (5.17) is used with $W_{mix,a} = [W_{p-1}\ \bar{P}_{a-p+1}]$, where $\bar{P}_{a-p+1}$ is the first $a-p+1$ columns of the PCA1 loading matrix when PCA is applied to R. This method for diagnosing faults will be referred to as the **FDA/PCA2 method.**

5.5 Comparison of PCA and FDA

Here the PCA and FDA dimensionality reduction techniques are compared via theoretical and graphical analyses for the case where PCA is applied to all the data in all the classes together (PCA1 in Section 4.6). This highlights the geometric differences between the two dimensionality reduction procedures. The way in which FDA can result in superior fault diagnosis to that achieved by PCA is also shown.

The optimization problems for PCA and FDA have been stated mathematically in (4.1) and (5.8), respectively. It can be shown that the PCA loading vectors and FDA vectors can also be calculated by computing the stationary points of the optimization problems

$$\max_{\mathbf{v}\neq 0} \frac{\mathbf{v}^T S_t \mathbf{v}}{\mathbf{v}^T \mathbf{v}} \tag{5.19}$$

and

$$\max_{\mathbf{v}\neq 0} \frac{\mathbf{v}^T S_t \mathbf{v}}{\mathbf{v}^T S_w \mathbf{v}}, \tag{5.20}$$

respectively. Equations 5.19 and 5.20 indicate that the PCA and FDA vectors are identical for the case when $S_w = \sigma I$ where $\sigma > 0$. One case in which this situation occurs if the data in each class can be described by a uniformly distributed ball (*i.e.*, circle in 2-D space and sphere in 3-D space), even if the balls are of distinct sizes. Differences between the two techniques can occur only if there is elongation in the data used to describe any one of the classes. These elongated shapes occur for highly correlated data sets (see Figure 4.3), typical for data collected from industrial processes. Therefore, when PCA and FDA are applied in the same manner to process data, the PCA loading vectors and FDA vectors are expected to be significantly different, and the differing objectives, (5.19) and (5.20), suggest that FDA will be significantly better for discriminating among classes of faults.

Figure 5.1 illustrates a difference between PCA and FDA that can occur when the distribution of the data in the classes is somewhat elongated. The first FDA vector and PCA loading vector are nearly perpendicular, and the linear transformation of the data onto the first FDA vector is much better able to separate the data in the two classes than the linear transformation of the data onto the first PCA loading vector.

The linear transformations of Fisher's data (introduced in Chapter 4) onto the first two PCA and FDA loading vectors are shown in Figure 5.2. The within-class-scatter matrix and between-class-scatter matrix are calculated as

$$S_w = \begin{bmatrix} 56.8 & 37.3 & 16.4 & 9.17 \\ 37.3 & 88.4 & 10.1 & 17.1 \\ 16.4 & 10.1 & 8.75 & 4.64 \\ 9.17 & 17.1 & 4.64 & 22.8 \end{bmatrix} \tag{5.21}$$

and

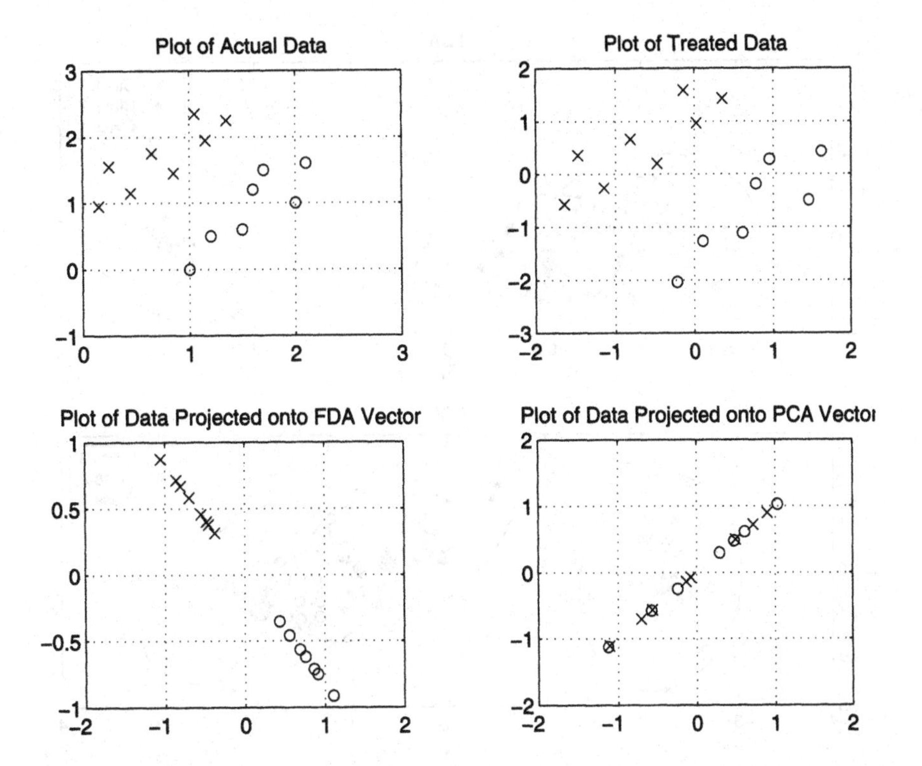

Fig. 5.1. A comparison of PCA and FDA for the linear transformation of the data in classes 'x' and 'o' onto the first FDA vector and PCA loading vector

$$S_b = \begin{bmatrix} 92.2 & -55.7 & 113 & 108 \\ -55.7 & 60.6 & -75.3 & -65.6 \\ 113 & -75.2 & 140 & 133 \\ 108 & -65.6 & 132 & 126 \end{bmatrix}, \tag{5.22}$$

respectively. Solving (5.9), we have $p - 1 = 2$ eigenvectors associated with non-zero eigenvalues, which are

$$\mathbf{w_1} = \begin{bmatrix} 0.15 \\ 0.12 \\ -0.96 \\ -0.18 \end{bmatrix} \tag{5.23}$$

and

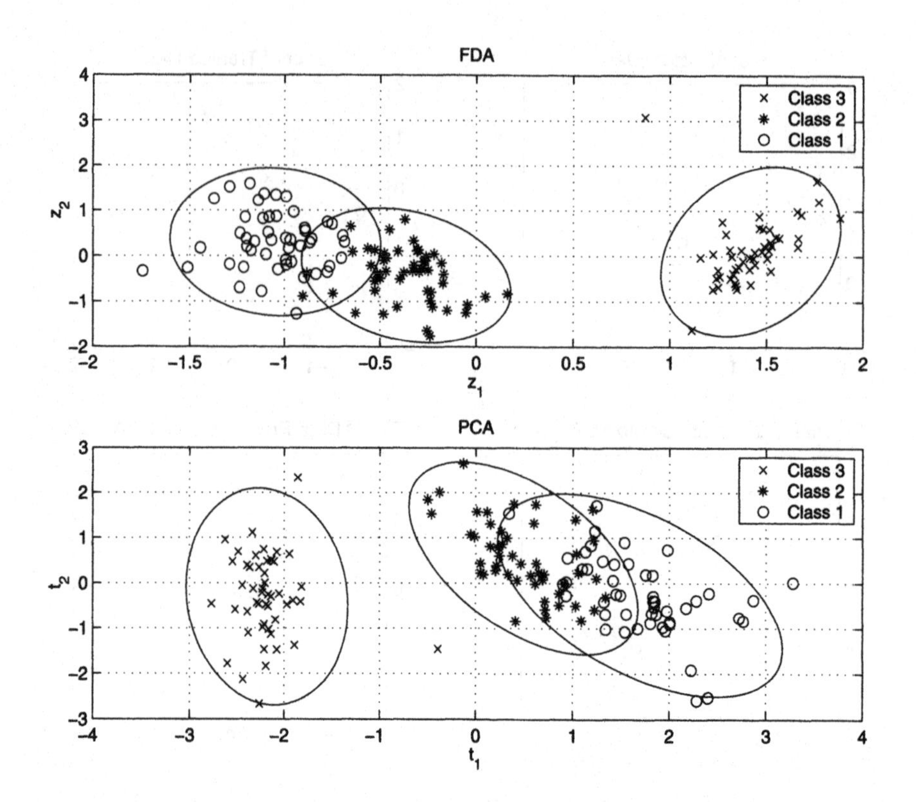

Fig. 5.2. The linear transformations of Fisher's data [45, 82] for three classes onto the first two FDA and PCA loading vectors, respectively

$$\mathbf{w_2} = \begin{bmatrix} -0.13 \\ -0.70 \\ -0.15 \\ 0.68 \end{bmatrix}, \tag{5.24}$$

and the corresponding eigenvalues are $\lambda_1 = 27$ and $\lambda_2 = 0.24$, respectively. The large λ_1 value indicates that there is a large separation of the class means relative to the class variances on z_1 (see Figure 5.2). Indeed the average values of z_1 for the 3 classes are -1.0, -0.37, and 1.42. The small λ_2 value indicates that the overall separation of the class means relative to the class variances is small in the z_2 direction. The average values of z_2 for the 3 classes are 0.30, -0.43, and 0.12.

The 95% elliptical confidence region for each class can be approximated by solving (3.8) with T_i^2 set to 6.64. The T_i^2 threshold is the same as in the example we showed in Chapter 4. Data falling in the intersection of the two elliptical confidence regions can result in misclassification. The degree of overlap between the confidence regions for Classes 1 and 2 is greater for PCA than for FDA (49 points vs. 17 points), indicating that the misclassification rates for PCA would be higher.

While the elliptical confidence region can be used to illustrate the qualitative classification performance for FDA, the discriminant function (5.16) can be used to determine the exact misclassification rates for the experiment data [82, 45]. The results are illustrated in Table 5.1 for different FDA reduction orders. Although Class 1 and Class 2 data overlap to some extent (see Figure 5.2), the discriminant function (5.16) is able to classify almost all of the data points correctly. Indeed, no more than 3 out of 50 data points are misclassified regardless of the order selection (see Table 5.1).

Table 5.1. The misclassification rates for the training data [45, 82] for FDA

Order (a)	1	2	3	4
Class 1 Misclassifications	0.06	0.02	0.02	0.02
Class 2 Misclassifications	0.06	0.06	0.04	0.06
Class 3 Misclassifications	0	0	0	0
Overall Misclassifications	0.04	0.027	0.02	0.027

This example is effective at illustrating the difference in the objectives between PCA and FDA. By comparing the limits of the horizontal and vertical axes and visually inspecting the data, it is clear that the span of the PCA linear transformation is larger than the FDA linear transformation. While PCA is better able to separate the data as a whole, FDA is better able to separate the data among the classes (*, o, x). This is evident in the degree of overlap between '*' and 'o' data regions in the two plots, in which the data points '*' and 'o' barely overlap for the FDA linear transformation, while there is a clear intermingling of data for the PCA linear transformation.

All of Fisher's data was used for training the FDA and PCA models in the previous example. A much more accurate comparison of PCA and FDA is to train the techniques with one data set (the training data), then apply them to a new data set (the testing data). In this example two fifth of Fisher's data (20 observations for each class, for a total of 60 observations) were used for training, while the rest of the data (30 observations for each class, for a total of 90 observations) were used for testing. The overall misclassification rates of the training data and testing data using the data-driven fault diagnosis methods are shown in Table 5.2 and 5.3, respectively.

The overall misclassification rates for FDA, FDA/PCA1, and FDA/PCA2 were the same at a given reduction order (Section 10.8 has an example where FDA/PCA1 and FDA/PCA2 produce lower overall misclassification rates). The FDA vectors corresponding to the two non-zero eigenvalues are very effective in discriminating the three classes. At $a = 2$, the overall misclassi-

Table 5.2. Overall misclassification rates for the training data [45, 82] using several data-driven fault diagnosis methods (60 observations in the training set)

Order (a)	1	2	3	4
FDA	0	0	0	0
FDA/PCA1	0	0	0	0
FDA/PCA2	0	0	0	0
PCA1	0.083	0.033	0.020	0
PCAm	0.28	0.20	0.15	0.13
MS	–	–	–	0

Table 5.3. Overall misclassification rates of the testing data [45, 82] using several data-driven fault diagnosis methods (60 observations in the training set)

Order (a)	1	2	3	4
FDA	0.067	0.067	0.078	0.033
FDA/PCA1	0.067	0.067	0.078	0.033
FDA/PCA2	0.067	0.067	0.078	0.033
PCA1	0.10	0.10	0.044	0.033
PCAm	0.17	0.18	0.11	0.11
MS	–	–	–	0.033

fication rate for the testing set is 0.0667 (*i.e.*, 84 out of 90 data points were correctly classified).

For $a < p$, the FDA methods had a lower overall misclassification rate than either PCA method. This agrees with earlier comments that FDA can do a much better job at diagnosing faults than PCA, especially at lower reduction orders. At any reduction order, PCA1 gave lower overall misclassification rates than PCAm. This supports our discussion in Section 4.6 that PCA1 will usually produce a better PCA representation for diagnosing faults. For $a = 4$, all of the methods, except for PCAm, gave the same overall misclassification rates. As discussed in Section 4.6, MS is the same as PCA1 when all orders are included. This does not always hold for the FDA methods.

To illustrate the dependence of the number of data points used in the training set on the proficiency of classification, another example was run using 120 observations in the training set and 30 observations in the testing set. The overall misclassification rates for the training data and testing data are shown in Table 5.4 and 5.5, respectively.

This example shows that, with more data points in the training set, the overall misclassification rates in the testing set for all methods are significantly lower. This example shows the same trends that all of the FDA methods outperforms the PCA methods, and that PCA1 outperforms PCAm.

Note that this data set is a relatively small-scale example, in which dimensionality reduction was not necessary for providing low misclassification rates. The benefit of dimensionality reduction is most apparent for the classification of new data from *large-scale systems*, in which training data are insufficient

Table 5.4. Overall misclassification rates for the training data [45, 82] using several data-driven fault diagnosis methods (120 observations in the training set)

Order (a)	1	2	3	4
FDA	0.05	0.033	0.033	0.033
FDA/PCA1	0.05	0.033	0.033	0.033
FDA/PCA2	0.05	0.033	0.033	0.033
PCA1	0.092	0.10	0.050	0.033
PCAm	0.20	0.19	0.067	0.067
MS	–	–	–	0.033

Table 5.5. Overall misclassification rates for the testing data [45, 82] using several data-driven fault diagnosis methods (120 observations in the training set)

Order (a)	1	2	3	4
FDA	0	0	0	0
FDA/PCA1	0	0	0	0
FDA/PCA2	0	0	0	0
PCA1	0.033	0.033	0	0
PCAm	0.067	0.067	0.067	0.067
MS	–	–	–	0

(practical case in industry). Applications of the methods to simulated plant data in Chapter 10 illustrate this point.

5.6 Dynamic FDA

As mentioned in Section 4.8, CUSUM and EWMA charts can be used to capture the serial correlations in the data for PCA. CUSUM and EWMA charts can also be generalized for FDA. The pattern classification method for fault diagnosis discussed in Chapter 3 and Section 5.4 can be extended to take into account the serial (temporal) correlations in the data, by augmenting the observation vector and stacking the data matrix in the same manner as (4.44). This method will be referred to as **dynamic FDA** (DFDA). This enables the pattern classification system to use more information in classifying the observations. Since the information contained in the augmented observation vector is a superset of the information contained in a single observation vector, it is expected from a theoretical point of view that the augmented vector approach can result in better performance. However, the dimensionality of the problem is increased by stacking the data, where the magnitude of the increase depends on the number of lags h. This implies that more data may be required to determine the mean vector and covariance matrix to the same level of accuracy for each class. In practice, augmenting the observation vector is expected to perform better when there is both significant serial correlation and there are enough data to justify the larger dimensionality. Since the amount of data n is usually fixed, performing dimensionality reduction using FDA becomes even more critical to the pattern classification system when the number of lags h is large. The application of FDA/PCA1 to (4.44) will be referred to as **DFDA/DPCA1**, and the developments in this chapter for FDA readily apply to DFDA and DFDA/DPCA1.

5.7 Homework Problems

1. An unknown fault is a fault that is not represented in the training set. Assume that the known fault classes are augmented with an additional class which contains normal operating data (see Section 5.4). It is possible that using (5.16) by itself can be unable to detect a fault which can be detected by the joint application of the PCA T^2 and Q statistics discussed in Chapter 4. Construct data sets (in which you apply both PCA and FDA) to illustrate the key reasoning underlying this conclusion.
2. Define a residual-based statistic for FDA similar to the Q statistic used in PCA. Would the FDA-based Q statistic be expected to outperform the PCA-based Q statistic for fault detection? Construct data sets (in which you apply both PCA and FDA) to illustrate the key reasoning underlying your conclusions. How does this answer depend on the reduction order for FDA?
3. Derive Equations 5.19 and 5.20.
4. Describe in detail how to blend FDA with CUSUM and EWMA, including the equations for the thresholds.
5. Write a one page technical summary of the classic paper by Fisher on discriminant analysis [82]. Compare the equations derived by Fisher to the equations in this chapter. Explain any significant differences.
6. Peterson and Mattson [256] consider more general criteria for dimensionality reduction. Compare their criteria to the Fisher criterion. What are the advantages and disadvantages of each? For what types of data would you expect one criterion to be preferable over the others?
7. Show that the FDA vectors are not necessarily orthogonal (hint: the easiest way to show this is by example). Compare FDA with PLS and PCA in this respect.

6. Partial Least Squares

6.1 Introduction

Partial least squares (PLS), also known as **projection to latent structures**, is a dimensionality reduction technique for maximizing the covariance between the predictor (independent) matrix X and the predicted (dependent) matrix Y for each component of the reduced space [98, 350]. A popular application of PLS is to select the matrix Y to contain only product quality data which can even include off-line measurement data, and the matrix X to contain all other process variables [207]. Such inferential models (also known as soft sensors) can be used for the on-line prediction of the product quality data [215, 222, 223], for incorporation into process control algorithms [158, 259, 260], as well as for process monitoring [207, 259, 260]. Discriminant PLS selects the matrix X to contain all process variables and selects the Y matrix to focus PLS on the task of fault diagnosis [46].

PLS computes loading and score vectors that are correlated with the predicted block while describing a large amount of the variation in the predictor block [343]. If the predicted block has only one variable, the PLS dimensionality reduction method is known as PLS1; if the predicted block has multiple variables, the dimensionality reduction method is known as PLS2. PLS requires calibration and prediction steps. The most popular algorithm used in PLS to compute the parameters in the calibration step is known as **non-iterative partial least squares** (NIPALS) [98, 343]. Another algorithm, known as SIMPLS, can also be used [62]. As mentioned, the predicted blocks used in discriminant PLS and in other applications of PLS are different. In chemometrics and process control applications, where PLS is most commonly applied, the predicted variables are usually measurements of product quality variables. In pattern classification, where discriminant PLS is used, the predicted variables are dummy variables (1 or 0) where '1' indicates an in-class member while '0' indicates a non-class member [9, 64, 244]. In the prediction step of discriminant PLS, discriminant analysis is used to determine the predicted class [244].

Section 6.2 defines the PLS1 and PLS2 algorithms in enough detail to allow the reader to implement these techniques. Section 6.3 discusses the selection of the reduction order. Section 6.4 discusses fault detection, identification, and diagnosis using PLS. The PLS and PCA techniques are compared in

Section 6.5. Section 6.6 summarizes several variations of the PLS algorithms for process monitoring.

6.2 PLS Algorithms

PLS requires a matrix $X \in \mathcal{R}^{n\times m}$ and a matrix $Y \in \mathcal{R}^{n\times p}$, where m is the number of predictor variables (the number of measurements in each observation), n is the total number of observations in the training set, and p is the number of observation variables in Y. When Y is selected to contain only the product quality variables, then p is the number of product quality variables. When Y is selected as in discriminant PLS, p is the number of fault classes.

In discriminant PLS, diagnosed data are needed in the calibration. To aid in the description of discriminant PLS, the data in X will be ordered in a particular way. With p fault classes, suppose that there are $n_1, n_2, \ldots, n_p$ observations for each variable in Classes 1, 2, ... , p respectively. Collect the training set data into the matrix $X \in \mathcal{R}^{n\times m}$, as shown in (2.5), so that the first n_1 rows contain data from Fault 1, the second n_2 rows contain data from Fault 2, and so on. Altogether, there are $n_1 + n_2 + \cdots + n_p = n$ rows. There are two methods, known as PLS1 and PLS2, to model the predicted block. In PLS1, each of the p predicted variables is modeled separately, resulting in one model for each class. In PLS2, all predicted variables are modeled simultaneously [217].

In PLS2, the predicted block $Y \in \mathcal{R}^{n\times p}$ contains p product quality variables; in discriminant PLS2, the predicted block $Y \in \mathcal{R}^{n\times p}$ is

$$Y = \underbrace{\begin{bmatrix} 1 & 0 & 0 & \cdots & 0 \\ \vdots & \vdots & \vdots & \cdots & \vdots \\ 1 & 0 & 0 & \cdots & 0 \\ 0 & 1 & 0 & \cdots & 0 \\ \vdots & \vdots & \vdots & \cdots & \vdots \\ 0 & 1 & 0 & \cdots & 0 \\ \vdots & \cdots & \ddots & \cdots & \vdots \\ \vdots & \cdots & \ddots & \cdots & \vdots \\ 0 & 0 & 0 & \cdots & 1 \\ \vdots & \vdots & \vdots & \cdots & \vdots \\ 0 & 0 & 0 & \cdots & 1 \end{bmatrix}}_{p \text{ columns}} \tag{6.1}$$

where each column in Y corresponds to a class. Each element of Y is filled with either *one* or *zero*. The first n_1 elements of Column 1 are filled with a '1', which indicates that the first n_1 rows of X are data from Fault 1. In discriminant PLS1, the algorithm is run p times, each with the same X, but for each individual column of Y in (6.1).

As mentioned in Section 2.2, data pretreatment is applied first, so that X and Y are mean-centered and scaled. The matrix X is decomposed into a score matrix $T \in \mathcal{R}^{n \times a}$ and a loading matrix $P \in \mathcal{R}^{m \times a}$, where a is the PLS component (reduction order), plus a residual matrix $E \in \mathcal{R}^{n \times m}$:

$$X = TP^T + E. \tag{6.2}$$

The matrix product TP^T can be expressed as the sum of the product of the **score vectors** $\mathbf{t_j}$ (the jth column of T) and the **loading vectors** $\mathbf{p_j}$ (the jth column of P) [98, 157, 343]:

$$X = \sum_{j=1}^{a} \mathbf{t_j} \mathbf{p_j}^T + \mathbf{E}. \tag{6.3}$$

Similarly, Y is decomposed into a score matrix $U \in \mathcal{R}^{n \times a}$, a loading matrix $Q \in \mathcal{R}^{p \times a}$, plus a residual matrix $\tilde{F} \in \mathcal{R}^{n \times p}$:

$$Y = UQ^T + \tilde{F}. \tag{6.4}$$

The matrix product UQ^T can be expressed as the sum of the product of the score vectors $\mathbf{u_j}$ (the jth column of U) and the loading vectors $\mathbf{q_j}$ (the jth column of Q):

$$Y = \sum_{j=1}^{a} \mathbf{u_j} \mathbf{q_j}^T + \tilde{\mathbf{F}}. \tag{6.5}$$

The decompositions in (6.3) and (6.5) have the same form as that used in PCA (see (4.5)). The matrices X and Y are represented as the sum of a series of rank one matrices. If a is set equal to $\min(m, n)$, then E and $\tilde{F}$ are zero and PLS reduces to ordinary least squares. Setting a less than $\min(m, n)$ reduces noise and collinearity. The goal of PLS is to determine the loading and score vectors which are correlated with Y while describing a large amount of the variation in X.

PLS regresses the estimated Y score vector $\widehat{\mathbf{u}}_\mathbf{j}$ to the X score vector $\mathbf{t_j}$ by

$$\widehat{\mathbf{u}}_\mathbf{j} = b_j \mathbf{t_j} \tag{6.6}$$

where b_j is the regression coefficient. In matrix form, this relationship can be written

$$\widehat{U} = TB \tag{6.7}$$

where $B \in \mathcal{R}^{a \times a}$ is the diagonal regression matrix with $B_{jj} = b_j$, and $\hat{U}$ has $\widehat{\mathbf{u}}_\mathbf{j}$ as its columns. Substituting $\widehat{U}$ from (6.7) in for U in (6.4), and taking into account that this will modify the residual matrix, gives

$$Y = TBQ^T + F \tag{6.8}$$

where F is the prediction error matrix. The matrix B is selected such that the induced 2-norm of F (the maximum singular value of F [104]), $\|F\|_2$, is minimized [157]. The score vectors $\mathbf{t_j}$ and $\mathbf{\hat{u}_j}$ are calculated for each PLS factor ($j = 1, 2, ..., a$) such that the covariance between X and Y is maximized at each factor. In PLS1, similar steps are performed, resulting in

$$\mathbf{y_i} = T_i B_i \mathbf{q_i}^T + \mathbf{f_i} \tag{6.9}$$

where $\mathbf{y_i} \in \mathcal{R}^n$ is the ith column of Y, $T_i \in \mathcal{R}^{n \times a}$ is the score matrix, $B_i \in \mathcal{R}^{a \times a}$ is the regression matrix, $\mathbf{q_i} \in \mathcal{R}^a$ is the loading vector, and $\mathbf{f_i} \in \mathcal{R}^n$ is the prediction error vector. Since there are p columns in Y, the range of i is from 1 to p.

Now if the score and loadings matrices for X and Y were calculated separately, then their successive score vectors could be weakly related to each other, so that the regression (6.6) which relates X and Y would result in a poor reduced dimension relationship. The NIPALS algorithm is an iterative approach to computing modified score vectors so that rotated components result which lead to an improved regression in (6.6). It does this by using the score vectors from Y in the calculation of the score vectors for X, and *vice versa*.

For the case of PLS2, the NIPALS algorithm computes the parameters using (6.10) to (6.20) [98, 157, 343]. The first step is the cross regression of X and Y, which are scaled so as to have zero mean and unit variance for each variable. Initialize the NIPALS algorithm using $E_0 = X$ and $F_0 = Y$, $j = 1$, and $\mathbf{u_j}$ equal to any column of F_{j-1}. Equations 6.10-6.13 are iteratively computed until convergence, which is determined by comparing $\mathbf{t_j}$ with its value from a previous iteration (the nomenclature $\|\cdot\|$ refers to the vector 2-norm, also known as the Euclidean norm).

$$\mathbf{w_j} = \frac{E_{j-1}^T \mathbf{u_j}}{\|E_{j-1}^T \mathbf{u_j}\|} \tag{6.10}$$

$$\mathbf{t_j} = E_{j-1} \mathbf{w_j} \tag{6.11}$$

$$\mathbf{q_j} = \frac{F_{j-1}^T \mathbf{t_j}}{\|F_{j-1}^T \mathbf{t_j}\|} \tag{6.12}$$

$$\mathbf{u_j} = F_{j-1} \mathbf{q_j} \tag{6.13}$$

Proceed to (6.14) if convergence; return to (6.10) if not. Mathematically, determining $\mathbf{t_1}$, $\mathbf{u_1}$, and $\mathbf{w_1}$ from (6.10) to (6.13) is the same as iteratively determining the eigenvectors of XX^TYY^T, YY^TXX^T, and X^TYY^TX associated with the largest eigenvalue, respectively [266, 344].

In the second step, $\mathbf{p_j}$ is calculated as

$$\mathbf{p_j} = \frac{E_{j-1}^T \mathbf{t_j}}{\mathbf{t_j}^T \mathbf{t_j}} \tag{6.14}$$

The final values for $\mathbf{p_j}$, $\mathbf{t_j}$, and $\mathbf{w_j}$ are scaled by the norm of $\mathbf{p_{j,old}}$:

$$\mathbf{p_{j,new}} = \frac{\mathbf{p_{j,old}}}{\|\mathbf{p_{j,old}}\|} \tag{6.15}$$

$$\mathbf{t_{j,new}} = \mathbf{t_{j,old}} \|\mathbf{p_{j,old}}\| \tag{6.16}$$

$$\mathbf{w_{j,new}} = \mathbf{w_{j,old}} \|\mathbf{p_{j,old}}\| \tag{6.17}$$

Although it is common to apply the scalings (6.15) to (6.17) in the algorithm [98, 343, 344], the scalings are not absolutely necessary [215]. In particular, the score vectors $\mathbf{t_j}$ used to relate X to Y in (6.6) are orthogonal in either case.

Now that $\mathbf{u_j}$ and $\mathbf{t_j}$ are computed using the above expressions, the regression coefficient b_j that relates the two vectors can be computed from

$$b_j = \frac{\mathbf{u_j}^T \mathbf{t_j}}{\mathbf{t_j}^T \mathbf{t_j}} \tag{6.18}$$

The residual matrices E_j and F_j needed for the next iteration are calculated from

$$E_j = E_{j-1} - \mathbf{t_j} \mathbf{p_j}^T \tag{6.19}$$

and

$$F_j = F_{j-1} - b_j \mathbf{t_j} \mathbf{q_j}^T. \tag{6.20}$$

This removes the variance associated with the already calculated score and loading vectors before computing the score and loading vectors for the next iteration. The entire procedure is repeated for the next factor (commonly called as latent variable [343, 344]) $(j+1)$ starting from (6.10) until $j = \min(m, n)$.

As discussed in the next section, predictions based on the PLS model can be computed directly from the observation vector and $\mathbf{p_j}, \mathbf{q_j}, \mathbf{w_j}$, and b_j for $j = 1, 2, \ldots, \min(m, n)$. We will also see an alternative approach where the predictions are obtained from the regression matrix $B2_j$ [217, 344]

$$B2_j = W_j(P_j^T W_j)^{-1}(T_j^T T_j)^{-1} T_j^T F_0 \tag{6.21}$$

where the matrices $P_j \in \mathcal{R}^{\min(m,n)\times j}$, $T_j \in \mathcal{R}^{n\times j}$, and $W_j \in \mathcal{R}^{\min(m,n)\times j}$ are formed by stacking the vectors $\mathbf{p_j}, \mathbf{t_j}$, and $\mathbf{w_j}$, respectively. This matrix is saved for $j = 1, 2, \ldots, \min(m, n)$.

The NIPALS algorithm for PLS1 is calculated using (6.22) to (6.27). Initialize the NIPALS algorithm using $E_0 = X$, $j = 1$, and set $i = 1$. The following equations are used:

$$\mathbf{w_{i,j}} = \frac{E_{j-1}^T \mathbf{y_i}}{\|E_{j-1}^T \mathbf{y_i}\|} \tag{6.22}$$

$$\mathbf{t_{i,j}} = E_{j-1} \mathbf{w_{i,j}} \tag{6.23}$$

$$\mathbf{p_{i,j}} = \frac{E_{j-1}^T \mathbf{t_{i,j}}}{\mathbf{t_{i,j}}^T \mathbf{t_{i,j}}} \tag{6.24}$$

After rescaling of $\mathbf{p_{i,j}}, \mathbf{t_{i,j}}$, and $\mathbf{w_{i,j}}$ in a manner similar to that used in (6.15) to (6.17), the regression coefficient $b_{i,j}$ is computed from

$$b_{i,j} = \frac{\mathbf{y_i}^T \mathbf{t_{i,j}}}{\mathbf{t_{i,j}}^T \mathbf{t_{i,j}}} \tag{6.25}$$

The residuals for the next iteration are calculated as follows

$$E_j = E_{j-1} - \mathbf{t_{i,j}} \mathbf{p_{i,j}^T} \tag{6.26}$$

$$\mathbf{f_{i,j}} = \mathbf{f_{i,j-1}} - b_{i,j} \mathbf{t_{i,j}} q_{i,j} \tag{6.27}$$

where $\mathbf{f_{0,i}} = \mathbf{y_i}$ and $q_{i,j} = 1$. The entire procedure is repeated for the next latent variable $(j+1)$ starting from (6.22) until $j = \min(m, n)$. After all the parameters for $i = 1$ are calculated, the algorithm is repeated for $i = 2, 3, \ldots, p$.

As discussed in the next section, predictions based on the PLS model can be computed directly from the observation vector and the $\mathbf{p_{i,j}}$, $\mathbf{w_{i,j}}$, and $b_{i,j}$. Alternatively, the predictions are obtained from the regression matrix $B1_j$ [9, 217]

$$B1_j = [\,\mathbf{b_{1,j}}\, \mathbf{b_{2,j}} \cdots \mathbf{b_{p,j}}\,] \tag{6.28}$$

where

$$\mathbf{b_{i,j}} = W_{i,j} \left(P_{i,j}^T W_{i,j}\right)^{-1} (T_{i,j}^T T_{i,j})^{-1} T_{i,j}^T \mathbf{f_{0,j}} \tag{6.29}$$

the matrices $P_{i,j} \in \mathcal{R}^{\min(m,n)\times j}$, $W_{i,j} \in \mathcal{R}^{\min(m,n)\times j}$, and $T_{i,j} \in \mathcal{R}^{n\times j}$ are formed by stacking the vectors $\mathbf{p_{i,j}}$, $\mathbf{w_{i,j}}$, and $\mathbf{t_{i,j}}$, respectively.

6.3 Reduction Order and PLS Prediction

It is important to have a proper number a of PLS factors selected in order to obtain a good prediction, since too high a number (the maximum theoretical value for a is the rank of X) will cause a magnification of noise and poor process monitoring performance. A standard way to determine the proper reduction order, denoted as c, is to apply cross-validation using the **prediction residual sum of squares** (PRESS). The order c is set to be the order at which PRESS is minimum [98]. As discussed previously, the weakness of this approach is that it requires that the data be split into two parts (the training and the testing sets), with the PLS vectors computed based only on the data from the testing set.

In the case of fault diagnosis, an alternative approach is to select the value of c which minimizes the information criterion (5.12). To determine c, the PLS vectors are constructed using all of the data, and then the PLS vectors are applied to all of the data to calculate the misclassification rates for each choice of the reduction order, where the misclassification rate is defined to be the ratio of the number of incorrectly assigned classes to the total number of classifications made (the number of observations in the training set).

For each factor $j = 1, 2, \ldots, \min(m, n)$, the estimated score vector $\widehat{\mathbf{t}}_{\mathbf{j}}$ and matrix residual E_j are

$$\widehat{\mathbf{t}}_{\mathbf{j}} = E_{j-1}\mathbf{w}_{\mathbf{j}} \tag{6.30}$$

$$E_j = E_{j-1} - \widehat{\mathbf{t}}_{\mathbf{j}}\mathbf{p}_{\mathbf{j}}^T \tag{6.31}$$

where $E_0 = X$. To compute a prediction of the predicted block $Y_{train2,a}$ of the training set using PLS2 with a PLS components:

$$Y_{train2,a} = F_j = \sum_{j=1}^{a} b_j \widehat{\mathbf{t}}_{\mathbf{j}}\mathbf{q}_{\mathbf{j}}^T. \tag{6.32}$$

For PLS1, the prediction of the predicted block $Y_{train1,a}$ of the training set using PLS1 with a PLS components is computed by

$$Y_{train1,a} = \left[\mathbf{y}_{\mathbf{train1},a}\ \mathbf{y}_{\mathbf{train2},a} \cdots \mathbf{y}_{\mathbf{trainp},a}\right] \tag{6.33}$$

where

$$\mathbf{y}_{\mathbf{traini},a} = \mathbf{f}_{\mathbf{i},\mathbf{j}} = \sum_{j=1}^{a} \mathbf{b}_{\mathbf{i},\mathbf{j}} \widehat{\mathbf{t}}_{\mathbf{i},\mathbf{j}} q_{i,j} \tag{6.34}$$

Alternatively, a prediction of PLS2 with a PLS components is given by the regression equation [9]:

$$Y_{train2,a} = XB2_a \tag{6.35}$$

The above equation is also used for the alternative prediction of PLS1 by replacing $B2_a$ with $B1_a$.

6.4 Fault Detection, Identification, and Diagnosis

One common approach to using PLS is to apply it in the same manner as PCA, selecting the Y matrix to be the product quality variables. Monitoring the PLS scores in this way has the advantage over the PCA scores in that the PLS scores will only monitor variations in X which are known to be related to the product quality variables. All the fault detection, identification, and diagnosis techniques for PCA can be applied in exactly the same way for PLS (e.g., including the Q and T^2 statistics, contribution plots, and discriminant analysis) [171, 343].

The use of discriminant PLS for fault diagnosis requires significantly more explanation. In discriminant PLS, the rows of Y_{train} will not have the form $[0, 0, 0, \ldots, 1, \ldots, 0, 0]$, which requires a method for assigning the class c_k to each observation k. One method is to assign c_k to correspond to the column index whose element is the closest to one [244]. A second method is to assign c_k to correspond to the column whose element has the maximum value.

The term **overestimation** refers to the case where the element of Y_{train} for an in-class member > 1 or the element of Y_{train} for a non-class member > 0. **Underestimation** is where the element of Y_{train} for an in-class member < 1 or the element of Y_{train} for a non-class member < 0. Both assignment methods give accurate classifications in the ideal case, that is, when none of the elements of Y_{train} are overestimated nor underestimated, and in the case where all of the elements of Y_{train} are underestimated. If all of the elements of Y_{train} are overestimated, then the first assignment method can give high misclassification rates, while the second assignment method will still tend to give good classifications [244]. The second assignment method is preferred because of this wider usefulness.

If some of the elements of Y_{train} are underestimated while others are overestimated, either of the above assignment methods can perform poorly. A method to resolve this problem is to take account of the underestimation and overestimation of Y into a second cycle of PLS algorithm [244]. The NIPALS algorithm is run for the second time for PLS1 and PLS2 by replacing $\mathbf{y}_i$ by $\mathbf{y}_{\mathbf{train1},i}$ and Y by Y_{train2}, respectively. To distinguish between the *normal* PLS method and this *adjusted* method, PLS1 and PLS2 are denoted as $\mathrm{PLS1_{adj}}$ and $\mathrm{PLS2_{adj}}$, respectively. The predicted Y of the training set using $\mathrm{PLS1_{adj}}$ and $\mathrm{PLS2_{adj}}$, denoted as $Y_{train1,adj}$ and $Y_{train2,adj}$, are obtained in similar fashion as PLS1 and PLS2, respectively.

The effectiveness of the algorithm can be determined by applying it to a testing set $X_{test} \in \mathcal{R}^{r \times m}$. The predicted block Y_{test1} of the testing set using PLS1 is calculated using (6.30) to (6.31) and (6.33) to (6.34) by replacing X with X_{test} while the predicted block Y_{test2} of the testing set using PLS2 is calculated using (6.30) to (6.32) by replacing X with X_{test}. The predicted blocks $Y_{test1,adj}$ and $Y_{test2,adj}$ using $\mathrm{PLS1_{adj}}$ and $\mathrm{PLS2_{adj}}$, respectively, are obtained similarly.

To illustrate the application of discriminant PLS2, the same experimental data set [82, 45] is used as in Chapter 4. The predictor matrix X is formed by using data from all three classes, where $n = 150$ and $m = 4$; the corresponding predicted matrix Y is formed as in (6.1), where $p = 3$. The matrices X and Y are first autoscaled. The NIPALS algorithm is initialized using $E_0 = X$, $F_0 = Y$, and $\mathbf{u_1}$ arbitrarily set to the third column of Y. After 12 iterations of (6.10)-(6.13), the score vector $\mathbf{t_1}$ converges with an error of less than 10^{-10}. The following vectors are then obtained:

$$\begin{aligned} \mathbf{w_1} &= [0.48\ -0.32\ 0.60\ 0.56]^T, \\ \mathbf{p_1} &= [0.52\ -0.29\ 0.58\ 0.56]^T. \end{aligned} \tag{6.36}$$

The same procedure is done for E_1 and F_1, which results in

$$\begin{aligned} \mathbf{w_2} &= [-0.28\ -0.93\ 0.023\ -0.28]^T, \\ \mathbf{p_2} &= [-0.37\ -0.91\ -0.045\ -0.16]^T. \end{aligned} \tag{6.37}$$

Since the rank of X is four, the procedure can be repeated until $j = 4$. Since only two factors are retained in the example as shown in Chapter 4, we will stop the calibration here and form the regression matrix $B2_2$ as

$$B2_2 = \begin{bmatrix} -0.21 & -0.051 & 0.26 \\ 0.36 & -0.46 & 0.096 \\ -0.33 & 0.078 & 0.25 \\ -0.26 & -0.038 & 0.30 \end{bmatrix}. \tag{6.38}$$

The matrix $Y_{train2,2}$ is formed using (6.35). With the ith column of $Y_{train2,2}$ denoted by y_i, the three-dimensional plot of y_1 vs. y_2 vs. y_3 is illustrated in Figure 6.1. The data are reasonably well separated. Notice that all the 'x' points have large y_3 values and small y_2 and y_1 values, so all Class 3 data would be correctly assigned. Some of the 'o' and '*' points overlap, which indicates that a small portion of the Class 2 data may be misclassified as Class 1 and *vice versa.*

The way to diagnose faults, based on the rows of Y, was discussed above. An alternative fault diagnosis approach based on discriminant PLS is to apply discriminant analysis to the PLS scores for classification [160]. In the terminology introduced in Chapter 5, for classifying p classes, the $p-1$ PLS directions can have substantially non-zero between-groups variance. This method can also provide substantially improved fault diagnosis over PCA [160].

6.5 Comparison of PCA and PLS

For fault diagnosis, a predicted block Y is not used in PCA, instead a linear transformation is performed in X such that the highest ranked PCA vectors

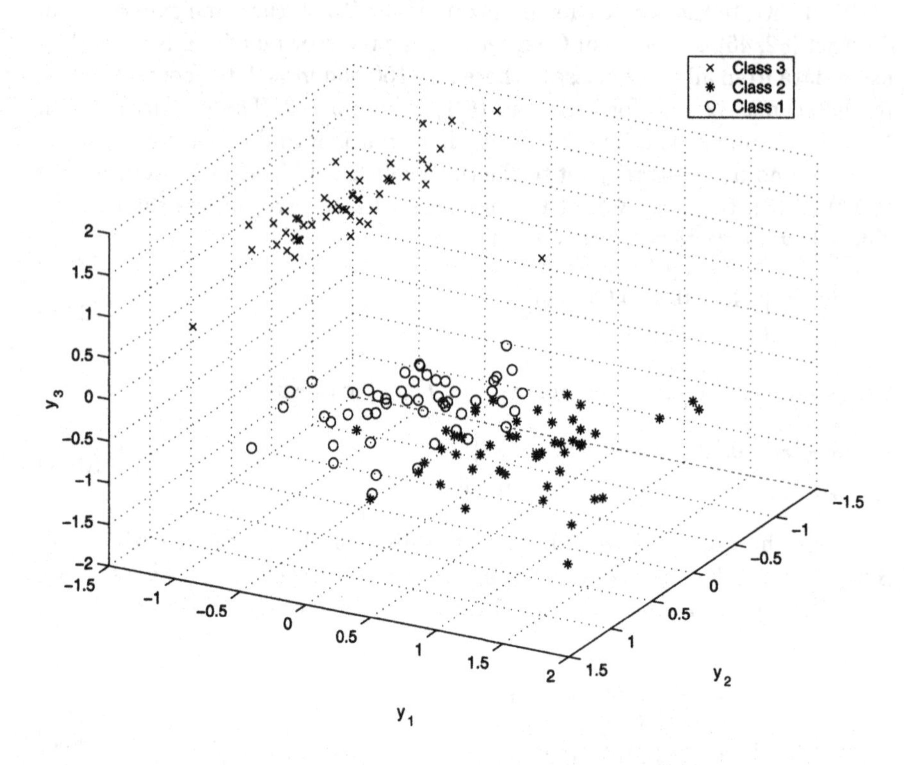

Fig. 6.1. The discriminant PLS predicted matrix plot for the data from [82, 45]

retain most of the variation in X. As described in Chapter 4, the retained scores can be used with discriminant analysis for classification. The disadvantage of the PCA approach is that the highest ranked PCA vectors may not contain the discriminatory power needed to diagnose faults.

PCA maximizes the variance in X while PLS maximizes the covariance between X and Y. By specifying Y to include the fault information as in discriminant PLS, the PLS vectors are computed so as to provide a lower-dimensional representation which is correlated with differences in fault class. Thus, fewer of the discriminant PLS vectors should be required and lower misclassification rates obtained. As discriminant PLS exploits fault information when constructing its lower-dimensional model, it would be expected that discriminant PLS can provide better fault diagnosis than PCA. However, this is not always true, as will be demonstrated in application in Chapter 10.

The projection of the experimental data taken from [82, 45] onto the first two PCA and discriminant PLS loading vectors is shown in Figure 6.2. Recall that the PCA model is built based on the data from all three classes. The two plots look similar indicating that PCA and discriminant PLS give similar separability of the data when two score vectors are used. For data of high dimension, our experience is that similarity between the first few PCA and PLS score vectors is often observed [157]. For score vectors of higher orders, the difference between PCA and discriminant PLS usually becomes more apparent. In this example, the loading matrices corresponding to all four loading vectors for PCA and discriminant PLS are

$$P_{PCA} = \begin{bmatrix} 0.5255 & -0.3634 & 0.6686 & -0.3804 \\ -0.2695 & -0.9266 & -0.1869 & 0.1842 \\ 0.5837 & -0.0081 & -0.0013 & 0.8119 \\ 0.5572 & -0.0969 & -0.7197 & -0.4027 \end{bmatrix} \tag{6.39}$$

and

$$P_{PLS} = \begin{bmatrix} 0.5167 & -0.3709 & 0.7510 & -0.2896 \\ -0.2885 & -0.9136 & -0.0275 & 0.2084 \\ 0.5836 & -0.0449 & 0.0024 & 0.8001 \\ 0.5561 & -0.1607 & -0.6597 & -0.4823 \end{bmatrix}, \tag{6.40}$$

respectively.

Note that the first PCA and discriminant PLS loading vectors are very closely aligned and the fourth loading vectors are much less so. Recall that the loading vectors for PCA are orthogonal. In PLS, the loading vectors are rotated slightly in order to capture a better relationship between the predicted and predictor blocks (*i.e.*, maximize the covariance between X and Y) [157]. As a result of this rotation, the PLS loading vectors are rarely orthogonal. In general, the rotation for the first PLS loading vector is usually small. As the order increases, the deviation from orthogonality for the discriminant PLS loading vectors usually increases. Although the discriminant PLS loading vectors are not orthogonal, their score vectors are indeed orthogonal (see Homework Problem 4).

6.6 Other PLS Methods

The PLS methods described in this chapter can be extended to take into account the serial correlations in the data, by augmenting the observation vector and stacking the data matrix in the same manner as (4.44). The matrix Y has to be changed correspondingly. Implementation of this approach is left as an exercise for the readers (see Homework Problem 5).

The PLS approaches can be generalized to nonlinear systems using **nonlinear partial least squares** (NPLS) algorithms [83, 213, 349]. In NPLS, the relationship between $\hat{\mathbf{u}}_j$ and $\mathbf{t}_j$ in (6.6) is replaced by

$$\hat{\mathbf{u}}_j = f(\mathbf{t}_j) \tag{6.41}$$

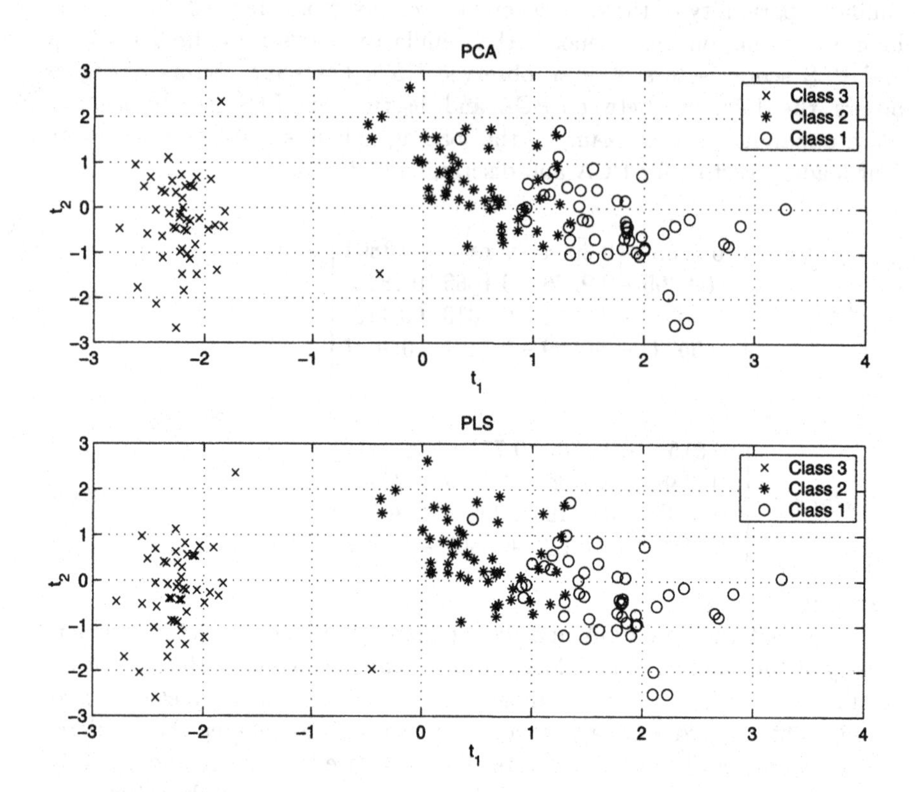

Fig. 6.2. The projections of experimental data [82, 45] for three classes onto the first two discriminant PLS and PCA loading vectors, respectively

where $f(\mathbf{t_j})$ is a nonlinear, continuous, and differentiable function in $\mathbf{t_j}$. The simplest nonlinear relationship for NPLS is a quadratic function

$$f(t_{j,k}) = a_j + b_j t_{j,k} + c_j t_{j,k}^2 \tag{6.42}$$

and $f(\mathbf{t_j}) = [f(t_{j,1})\ f(t_{j,2}) \ldots f(t_{j,n})]^T$. This NPLS model is commonly known as **quadratic partial least squares** (QPLS). At each iteration of QPLS, the ordinary PLS steps are applied to $\mathbf{t_j}$, $\mathbf{q_j}$, and $\mathbf{u_j}$, and ordinary least squares are used to estimate the coefficients a_j, b_j, and c_j (see [349] for the detailed procedure). The nonlinearities can also be based on sigmoidal functions as used in artificial neural networks [122, 268].

For systems with mild nonlinearities, the same degree of fit can usually be obtained by a linear model with several factors, or by a nonlinear model with fewer dimensions [349]. In cases where the systems display strong nonlinearities (*i.e.*, if the nonlinearities have maxima, minima, or have significant curvature), a nonlinear model is appropriate and NPLS can perform better than linear PLS especially when the systems are well-determined and with high observation/variable ratio. However, for an underdetermined system, the models cannot be fitted with acceptable variance using NPLS because of the small number of degrees of freedom in the data sets [83].

Other PLS methods in the literature that have been applied to either simulations or actual process applications are **recursive partial least squares** (RPLS) [266], multiblock partial least squares [208, 343], and multiway partial least squares [243, 343]. The multiway technique is especially useful for the monitoring of batch processes, in which the predictor X is usually selected to be a three-dimensional array $(i \times j \times k)$. A straightforward generalization of the PLS technique to the multiway technique provides a strategy for the detection and diagnosis of faults in batch processes.

6.7 Homework Problems

1. Describe in some detail how to formulate the Q and T^2 statistics for detecting faults using PLS, where Y is the matrix of product quality variables. Compare and contrast this fault detection approach with the PCA-based Q and T^2 statistics. Describe in detail how to generalize the discriminant-based PCA methods for fault diagnosis to PLS, where Y is the matrix of product quality variables. How would you expect the performance of this approach to compare with the performance of discriminant PLS?
2. Generalize PLS as described in Problem 1 to EWMA and CUSUM versions, and to dynamic PLS.
3. Show that the PCA loading vectors for the experimental data from [45, 82] are orthogonal (hint: compute $P_{PCA}^T P_{PCA}$ using P_{PCA} in (6.39)). Show that the PLS loading vectors for the data are not orthogonal. Calculate the angle between the jth PCA and j^{th} PLS loading vector for the data for $j = 1, \ldots, 4$. How does the angle change as a function of j?
4. Show that the discriminant PLS loading vectors are not orthogonal, and their score vectors are orthogonal for the experimental data from [45, 82].
5. Generalize discriminant PLS to dynamic discriminant PLS.
6. Provide a detailed comparison of FDA and discriminant PLS. Which method would be expected to do a better job diagnosing faults? Why?
7. Read an article on the use of multiway PLS (e.g., [170, 243]) and write a report describing in detail how the technique is implemented and applied. Describe how the computations are performed and how the statistics are computed. Formulate a discriminant multiway PLS algorithm. For what types of processes are these algorithms suited? Provide some hypothetical examples.
8. Read an article on the application of multiblock PLS (e.g., [84, 208]) and write a report describing in detail how the technique is implemented and applied. Describe how the computations are performed and how the

statistics are computed. Formulate a discriminant multiblock PLS algorithm. For what types of processes are these algorithms suited? Provide some hypothetical examples.

9. Read an article on the application of nonlinear PLS (e.g., [83, 213, 349]) and write a report describing in detail how the technique is implemented and applied. Describe how the computations are performed and how the statistics are computed. For what types of processes are these algorithms suited? Provide some hypothetical examples.

7. Canonical Variate Analysis

7.1 Introduction

In Section 4.7, it was shown how DPCA can be applied to develop an autoregressive with input ARX model and to monitor the process using the ARX model. The weakness of this approach is the inflexibility of the ARX model for representing linear dynamical systems. For instance, a low order **autoregressive moving average** ARMA (or **autoregressive moving average with input** ARMAX) model with relatively few estimated parameters can accurately represent a high order ARX model containing a large number of parameters [199]. For a **single-input-single-output** (SISO) process, an ARMAX(h) model is:

$$y_t = \sum_{i=1}^{h} \alpha_i y_{t-i} + \sum_{i=0}^{h} \beta_i u_{t-i} + \sum_{i=1}^{h} \gamma_i e_{t-i} + e_t \tag{7.1}$$

where y_t and u_t are the output and input at time t, respectively, $\alpha_1, \ldots, \alpha_h$, $\beta_1, \ldots, \beta_h$, and $\gamma_1, \ldots, \gamma_h$ are constant coefficients, and e_t is a white noise process with zero mean [336]. For an invertible process, the ARMAX(h) model can be written as an infinite-order ARX model [336]:

$$y_t = \sum_{i=1}^{\infty} \pi_i y_{t-i} + \sum_{i=0}^{\infty} \rho_i u_{t-i} + e_t. \tag{7.2}$$

The constant coefficients $\pi_1, \pi_2, \ldots$ and $\rho_1, \rho_2, \ldots$ are determined from the coefficients in (7.1) via the backshift and division operations [336].

The classical approach to identifying ARMAX processes requires the *a priori* parameterization of the ARMAX model and the subsequent estimation of the parameters via the solution of a least squares problem [199]. To avoid over-parameterization and identifiability problems, the structure of the ARMAX model needs to be properly specified; this is especially important for multivariable systems with a large number of inputs and outputs. This structure specification for ARMAX models is analogous to specifying the observability (or controllability) indices and the state order for state-space models, and is not trivial for higher-order multivariable systems [317]. Another problem with the classical approach is that the least squares problem requires the solution of a nonlinear optimization problem. The solution of the nonlinear optimization problem is iterative, can suffer from convergence problems, can be overly sensitive to small data fluctuations, and the required amount of computation to solve the optimization problem cannot be bounded [189].

To avoid the problems of the classical approach, a class of system identification methods for generating state-space models called **subspace algorithms** has been developed in the past few years. The class of state-space models is equivalent to the class of ARMAX models [12, 199]. That is, given a state-space model, an ARMAX model with an identical input-output mapping can be determined, and *vice versa.* The subspace algorithms avoid *a priori* parameterization of the state-space model by determining the states of the system directly from the data, and the states along with the input-output data allow the state-space and covariance matrices to be solved directly via *linear* least squares [317] (see Figure 7.1). These algorithms rely mostly on the singular value decomposition (SVD) for the computations, and therefore do not suffer from the numerical difficulties associated with the classical approach.

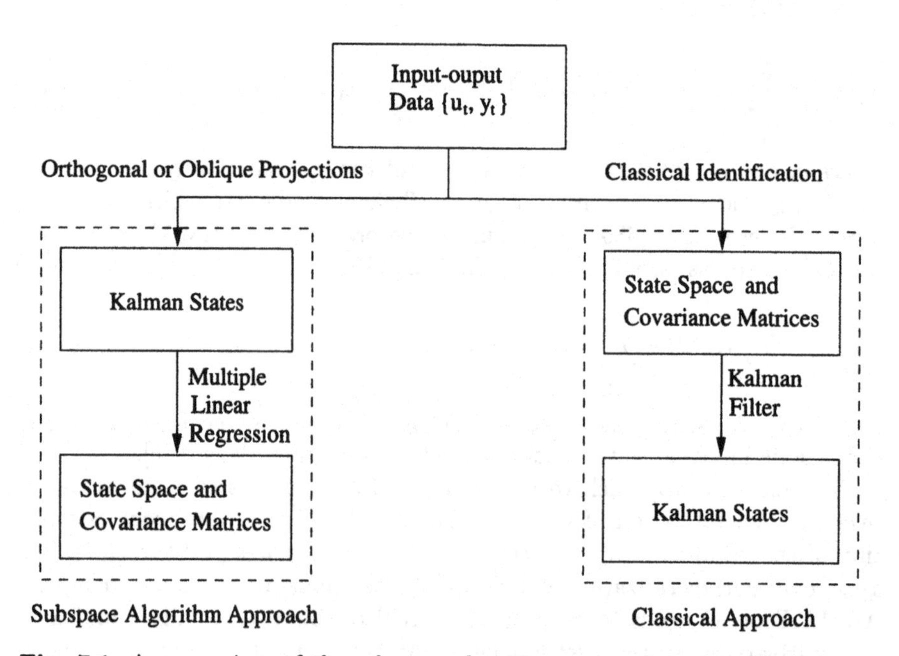

Fig. 7.1. A comparison of the subspace algorithm approach to the classical approach for identifying the state-space model and extracting the Kalman states [318]

Three popular subspace algorithms are **numerical algorithms for subspace state space system identification** (N4SID), **multivariable output-error state space** (MOESP), and **canonical variate analysis** (CVA) [318]. Although the subspace algorithm based on CVA is often referred to as "CVA", CVA is actually a dimensionality reduction technique in multivariate statistical analysis involving the selection of pairs of variables from the *inputs* and *outputs* that maximize a correlation measure [189]. For clarity of presentation, "CVA" in this book refers to the dimensionality reduction technique, and the subspace algorithm based on CVA is called the **CVA algorithm**. The philosophy of CVA shares many common features to PCA, FDA, and PLS (see Section 7.2), which makes it a natural subspace identification technique for use in developing process monitoring statistics. The CVA-based statistics described in in this chapter can be readily generalized to the other subspace identification algorithms.

To fully understand all aspects of CVA requires knowledge associated with materials outside of the scope of this book. Enough information is given in this chapter for the readers to gain some intuitive understanding of how CVA works and to implement the process monitoring techniques. Section 7.2 describes the CVA Theorem and an interpretation of the theorem indicating the optimality of CVA for dimensionality reduction. Section 7.3 describes the CVA algorithm with a statistical emphasis. Determination of the state-space model and the issues of system identifiability are discussed in Section 7.4. Section 7.5 addresses the computational issues of CVA. A procedure for automatically and optimally selecting the state order of the state-space model is presented in Section 7.6. Section 7.7 presents a systems theory interpretation for the CVA algorithm and the other subspace algorithms. Section 7.8 discusses the process monitoring measures developed for the states extracted by the CVA algorithm.

7.2 CVA Theorem

CVA is a linear dimensionality reduction technique, optimal in terms of maximizing a correlation measure between two sets of variables. The **CVA Theorem** states that given a vector of variables $\mathbf{x} \in \mathcal{R}^m$ and another vector of variables $\mathbf{y} \in \mathcal{R}^n$ with covariance matrices Σ_{xx} and Σ_{yy}, respectively, and cross covariance matrix Σ_{xy}, there exist matrices $J \in \mathcal{R}^{m \times m}$ and $L \in \mathcal{R}^{n \times n}$ such that

$$J\Sigma_{xx}J^T = I_{\bar{m}}, \qquad L\Sigma_{yy}L^T = I_{\bar{n}}, \tag{7.3}$$

and

$$J\Sigma_{xy}L^T = D = \mathrm{diag}(\gamma_1, \cdots, \gamma_r, 0, \cdots, 0), \tag{7.4}$$

where $\gamma_1 \geq \cdots \geq \gamma_r$, $\bar{m} = \mathrm{rank}(\Sigma_{xx})$, $\bar{n} = \mathrm{rank}(\Sigma_{yy})$, D contains the **canonical correlations** γ_i, $I_{\bar{m}} \in \mathcal{R}^{m \times m}$ is a diagonal matrix containing the first $\bar{m}$ diagonal elements as one and the rest of the diagonal elements as zero, and $I_{\bar{n}} \in \mathcal{R}^{n \times n}$ is the diagonal matrix containing the first $\bar{n}$ diagonal elements as one and the rest of the diagonal elements as zero [189]. The vector of **canonical variables** $\mathbf{c} = J\mathbf{x}$ contains a set of uncorrelated random variables and has the covariance matrix

$$\Sigma_{cc} = J\Sigma_{xx}J^T = I_{\bar{m}}, \tag{7.5}$$

and the vector of **canonical variables** $\mathbf{d} = L\mathbf{y}$ contains a set of uncorrelated random variables and has the covariance matrix

$$\Sigma_{dd} = L\Sigma_{yy}L^T = I_{\bar{n}}. \tag{7.6}$$

The cross covariance matrix between $\mathbf{c}$ and $\mathbf{d}$ is diagonal

$$\Sigma_{cd} = J\Sigma_{xy}L^T = D = \mathrm{diag}(\gamma_1, \cdots, \gamma_r, 0, \cdots, 0), \tag{7.7}$$

which indicates that the two vectors are only pairwise correlated. The degree of the pairwise correlations is indicated and can be ordered by the canonical correlations γ_i.

CVA is equivalent to a **generalized singular value decomposition** (GSVD) [184, 189]. When Σ_{xx} and Σ_{yy} are invertible, the projection matrices J and L and the matrix of canonical correlations D can be computed by solving the SVD

$$\Sigma_{xx}^{-1/2}\Sigma_{xy}\Sigma_{yy}^{-1/2} = U\Sigma V^T \tag{7.8}$$

where $J = U^T\Sigma_{xx}^{-1/2}$, $L = V^T\Sigma_{yy}^{-1/2}$, and $D = \Sigma$ [185]. It is easy to verify that J, L, and D computed from (7.8) satisfy (7.3) and (7.4). The weightings $\Sigma_{xx}^{-1/2}$ and $\Sigma_{yy}^{-1/2}$ ensure that the canonical variables are uncorrelated and have unit variance, and the matrices U^T and V^T rotate the canonical variables so that $\mathbf{c}$ and $\mathbf{d}$ are only pairwise correlated. The degree of the pairwise correlations is indicated by the diagonal elements of Σ. Note that the GSVD mentioned above is not the same as the GSVD described in most of the mathematics literature [104, 316].

A CVA-related approach in the multivariate statistics literature [67, 180, 193, 234, 300] is known as **canonical correlation analysis** (CCA), which can be generalized into the CVA Theorem [180, 234]. While both CCA and CVA are suitable for correlating two sets of variables, CVA has been applied on time series data (see Section 7.3). To emphasize the application of the process monitoring algorithm on time series data, we prefer to use the terminology CVA over CCA.

Several dimensionality reduction techniques have been interpreted in the framework of the GSVD [193, 189]. For example, consider the case where the left hand side of (7.8) is replaced by $\Sigma_{xx}^{1/2}$. Then

$$\Sigma_{xx}^{1/2} = U\Sigma V^T. \tag{7.9}$$

Using the fact that $U = V$ (Since $\Sigma_{xx}^{1/2}$ is symmetric), squaring both sides give

$$\Sigma_{xx} = U\Sigma^2 V^T. \tag{7.10}$$

The corresponding equation (4.3) for PCA is

$$\Sigma_{xx} = U\Lambda V^T. \tag{7.11}$$

We see that the diagonal elements of Σ in (7.9) are equal to the diagonal elements of Σ in (4.2).

CVA can be reduced to FDA. The generalized eigenvalue problem for FDA (5.9) can be written as a function of x and y as defined in (7.3), where x contains the measurement variables and y contains dummy variables which represent class membership similarly to (6.1) [193].

PLS is also related with CVA, where both methods are equivalent to a GSVD on the covariance matrix. The difference is that CVA uses a weighting so as to maximize *correlation*, whereas PLS maximizes *covariance* [283]. CVA simultaneously obtains all components ($J, L,$ and D) in one GSVD, whereas the PLS algorithm is sequential in selecting the important components, working with the residuals from the previous step.

7.3 CVA Algorithm

In Section 7.2, the optimality and the structure abstraction of CVA were presented via the CVA Theorem. While the CVA concept for multivariate statistical analysis was developed by Hotelling [125], it was not applied to system identification until Akaike's work on the ARMA model [3, 4, 5, 189]. Larimore developed CVA for state-space models [184, 185, 189]. This section describes the linear state-space model and the CVA algorithm for identifying state space models directly from the data.

Given time series input data $\mathbf{u_t} \in \mathcal{R}^{m_u}$ and output data $\mathbf{y_t} \in \mathcal{R}^{m_y}$, the linear state-space model is given by [187]

$$\mathbf{x_{t+1}} = \Phi \mathbf{x_t} + G\mathbf{u_t} + \mathbf{w_t} \tag{7.12}$$

$$\mathbf{y_t} = H\mathbf{x_t} + A\mathbf{u_t} + B\mathbf{w_t} + \mathbf{v_t} \tag{7.13}$$

where $\mathbf{x_t} \in \mathcal{R}^k$ is a k-order state vector and $\mathbf{w_t}$ and $\mathbf{v_t}$ are white noise processes that are independent with covariance matrices Q and R, respectively. The state-space matrices Φ, G, H, A, and B along with the covariance matrices Q and R specify the state-space model. It is assumed here that the state-space matrices are constant (*time-invariance*) and the covariance matrices are constant (*weakly stationary*). The term $B\mathbf{w_t}$ in (7.13) allows the noise in the *output equation* (7.13) to be correlated with the noise in the *state equation* (7.12). Omitting the term Bw_t, typically done for many state-space models, may result in a state order that is not minimal [185]. Time-varying trends in the data can be fitted by augmenting polynomial functions of time to the state-space model; a software package that implements this is *ADAPTx Version 3.03* [187].

An important aspect of the CVA algorithm is the separation of *past* and *future*. At a particular time instant $t \in (1, \cdots, n)$ the vector containing the information from the past is

$$\mathbf{p_t} = \left[\mathbf{y}_{t-1}^T, \mathbf{y}_{t-2}^T, \cdots, \mathbf{u}_{t-1}^T, \mathbf{u}_{t-2}^T, \cdots\right]^T, \tag{7.14}$$

and the vector containing the output information in the present and future is

$$\mathbf{f_t} = \left[\mathbf{y}_t^T, \mathbf{y}_{t+1}^T, \cdots\right]^T. \tag{7.15}$$

Assuming the data is generated from a linear state space model with a finite number of states k, the elements of the state vector $\mathbf{x_t}$ is equal to a set of k linear combinations of the past,

$$\mathbf{x_t} = J_k \mathbf{p_t} \tag{7.16}$$

where $J_k \in \mathcal{R}^{k \times m_p}$ is a constant matrix with $m_p < \infty$. The state vector $\mathbf{x_t}$ has the property that the conditional probability of the future $\mathbf{f_t}$ conditioned on the past $\mathbf{p_t}$ is equal to the conditional probability of the future $\mathbf{f_t}$ conditioned on the state $\mathbf{x_t}$

$$P(\mathbf{f_t}|\mathbf{p_t}) = P(\mathbf{f_t}|\mathbf{x_t}). \tag{7.17}$$

In other words, the state provides as much information as past data do as to the future values of the output. This also indicates that only a finite number of linear combinations of the past affects the future outputs. This property of the state vector can be extended to include future inputs [187]

$$P((\mathbf{f_t}|\mathbf{q_t})|\mathbf{p_t}) = P((\mathbf{f_t}|\mathbf{q_t})|\mathbf{x_t}) \tag{7.18}$$

where $\mathbf{q_t} = [\mathbf{u}_t^T, \mathbf{u}_{t+1}^T, \cdots]^T$. In the process identification literature, a process satisfying (7.18) is said to be a **controlled Markov process** of order k.

Let the k-order memory, $\mathbf{m_t} \in \mathcal{R}^k$, be a set of k linear combinations of the past $\mathbf{p_t}$

$$\mathbf{m_t} = C_k \mathbf{p_t} \tag{7.19}$$

where $C_k \in \mathcal{R}^{k \times m_p}$. The term "memory" is used here instead of "state" because the vector $\mathbf{m_t}$ may not necessarily contain all the information in the past (for instance, the dimensionality of k may not be sufficient to capture all the information in the past). The goal of process identification is to provide the optimal prediction of the future outputs based on the past and current state. Now in a real process the true state order k is unknown, so instead the future outputs are predicted based on the current memory:

$$\hat{f}_t(\mathbf{m_t}) = \Sigma_{fm}\Sigma_{mm}^{-1}\mathbf{m_t} \tag{7.20}$$

where $\hat{\mathbf{f}}_\mathbf{t}(\mathbf{m_t})$ is the optimal linear prediction of the future $\mathbf{f_t}$ based on the memory $\mathbf{m_t}$ [187]. The CVA algorithm computes the optimal matrix for C_k in (7.19), that is, the matrix C_k which minimizes the average prediction error:

$$E\{(\mathbf{f_t} - \hat{\mathbf{f}}_\mathbf{t})\Lambda^{\dagger}(\mathbf{f_t} - \hat{\mathbf{f}}_\mathbf{t})\} \tag{7.21}$$

where E is the expectation operator and $\Lambda^{\dagger}$ is the pseudo inverse of Λ, which is a positive semidefinite symmetric matrix used to weigh the relative importance of the output variables over time. The choice $\Lambda = \Sigma_{ff}$ results in nearly maximum likelihood estimates [184, 283].

The optimal value for C_k in (7.19) is computed via the GSVD by substituting the matrix Σ_{xx} with Σ_{pp}, Σ_{yy} with Σ_{ff}, and Σ_{xy} with Σ_{pf} in (7.3) and (7.4) [187]. The optimal estimate for matrix C_k is equal to J_k, where J_k is the first k rows of the matrix J in (7.3) [189]. The optimal k-order memory is

$$\mathbf{m}_\mathbf{t}^{opt} = J_k\mathbf{p_t}. \tag{7.22}$$

The structure of the solution indicates that the optimal memory for order k is a subset of the *optimal* memory for order $k+1$. The optimal memory for a given order k corresponds to the first k states of the system [187], and these states are referred to as the **CVA states.**

7.4 State Space Model and System Identifiability

The process monitoring statistics described in Section 7.8 are based on the matrix J which is used to construct the CVA states, and do not require the construction of an explicit state-space model (7.12)-(7.13). The calculation of the state space matrices in (7.12)-(7.13) is described here for completeness.

Assuming the order of the state space model, k, is chosen to be greater than or equal to the order of the minimal state space realization of the actual system, the state vectors $\mathbf{x_t}$ in (7.12) and (7.13) can be replaced by the state estimate $\mathbf{m_t}$:

$$\begin{bmatrix} \mathbf{m_{t+1}} \\ \mathbf{y_t} \end{bmatrix} = \begin{bmatrix} \Phi & G \\ H & A \end{bmatrix} \begin{bmatrix} \mathbf{m_t} \\ \mathbf{u_t} \end{bmatrix} + \begin{bmatrix} I & 0 \\ B & I \end{bmatrix} \begin{bmatrix} \mathbf{w_t} \\ \mathbf{v_t} \end{bmatrix} \tag{7.23}$$

Since $\mathbf{u_t}$ and $\mathbf{y_t}$ are known, and $\mathbf{m_t}$ can be computed once J_k in (7.22) is known, this equation's only unknowns (Φ, G, H, A, and B) are linear in the parameters. The state space matrices can be estimated by multiple linear regression (see Figure 7.1)

$$\begin{bmatrix} \hat{\Phi} & \hat{G} \\ \hat{H} & \hat{A} \end{bmatrix} = \hat{\Sigma}_{my,mu} \hat{\Sigma}^{-1}_{mu,mu} \tag{7.24}$$

where

$$\mathbf{mu} = \begin{bmatrix} \mathbf{m_t} \\ \mathbf{u_t} \end{bmatrix}, \ \mathbf{my} = \begin{bmatrix} \mathbf{m_{t+1}} \\ \mathbf{y_t} \end{bmatrix}, \tag{7.25}$$

and $\hat{\Sigma}_{i,j}$ represents the sample covariance matrix for variables i and j. The error of the multiple regression has the covariance matrix

$$\begin{bmatrix} S_{11} & S_{12} \\ S_{21} & S_{22} \end{bmatrix} = \hat{\Sigma}_{my,my} - \hat{\Sigma}_{my,mu} \hat{\Sigma}^{-1}_{mu,mu} \hat{\Sigma}^{T}_{my,mu}, \tag{7.26}$$

and the matrices $\hat{B} = S_{21} S^{\dagger}_{11}$, $\hat{Q} = S_{11}$, and $\hat{R} = S_{22} - S_{21} S^{\dagger}_{11} S_{12}$ where $\dagger$ signifies the pseudo-inverse [104]. With the matrices $\hat{A}$, $\hat{B}$, $\hat{H}$, $\hat{G}$, $\hat{\Phi}$, $\hat{Q}$, and $\hat{R}$ estimated, the state space model as shown in (7.12) and (7.13) can be used for various applications such as multistep predictions and forecasts, for example, as needed in model predictive control [159, 283].

There is a significant advantage in terms of identifiability of state space identification approaches over classical identification based on polynomial transfer functions. For polynomial transfer functions, it is always possible to find particular values of the parameters that produce arbitrarily poor conditioning [102, 187], and hence a loss in identifiability of the model [264, 325]. The simplest example of this is when a process pole nearly cancels a process zero.

The state space model estimated using (7.24) and (7.26) is globally identifiable, so that the method is statistically well-conditioned [189]. The CVA algorithm guarantees the choice of a well-conditioned parameterization.

7.5 Lag Order Selection and Computation

The discussion in Section 7.3 assumes that an infinite amount of data is available. For the computational problem, there is a finite amount of data available, and the vectors $\mathbf{p_t}$, $\mathbf{f_t}$, and $\mathbf{q_t}$ are truncated as

$$\mathbf{p_t} = \left[\mathbf{y}^T_{\mathbf{t-1}}, \mathbf{y}^T_{\mathbf{t-2}}, \cdots, \mathbf{y}^T_{\mathbf{t-h}}, \mathbf{u}^T_{\mathbf{t-1}}, \mathbf{u}^T_{\mathbf{t-2}}, \cdots, \mathbf{u}^T_{\mathbf{t-h}}\right]^T, \tag{7.27}$$

$$\mathbf{f_t} = \left[\mathbf{y}^T_{\mathbf{t}}, \mathbf{y}^T_{\mathbf{t+1}} \cdots, \mathbf{y}^T_{\mathbf{t+l-1}}\right]^T, \tag{7.28}$$

$$\mathbf{q_t} = \left[\mathbf{u}_t^T, \mathbf{u}_{t+1}^T, \cdots, \mathbf{u}_{t+l-1}^T\right]^T \tag{7.29}$$

where h and l are the number of lags included in the vectors. Note that $\mathbf{p_t}$ with h lags directly corresponds to the observation vector for (4.44) with $h-1$ lags. Theoretically, the CVA algorithm does not suffer when $h = l > k$, where k is the state order of the system generating the data (actually, h and l just need to be larger than the largest observability index [318]). However, the state order of the system is not known *a priori*. The first step of computing of CVA is to determine the number of lags h. Assuming there are n observations in the training set and the maximum number for the lag order is max, Larimore suggests fitting autoregressive models with several different numbers of lags to the *training data*:

$$Y = C_j X_j + E_j \tag{7.30}$$

where the predicted matrix $Y \in \mathcal{R}^{(m_u+m_y)\times(n-max)}$ is given as:

$$Y = \begin{bmatrix} \mathbf{y}_{\mathbf{max+1}} & \mathbf{y}_{\mathbf{max+2}} & \cdots & \mathbf{y}_{\mathbf{n}} \\ \mathbf{u}_{\mathbf{max+1}} & \mathbf{u}_{\mathbf{max+2}} & \cdots & \mathbf{u}_{\mathbf{n}} \end{bmatrix} \tag{7.31}$$

and the predictor matrix $X_j \in \mathcal{R}^{j(m_u+m_y)\times(n-max)}$ with j lags is given as the first $j(m_u + m_y)$ rows of

$$X = \begin{bmatrix} \mathbf{y}_{\mathbf{max}} & \mathbf{y}_{\mathbf{max+1}} & \cdots & \mathbf{y}_{\mathbf{n-1}} \\ \mathbf{u}_{\mathbf{max}} & \mathbf{u}_{\mathbf{max+1}} & \cdots & \mathbf{u}_{\mathbf{n-1}} \\ \mathbf{y}_{\mathbf{max-1}} & \mathbf{y}_{\mathbf{max}} & \cdots & \mathbf{y}_{\mathbf{n-2}} \\ \mathbf{u}_{\mathbf{max-1}} & \mathbf{u}_{\mathbf{max}} & \cdots & \mathbf{u}_{\mathbf{n-2}} \\ \vdots & \vdots & \vdots & \vdots \\ \mathbf{y_1} & \mathbf{y_2} & \cdots & \mathbf{y}_{\mathbf{n-max}} \\ \mathbf{u_1} & \mathbf{u_2} & \cdots & \mathbf{u}_{\mathbf{n-max}} \end{bmatrix} \tag{7.32}$$

and $E_h \in \mathcal{R}^{(m_u+m_y)\times(n-max)}$ is the residual matrix for lag order j. The regression matrix for C_j is determined via least squares:

$$C_j = \Sigma_{YX_j}\Sigma_{X_jX_j}^{-1} \tag{7.33}$$

where the covariance matrix Σ_{YX_j} is equal to $\frac{1}{n-max}YX_j^T$. The residual matrix E_j is calculated for $j = 1, 2, \ldots, max$. The lag order h is selected to be the lag minimizing the small sample AIC criterion (7.37) discussed in Section 7.6. This ensures that large enough lags are used to capture all the statistically significant information in the data. The selection of the state order k is described in the next section.

The computational requirements are known *a priori* for the GSVD computation. The number of flop counts grows by order $(nh + h^3)$, and the required storage space is on the order $(n + h^2)$ [189].

The near optimality of the state-space model produced by the CVA algorithm has been observed in Monte Carlo simulations. The estimated **Kullback-Leibler information distances** (see Section 7.6) for both open- and closed-loop simulations were close to the information distances, related to the Cramer-Rao bound, corresponding to the minimum possible parameter estimation error for any unbiased estimation procedure [189]. Simulations have also verified the robustness of the CVA algorithm for systems involving feedback [189].

7.6 State Order Selection and Akaike's Information Criterion

The selection of the state order is an important step in identifying a state-space model. The existence of a *true* state order is highly suspect when dealing with real process data; however, the state order can be utilized as a trade-off parameter for the model complexity, similar to the order of model reduction, a, described for PCA, FDA, and PLS in Chapters 4, 5, and 6, respectively. For instance, choosing the state order too large results in the model overfitting the data, and choosing the state order too small results in the model underfitting the data. This section presents a method for state order selection based on **Akaike's information criterion** (AIC).

The agreement between two probability density functions can be measured in terms of the **Kullback-Leibler information distance** (KLID) [199]

$$I(p_*(x), \hat{p}(x)) = \int p_*(x) \ln \frac{p_*(x)}{\hat{p}(x)} dx \tag{7.34}$$

where x contains the random variables, $p_*(x)$ is the true probability density function, and $\hat{p}(x)$ is the estimated probability density function. The KLID is based on the statistical principles of sufficiency and repeated sampling in a predictive inference setting, and is invariant to model reparameterization [188]. If the true probability density function of the process data is known, then the information distance (7.34) could be computed for various state orders and the optimal state order would correspond to the minimum information distance.

For large samples, the optimal estimator of the information distance (7.34) for a given order k is the AIC,

$$AIC(k) = -2 \ln p(y^n, u^n; \hat{\theta}_k) + 2M_k \tag{7.35}$$

where p is the likelihood function [13], the vectors u^n and y^n contain n observations for the input and output variables, respectively, and $\hat{\theta}_k$ are the M_k independent parameters estimated for state order k. The order k is selected such that the AIC criterion (7.35) is minimized. The number of independent parameters in the state-space model (7.12) and (7.13) is

$$M_k = k(2m_y + m_u) + m_u m_y + \frac{m_y(m_y+1)}{2}. \tag{7.36}$$

The number of independent parameters is far less than the actual number of parameters in the state-space model [199], and the result (7.36) was developed by considering the size of the equivalence class of state-space models having the same input-output and noise characteristics [187].

For small samples, the AIC can be an inaccurate estimate of the KLID. This has led to the development of the small sample correction to the AIC [187]

$$AIC_C(k) = -2\ln p(y^n, u^n; \hat{\theta}_k) + 2fM_k \tag{7.37}$$

where the correction factor for small samples is

$$f = \frac{\bar{n}}{\bar{n} - \left(\dfrac{M_k}{m_u + m_y} + \dfrac{m_u + m_y + 1}{2}\right)} \tag{7.38}$$

where $\bar{n}$ is the number of one-step ahead predictions used to develop the model. The small sample correction to the AIC approaches the AIC ($f \to 1$) as the sample size increases ($\bar{n} \to \infty$). It has been reported to produce state order selections that are close to the optimal prescribed by the KLID [189]. Within the context of Section 3.3, the selection of the optimal state order results in an optimal tradeoff between the bias and variance effects on the model error.

7.7 Subspace Algorithm Interpretations

The book *Subspace Identification of Linear Systems* by Van Overschee and De Moor [318] presents a unified approach to the subspace algorithms. It shows that the three subspace algorithms (N4SID, MOESP, and CVA) can be computed with essentially the same algorithm, differing only in the choice of weights. Larimore [189] states that the other algorithms differ from the CVA algorithm only in the choice of the matrices Σ_{xx} and Σ_{yy} used in (7.3), and claims accordingly that the other algorithms are statistically suboptimal.

It has been proven under certain assumptions that the subspace algorithms can be used to produce asymptotically unbiased estimates of the state-space matrices [318]. However, the state-space matrices estimated by the three algorithms can be significantly different when the amount of input and output data is relatively small.

Van Overschee and De Moor also show that the state sequences generated by the subspace algorithms are the outputs of non-steady-state Kalman filter banks. The basis for the states is determined by the weights used by the various algorithms, and the state-space realizations produced by the algorithms are balanced realizations under certain frequency-weightings. Therefore, reducing the dimensionality of the memory in the subspace algorithms can be interpreted in the framework of the frequency-weighted balanced truncation techniques developed by Enns [79], with the exception that the subspace algorithms truncate the state-space model before the model is estimated (see Figure 7.2). The amount of model error introduced by reducing the order is minimized by eliminating only those states with the smallest effect on the input-output mapping, and for the CVA algorithm, the amount of model error is proportional to the canonical correlations [187]. The model reduction approach of the CVA algorithm has the advantage in that truncating the memory vector prior to the estimation of the state-space model instead of truncating the state vector based on a full order state-space model is much more computationally and numerically robust (see Figures 7.1 and 7.2). The degree of model reduction, or equivalently the selection of the state order, is an important step in the identification process, and a statistically optimal method was discussed in Section 7.6.

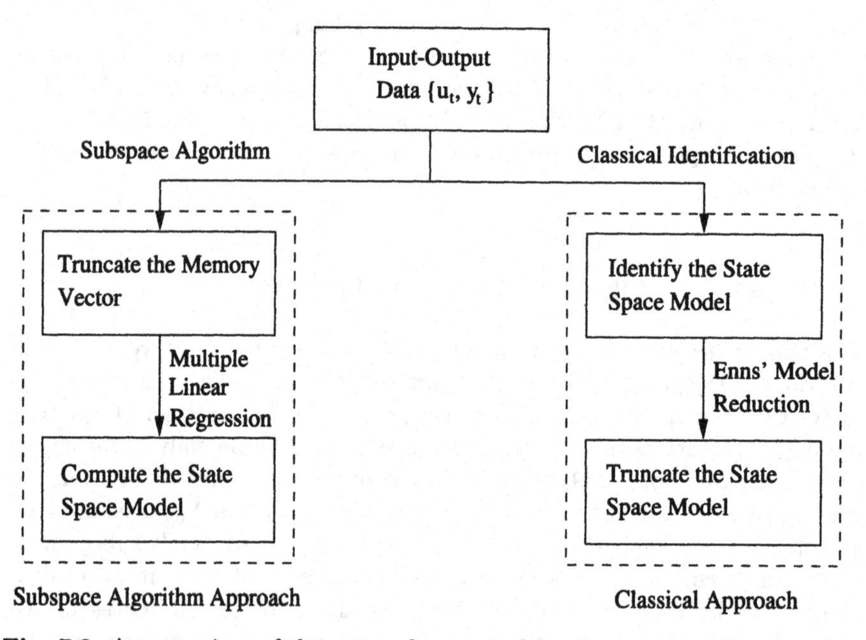

Fig. 7.2. A comparison of the approaches to model reduction using Enns' model reduction technique and the subspace algorithm [318]

7.8 Process Monitoring Statistics

The GSVD for the CVA algorithm produces a set of canonical variables, $\mathbf{c} = J\mathbf{p_t}$ (where $\mathbf{c} \in \mathcal{R}^{h(m_u+m_y)}$), that are uncorrelated and have unit variance. The T^2 statistic for the canonical variables is

$$T^2 = \mathbf{p_t}^T J^T J \mathbf{p_t}. \tag{7.39}$$

The T^2 statistic (7.39), however, may contain a large amount of noise and may not be very robust for monitoring the process. Reducing the order a for DPCA can increase the effectiveness of the T^2 statistic, and allows the process noise to be monitored separately via the Q statistic. An analogous approach is taken here for monitoring the process using the CVA states:

$$\mathbf{x_t} = J_k \mathbf{p_t} = U_k^T \hat{\Sigma}_{pp}^{-1/2} \mathbf{p_t} \tag{7.40}$$

where U_k contains the first k columns of U in (7.8)

A process monitoring statistic based on quantifying the variations of the CVA states has been applied by Negiz and Cinar to a milk pasteurization process [240, 241]. The measure is the T_s^2 statistic

$$T_s^2 = \mathbf{p_t}^T J_k^T J_k \mathbf{p_t}, \tag{7.41}$$

and assuming normality, the T_s^2 statistic follows the distribution

$$T_{s,\alpha}^2 = \frac{k(n^2-1)}{n(n-k)} F_\alpha(k, n-k) \tag{7.42}$$

where n is the number of observations (see 2.11). The T_s^2 statistic measures the variations *inside* the state space, and the process faults can be detected, as shown in Section 2.4, by choosing a level of significance and solving the appropriate threshold using $T_{s,\alpha}^2$.

The variations *outside* the state space can be measured using the statistic [279]

$$T_r^2 = \mathbf{p_t}^T J_q^T J_q \mathbf{p_t} \tag{7.43}$$

where J_q contains the last $q = h(m_u + m_y) - k$ rows of J in (7.8). Assuming normality, the T^2 statistic (7.43) follows the distribution

$$T_{r,\alpha}^2 = \frac{q(n^2-1)}{n(n-q)} F_\alpha(q, n-q). \tag{7.44}$$

A weakness of this approach is that T_r^2 can be overly sensitive because of the inversion of the small values of Σ_{xx} in (7.8) [145, 279]. This can result in a high false alarm rate. To address this concern, the threshold should be readjusted before applying the statistics for process monitoring (see Section 10.6 for an example).

The residual vector of the state-space model in terms of the past $\mathbf{p_t}$ can be calculated

$$\mathbf{r_t} = (I - J_k^T J_k)\mathbf{p_t}, \tag{7.45}$$

and the variation in the residual space can be monitored using the Q statistic similar to the (D)PCA approaches

$$Q = \mathbf{r_t}^T \mathbf{r_t}. \tag{7.46}$$

The statistics of T_r^2 and Q essentially measure the noise of the process. The T^2 statistic (7.39) is equal to $T_s^2 + T_r^2$, and by extracting the CVA states from the data, the variations in the state and measurement noise space can be decoupled and measured separately using T_s^2 and T_r^2, respectively. A violation of the T_s^2 statistic indicates that the states are out of control, and a violation of the T_r^2 statistic indicates that the characteristic of the measurement noise has changed and/or new states have been created in the process. This is similar to the PCA approach to fault detection outlined in Section 4.4, with the exception that the states of the system are extracted in a different manner. The flexibility of the state-space model and the near optimality of the CVA approach suggest that the CVA states more accurately represent the status of the operations compared to the scores using PCA or DPCA. Other CVA-based fault detection statistics are reported in the literature [190, 328].

The correlation structure of the CVA states allows the PCA-based statistics in Chapter 4 for fault identification and diagnosis to be applicable to the CVA model. It is straightforward to extend the PCA-based statistics to CVA. The total contribution statistic (4.25) can be computed for the CVA model by replacing the scores with the CVA estimated states, $\mathbf{m_t} = J_k \mathbf{p_t}$. The statistic (4.32) can be applied for fault identification using the residual vector in (7.45). A pattern classification system for fault diagnosis can be employed using the discriminant function (3.6) based on $(T_s^2)_i$, $(T_r^2)_i$, or Q_i for each class i. These discriminant functions can improve the classification system upon using the discriminant function (3.6) based on the entire observation space, $\mathbf{p_t}$, when most of the discriminatory power is contained in the state space or the residual space.

7.9 Homework Problems

1. Verify that the matrices J, L, and D computed from (7.8) satisfy (7.3) and (7.4).
2. Describe in some detail how to formulate the $CONT$ and RES statistics for identifying faults using CVA. Name advantages and disadvantages of this approach to alternative methods for identifying faults. Would $CONT$ or RES expected to perform better? Why?

3. Describe in detail how to formulate CVA for fault diagnosis. Name advantages and disadvantages of this approach to alternative methods for diagnosing faults.
4. Compare and contrast the CVA-based Q and T_r^2 statistics. Which statistic would you expect to perform better for fault detection? Why?
5. Read the following materials [189, 193, 283] and formulate PCA, PLS, FDA, and CVA in the framework of the generalized singular value decomposition. Based on the differences between the methods as represented in this framework, state the strengths and weaknesses of each method for applying process monitoring statistics.
6. Read a chapter in a book on the application of canonical correlation analysis (CCA) [67, 193, 234]. Compare and contrast CCA with FDA and CVA.
7. Compare and contrast the CVA-based statistics described in this chapter with the CVA-based process monitoring statistics reported in these papers [190, 328].
8. Read an article on the application of nonlinear CVA (e.g., [186]) and write a report describing in detail how the technique is implemented and applied. Describe how the computations are performed and how process monitoring statistics can be computed. For what types of processes are these algorithms suited? Provide some hypothetical examples.

Part IV

Application

8. Tennessee Eastman Process

8.1 Introduction

In Part IV the various data-driven process monitoring statistics are compared through application to a simulation of an industrial plant. The methods would ideally be illustrated on data collected during specific known faults from an actual industrial process, but this type of data is not publicly available for any large-scale industrial plant. Instead, many academics in process monitoring perform studies based on data collected from computer simulations of an industrial process. The process monitoring methods in this book are tested on the data collected from the process simulation for the **Tennessee Eastman process** (TEP). The TEP has been widely used by the process monitoring community as a source of data for comparing various approaches [16, 39, 40, 46, 99, 100, 113, 117, 183, 191, 270, 272, 271, 278, 279].

The TEP was created by the Eastman Chemical Company to provide a realistic industrial process for evaluating process control and monitoring methods [72]. The test process is based on a simulation of an actual industrial process where the components, kinetics, and operating conditions have been modified for proprietary reasons. The process consists of five major units: a reactor, condenser, compressor, separator, and stripper; and, it contains eight components: A, B, C, D, E, F, G, and H.

Chapter 8 describes the TEP in enough detail to interpret the application of the process monitoring statistics in Chapters 9 and 10. Sections 8.2 to 8.6 describe the process flowsheet, variables, faults, and simulation program. In reality, processes are operated under closed-loop control. To simulate realistic conditions, the second plant-wide control structure described in [205] was implemented to generate the data for demonstrating and comparing the various process monitoring methods. The control structure is described in Section 8.6. Detailed discussions on control structures for the TEP are available [219, 218, 237, 321].

8.2 Process Flowsheet

Figure 8.1 is a flowsheet for the industrial plant. The gaseous reactants A, C, D, and E and the inert B are fed to the reactor where the liquid products G and H are formed. The reactions in the reactor are:

$$\begin{aligned} A(g) + C(g) + D(g) &\rightarrow G(liq), \\ A(g) + C(g) + E(g) &\rightarrow H(liq), \\ A(g) + E(g) &\rightarrow F(liq), \\ 3D(g) &\rightarrow 2F(liq). \end{aligned} \tag{8.1}$$

The species F is a by-product of the reactions. The reactions are irreversible, exothermic, and approximately first-order with respect to the reactant concentrations. The reaction rates are Arrhenius functions of temperature where the reaction for G has a higher activation energy than the reaction for H, resulting in a higher sensitivity to temperature.

The reactor product stream is cooled through a condenser and then fed to a vapor-liquid separator. The vapor exiting the separator is recycled to the reactor feed through a compressor. A portion of the recycle stream is purged to keep the inert and byproduct from accumulating in the process. The condensed components from the separator (Stream 10) is pumped to a stripper. Stream 4 is used to strip the remaining reactants from Stream 10, which are combined with the recycle stream via Stream 5. The products G and H exiting the base of the stripper are sent to a downstream process which is not included in the diagram.

8.3 Process Variables

The process contains 41 measured and 12 manipulated variables. The manipulated variables are listed in Table 8.1. The 22 measured variables which are sampled every 3 minutes, XMEAS(1) through XMEAS(22), are listed in Table 8.2. The 19 composition measurements, XMEAS(23) through XMEAS(41), are described in Table 8.3. The composition measurements are taken from Streams 6, 9, and 11. The sampling interval and time delay for Streams 6 and 9 are both equal to 6 minutes, and for Stream 11 are equal to 15 minutes. All the process measurements include Gaussian noise.

8.4 Process Faults

The Tennessee Eastman process simulation contains 21 preprogrammed faults (see Table 8.4). Sixteen of these faults are known, and five are unknown. Faults 1-7 are associated with a step change in a process variable, e.g., in the cooling water inlet temperature or in feed composition. Faults 8-12 are

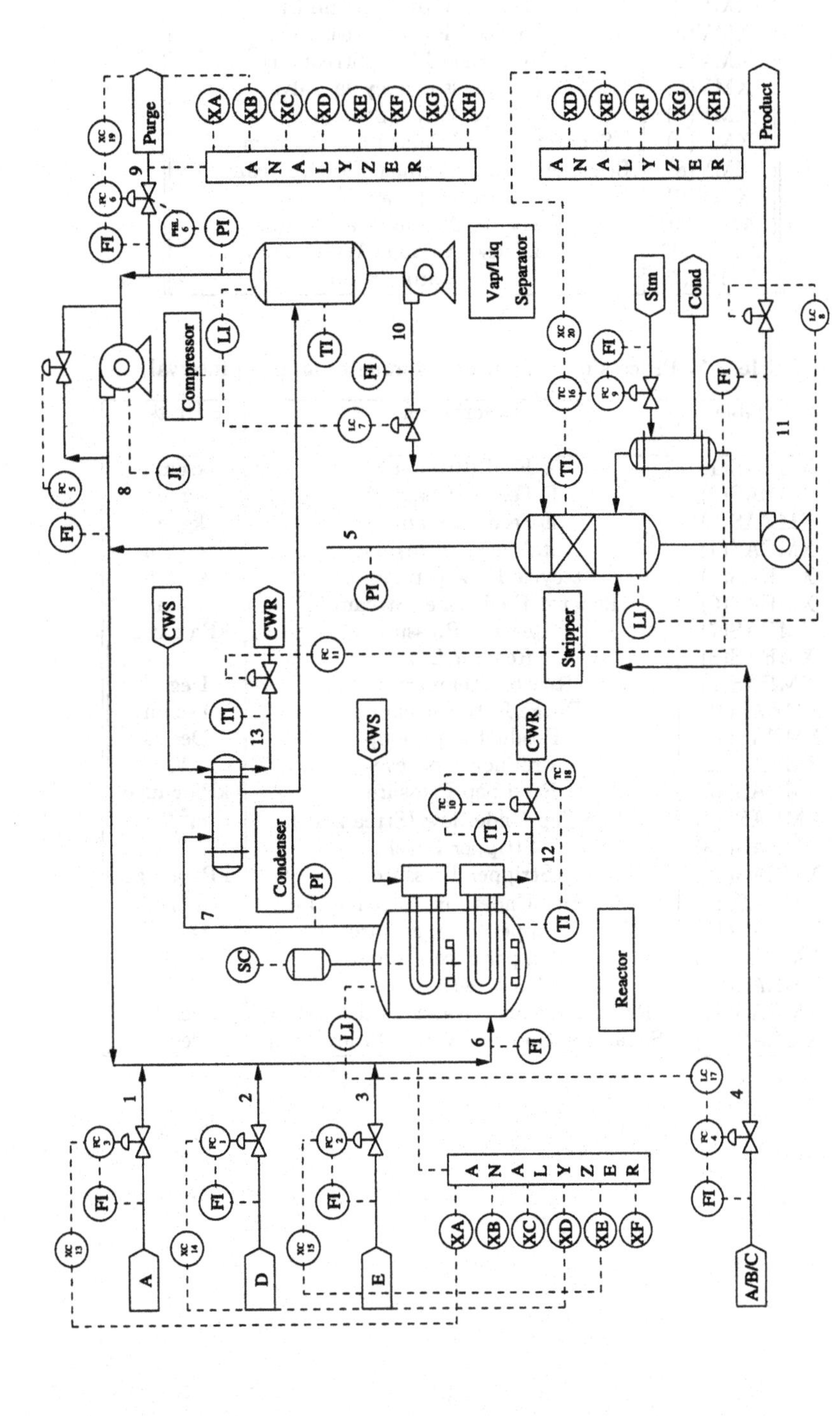

Fig. 8.1. A process flowsheet for the TEP with the second control structure in [205]

Table 8.1. Manipulated variables

Variable	Description
XMV(1)	D Feed Flow (Stream 2)
XMV(2)	E Feed Flow (Stream 3)
XMV(3)	A Feed Flow (Stream 1)
XMV(4)	Total Feed Flow (Stream 4)
XMV(5)	Compressor Recycle Valve
XMV(6)	Purge Valve (Stream 9)
XMV(7)	Separator Pot Liquid Flow (Stream 10)
XMV(8)	Stripper Liquid Product Flow (Stream 11)
XMV(9)	Stripper Steam Valve
XMV(10)	Reactor Cooling Water Flow
XMV(11)	Condenser Cooling Water Flow
XMV(12)	Agitator Speed

Table 8.2. Process measurements (3 minute sampling interval)

Variable	Description	Units
XMEAS(1)	A Feed (Stream 1)	kscmh
XMEAS(2)	D Feed (Stream 2)	kg/hr
XMEAS(3)	E Feed (Stream 3)	kg/hr
XMEAS(4)	Total Feed (Stream 4)	kscmh
XMEAS(5)	Recycle Flow (Stream 8)	kscmh
XMEAS(6)	Reactor Feed Rate (Stream 6)	kscmh
XMEAS(7)	Reactor Pressure	kPa gauge
XMEAS(8)	Reactor Level	%
XMEAS(9)	Reactor Temperature	Deg C
XMEAS(10)	Purge Rate (Stream 9)	kscmh
XMEAS(11)	Product Sep Temp	Deg C
XMEAS(12)	Product Sep Level	%
XMEAS(13)	Prod Sep Pressure	kPa gauge
XMEAS(14)	Prod Sep Underflow (Stream 10)	m^3/hr
XMEAS(15)	Stripper Level	%
XMEAS(16)	Stripper Pressure	kPa gauge
XMEAS(17)	Stripper Underflow (Stream 11)	m^3/hr
XMEAS(18)	Stripper Temperature	Deg C
XMEAS(19)	Stripper Steam Flow	kg/hr
XMEAS(20)	Compressor Work	kW
XMEAS(21)	Reactor Cooling Water Outlet Temp	Deg C
XMEAS(22)	Separator Cooling Water Outlet Temp	Deg C

Table 8.3. Composition measurements

Variable	Description	Stream	Sampling Interval (min.)
XMEAS(23)	Component A	6	6
XMEAS(24)	Component B	6	6
XMEAS(25)	Component C	6	6
XMEAS(26)	Component D	6	6
XMEAS(27)	Component E	6	6
XMEAS(28)	Component F	6	6
XMEAS(29)	Component A	9	6
XMEAS(30)	Component B	9	6
XMEAS(31)	Component C	9	6
XMEAS(32)	Component D	9	6
XMEAS(33)	Component E	9	6
XMEAS(34)	Component F	9	6
XMEAS(35)	Component G	9	6
XMEAS(36)	Component H	9	6
XMEAS(37)	Component D	11	15
XMEAS(38)	Component E	11	15
XMEAS(39)	Component F	11	15
XMEAS(40)	Component G	11	15
XMEAS(41)	Component H	11	15

Units are mole %. Dead time is equal to the sampling interval

associated with an increase in the variability of some process variables. Fault 13 is a slow drift in the reaction kinetics, and Faults 14, 15, and 21 are associated with sticking valves.

The sensitivity and robustness of the various process monitoring methods will be investigated in Chapter 10 by simulating the process under various fault conditions. The simulation program allows the faults to be implemented either individually or in combination with one another.

8.5 Simulation Program

The simulation code for the process is available in FORTRAN, and a detailed description of the process and simulation is available [72]. There are six modes to the process operation corresponding to various G/H mass ratios and production rates of Stream 11. Only the base case will be used here. The program is implemented with 50 states in open loop and a 1 second interval for integration. This integration interval is reasonable since the largest negative eigenvalue of the process is about 1.8 seconds. The simulation code for the process in open loop can be downloaded from `http://brahms.scs.uiuc.edu`.

Table 8.4. Process faults

Variable	Description	Type
IDV(1)	A/C Feed Ratio, B Composition Constant (Stream 4)	Step
IDV(2)	B Composition, A/C Ratio Constant (Stream 4)	Step
IDV(3)	D Feed Temperature (Stream 2)	Step
IDV(4)	Reactor Cooling Water Inlet Temperature	Step
IDV(5)	Condenser Cooling Water Inlet Temperature	Step
IDV(6)	A Feed Loss (Stream 1)	Step
IDV(7)	C Header Pressure Loss - Reduced Availability (Stream 4)	Step
IDV(8)	A, B, C Feed Composition (Stream 4)	Random Variation
IDV(9)	D Feed Temperature (Stream 2)	Random Variation
IDV(10)	C Feed Temperature (Stream 4)	Random Variation
IDV(11)	Reactor Cooling Water Inlet Temperature	Random Variation
IDV(12)	Condenser Cooling Water Inlet Temperature	Random Variation
IDV(13)	Reaction Kinetics	Slow Drift
IDV(14)	Reactor Cooling Water Valve	Sticking
IDV(15)	Condenser Cooling Water Valve	Sticking
IDV(16)	Unknown	
IDV(17)	Unknown	
IDV(18)	Unknown	
IDV(19)	Unknown	
IDV(20)	Unknown	
IDV(21)	The valve for Stream 4 was fixed at the steady state position	Constant Position

8.6 Control Structure

The simulation of the TEP is made available by the Eastman Chemical Company in open-loop operation. Since the process is open-loop unstable and industrial processes in reality are operated under closed loop, a plant-wide control scheme was employed when applying the process monitoring methods in Chapter 10. In [205, 206], four different plant-wide control structures using only Proportional (P) and Proportional-Integral (PI) controllers were investigated for the TEP. The second control structure listed in [205, 206] was chosen for this book because this structure provided the best performance according to the authors.

The control structure implemented to obtain the results in Chapter 10 is shown schematically in Figure 8.1. The control structure consists of nineteen loops, and the values of the control parameters and other details of the control structure are listed in Table 8.5. The exact values for the controller gains implemented by the author of [205] could not be determined because the controller gains were scaled to be dimensionless and the scalings on the controller inputs and outputs were not presented. However, we estimated the controller parameters based on the values from [205], and these parameters are reported in Table 8.5 with units consistent with the manipulated and measurement variables [72]. Some closed-loop simulations with the control parameters from Table 8.5 are shown in Figures 8.2 and 8.3. A comparison of these plots with those in [205] indicates that relatively similar values for the control parameters were employed for both sets of simulations. The simulation code for the process in closed loop can be downloaded from `http://brahms.scs.uiuc.edu`.

8.7 Homework Problems

1. Plot the manipulated and measured variables over time for one of the process faults in Table 8.4 using the closed-loop controllers (the code can be downloaded from `http://brahms.scs.uiuc.edu`). Explain how the effect of the process fault propagates through the plant, as indicated by the process variables. What is the physical mechanism for each of the process variable changes? Does each variable change in the way you would expect? Explain. For each variable, explain how its time history is affected by the closed-loop controllers. Which controllers mask the effect of the fault on the process variables? [Note to instructor: consider assigning a different fault to each student in the class.]
2. Describe the step-by-step procedure used to arrive at the plant-wide control structure used in this chapter (hint: read [206]).

Table 8.5. Control structure and parameter description

Loop Number	Manipulated Variable	Control Variable	Primary (P) Secondary (S)	Gain	Integral Time (sec.)	Sampling Interval (sec.)
1	XMV(1)	XMEAS(2)	S	0.172	∞	3
2	XMV(2)	XMEAS(3)	S	0.120	∞	3
3	XMV(3)	XMEAS(1)	S	98.3	∞	3
4	XMV(4)	XMEAS(4)	S	6.56	∞	3
5	XMV(5)	XMEAS(5)	–	−0.157	1	3
6*	XMV(6)	XMEAS(10)	S	122	∞	3
7	XMV(7)	XMEAS(12)	–	−2.94	∞	3
8	XMV(8)	XMEAS(15)	–	−2.31	∞	3
9	XMV(9)	XMEAS(19)	S	0.0891	∞	3
10	XMV(10)	XMEAS(21)	S	−1.04	1452	3
11	XMV(11)	XMEAS(17)	–	2.37	2600	3
12	–	–	–	–	–	–
13	XMEAS(1)	XMEAS(23)	P	1770	3168	360
14	XMEAS(2)	XMEAS(26)	P	0.143	3168	360
15	XMEAS(3)	XMEAS(27)	P	0.0284	5069	360
16	XMEAS(19)	XMEAS(18)	P/S	0.0283	236	3
17	XMEAS(4)	XMEAS(8)	P	14.6	3168	3
18	XMEAS(21)	XMEAS(9)	P	12.6	982	3
19	XMEAS(10)	XMEAS(30)	P	−2133	6336	360
20	XMEAS(18)	XMEAS(38)	P	−157	12408	900
21	–	–	–	–	–	–

The valve for Control Loop 6 is completely open for pressures above 2950 kPa gauge and closed for pressures below 2300 kPa gauge

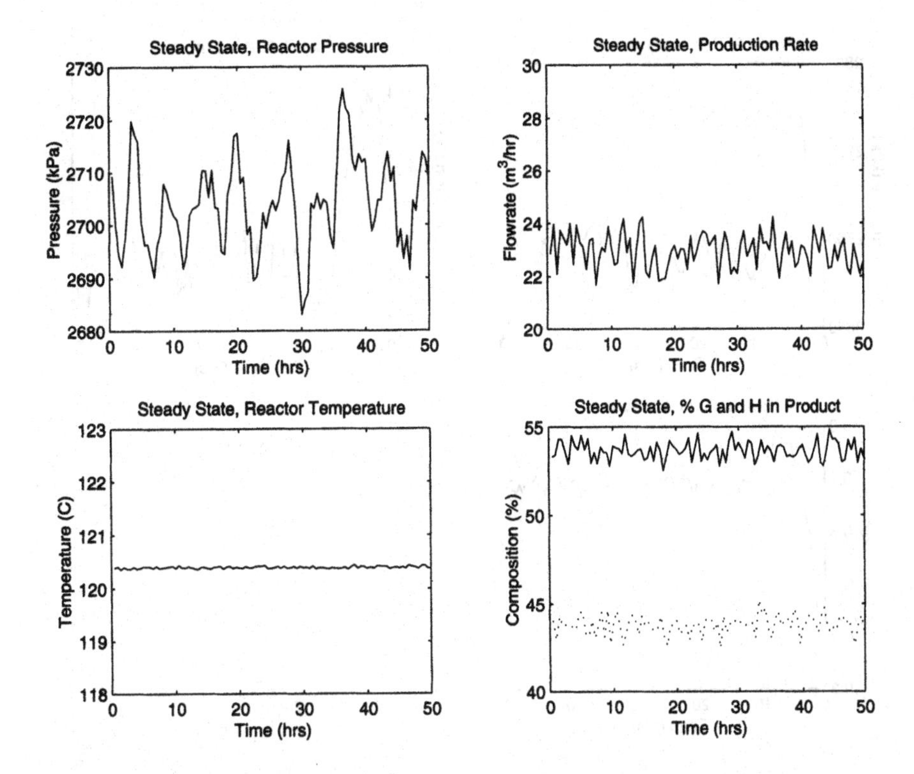

Fig. 8.2. Closed-loop simulation for the steady state case with no faults. The solid and dotted lines in the lower right plot represent the compositions of G and H, respectively.

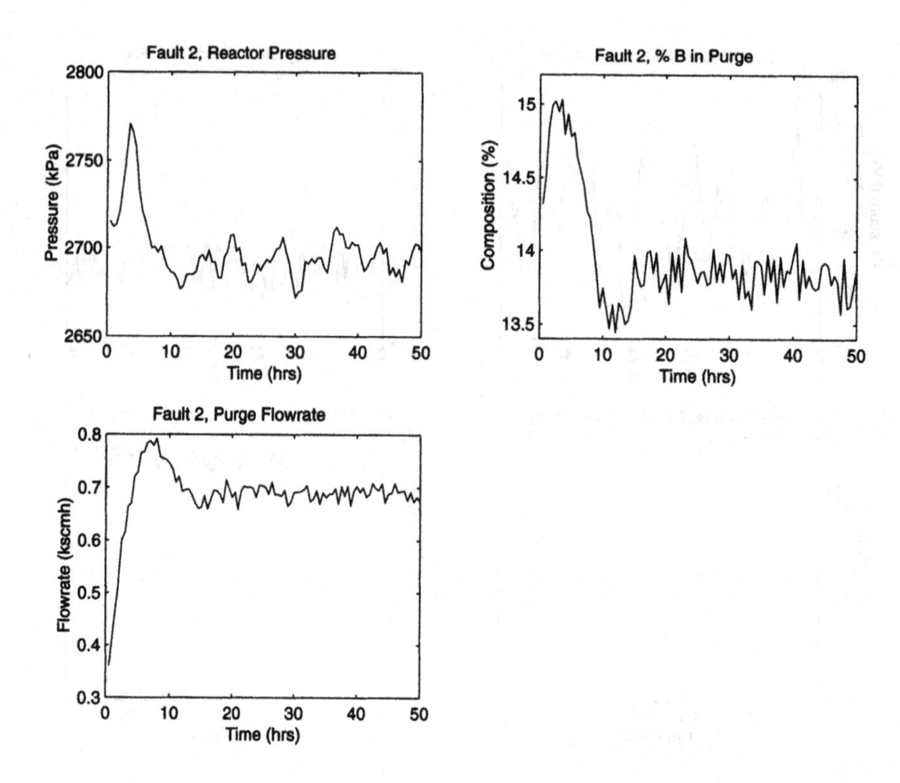

Fig. 8.3. Closed-loop simulation for a step change in the composition of the inert B (IDV(2) in Table 8.4)

9. Application Description

9.1 Introduction

Chapter 8 describes the process, the control system, and the type of faults for the Tennessee Eastman plant simulator. In Chapter 10, this simulator will be used to demonstrate and compare the various process monitoring methods presented in Part III. The process monitoring methods are tested on data generated by the TEP simulation code, operating under closed loop with the plant-wide control structure discussed in Section 8.6. The original simulation code allows 20 preprogrammed faults to be selectively introduced to the process [72]. We have added an additional fault simulation, which results in a total of 21 faults as shown in Table 8.4. In addition to the aforementioned aspects of the process, the process monitoring performance is dependent on the way in which the data are collected, such as the sampling interval and the size of the data sets.

The purpose of this chapter is to describe the data sets and to present the process monitoring measures employed for comparing the process monitoring methods. Section 9.2 describes how the data in the training and testing sets were generated by the TEP. A discussion on how the selection of the sampling interval and sample size of the data sets affects the process monitoring methods follows in Sections 9.3 and 9.4, respectively. Section 9.5 discusses the selection of the lag and order for each method. Sections 9.6, 9.7, and 9.8 present the measures investigated for fault detection, identification, and diagnosis, respectively. The process monitoring methods (covered in Parts II and III) used for these purpose are collected into Tables 9.2-9.4 which show how the methods are related.

9.2 Data Sets

The data in the training and testing sets included all the manipulated and measured variables (see Tables 8.1-8.3), except the agitation speed of the reactor's stirrer for a total of $m = 52$ observation variables. (The agitation speed was not included because it was not manipulated.) An observation vector at a particular time instant is given by

$$\mathbf{x} = [\text{XMEAS}(1), \cdots, \text{XMEAS}(41), \text{XMV}(1), \cdots, \text{XMV}(11)]^T. \qquad (9.1)$$

The observations were simulated with an integration step size of 1 second, and this did not produce any numerical inaccuracies. Although some of the observations are sampled continuously while other variables contain time delays (see Section 8.3), it simplifies the implementation to employ the same sampling interval for each variable when the data are collected for calculating multivariate process monitoring measures. A sampling interval of 3 minutes was used to collect the simulated data for the training and testing sets.

The data in the training set consisted of 22 different simulation runs, where the random seed was changed between each run. One simulation run (Fault 0) was generated with no faults; another simulation run (Fault 21) was generated by fixing the position of the valve for Stream 4 at the steady state position; and, each of the other 20 simulation runs (Faults 1-20) was generated under a different fault, each corresponding to a fault listed in Table 8.4. The simulation time for each run was 25 hours. The simulations started with no faults, and the faults were introduced 1 simulation hour into the run. The total number of observations generated for each run was $n = 500$, but only 480 observations were collected after the introduction of the fault. It is only these 480 observations actually used to construct the process monitoring measures.

The data in the testing set also consisted of 22 different simulation runs, where the random seed was changed between each run. These simulation runs directly correspond to the runs in the training set (Faults 0-21). The simulation time for each run was 48 hours. The simulation started with no faults, and the faults were introduced 8 simulation hours into the run. The total number of observations generated for each run was $n = 960$.

9.3 Sampling Interval

The amount of time in which quality data are collected from industrial processes during either in-control or out-of-control operations is usually limited in practice. Typically, only a small portion of the operation time exists where it can be determined with confidence that the data were not somehow corrupted and no faults occurred in the process. Also, the process supervisors do not generally allow faults to remain in the process for long periods of time for the purpose of producing data used in fault diagnosis algorithms.

Typically data collected during faulty operations are stored in historical databases in which engineers or operators diagnose the faults sometime after the fault occurs, and then enter that information into the historical database. The amount of such data available in the historical database is typically fixed and the sampling interval for the process monitoring methods needs to be determined.

It is desirable to detect, identify, and diagnose faults as soon as possible. This suggests a high sampling rate. Also, given a fixed time $T = n\Delta t$, it is beneficial from an information point-of-view to sample as fast as possible ($\Delta t \to 0$, $n \to \infty$). In terms of process monitoring, however, there are three possible problems with sampling as fast as possible. For the amount of data produced, the computational requirements may exceed the computational power available. Additionally, the model fit may be concentrated to the higher frequencies, where measurement noise is predominant. When identifying an ARX model via a least-squares approach, Ljung [199] shows how the bias is shifted when sampling with higher frequencies. This bias shift for fast sampling rates may be undesirable, especially if the faults primarily affect the lower frequency dynamics of the process. Finally, statistics that ignore serial correlation will generally perform more poorly for short sampling times.

The choice of the sampling interval for process monitoring is usually selected based on engineering judgment. For system identification, a rule of thumb is to set the sampling interval to one-tenth the time constant of the process [199]. Considering that many of the time constants of the Tennessee Eastman problem under closed loop appear to be about 2 hours (see Figure 8.2), it is advisable from a system identification point of view to sample at an interval of 12 minutes. This does not, however, take advantage of the instrumentation of the process, which allows much faster sampling rates (see Section 8.3). A sampling interval of 3 minutes was selected here to allow fast fault detection, identification, and diagnosis, and to allow a good comparison between techniques that either take into account or ignore serial correlations. In addition, the same sampling interval has been used in other applications of process monitoring to the TEP [46, 113, 183, 279].

An alternative approach would be to average each measurement over a period of time before using the data in the process monitoring algorithms. This and similar "moving window" techniques will generally reduce normal process variability and hence produce a more sensitive process monitoring method. However, this comes at a cost of delaying fault detection. Wise and co-workers [345] pointed out that the width of the windows (*i.e.*, the number of data points used to compute the average) had an important effect on the performance. In general, a "wide" window allows the detection of smaller changes, but does not respond as quickly to changes as "narrow" windows.

9.4 Sample Size

As mentioned in the previous section, the total time spanned by the training set is generally limited. In the cases where the total time $T = n\Delta t$ is fixed, the selection of the sampling interval Δt and the sample size n cannot be decoupled. Therefore, the effect of the sampling interval on the sample size should be considered when selecting the sampling interval, and *vice versa.*

An important consideration for the sample size is the total number of independent parameters contained in the model being identified. It is desirable to have the number of model parameters be much smaller than the total number of process variables m multiplied by the total number of observations n.

Because the data for this book are simulated by the TEP, the sample size is not limited by T and can be considered separately from the sampling interval. Downs and Vogel [72] recommend a simulation time between 24 and 48 hours to realize the full effect of the faults. With a sampling interval equal to 3 minutes, 24 to 48 hours of simulation time contain $n = 480$ to 960 observations. Simulations (see Figure 8.3) suggest that a run containing 24 simulation hours sufficiently captures the significant shifts in the data produced by the fault.

The sufficiency of the sample size for the training set $n = 480$ can be determined by examining the total number of independent parameters associated with the orders of the various process monitoring methods (see Table 9.1). The total number of states in the closed-loop process is $k = 61$; 50 states from the open-loop process plus 11 states from the PI controllers. For a state-space model of state order $k = 61$ with 11 inputs and 41 outputs, the number of independent parameters M_k is equal to 6985 according to (7.36). For fault detection using the PCA-based T^2 statistic (4.12), the number of estimated parameters M_a is equal to the number of independent degrees of freedom of the matrix product of $P\Sigma_a^{-2}P^T$ in (4.12), which is calculated from

$$M_a = \frac{a + 2am - a^2}{2}. \tag{9.2}$$

For $a = 51$, the number of independent parameters is 1377. For fault detection using the CVA-based T_s^2 statistic (7.41), the number of estimated parameters M_k is equal to the number of independent degrees of freedom of $J_k^T J_k$ in (7.41), which is calculated from

$$M_k = \frac{k + 2kmh - k^2}{2}. \tag{9.3}$$

For $h = 2$ and $k = 61$, the number of independent parameters is 4029. The total number of data points in the training set is equal to $nm = (480)(52) = 24,960$. The absolute minimum requirement to apply the PCA, CVA, or state-space model at a given order is that the number of data points is greater than the number of independent parameters in the model. The ratio of the number of data points to the number of independent parameters is $nm/M_k = (480)(52)/6985 = 3.57$ for the state-space model, $nm/M_a = 18.1$ for the PCA-based model, and $nm/M_k = 5.53$ for the CVA-based model. With all other variables being equal (e.g., the noise level), the larger the ratio is greater than one, the higher the accuracy of the model. For this data set, all ratios are greater than one, indicating that the size of the training set ($n = 480$) is sufficient to apply the PCA, CVA, and state-space model. Reducing the order may still result in a higher quality model, depending on the noise level. As shown in Table 9.1, the state-space model requires the largest number of independent parameters, followed by CVA, and PCA. A PCA model of a given order has significantly less independent parameters, but does not take into account serial correlations.

Table 9.1. The number of independent parameters estimated for the various models and orders

Order†	Inputs	Outputs	Parameters State Space††	Parameters PCA†††	Parameters CVA††††
1	11	41	1405	52	104
11	11	41	2335	517	1089
21	11	41	3265	882	1974
31	11	41	4195	1147	2759
41	11	41	5125	1312	3444
51	11	41	6055	1377	4029
61	11	41	6985	–	4514

† The order is equal to a for PCA and the state order k for the state-space model and CVA
†† The number of parameters is based on (7.36)
††† The number of parameters is based on (9.2)
†††† The number of parameters is based on (9.3), using $h = 2$ lags

9.5 Lag and Order Selection

The number of lags included in the DPCA, DFDA, and CVA process monitoring methods can substantially affect the monitoring performance. It is best to choose the number of lags as the minimum needed to capture the dynamics of the process accurately. Choosing the number of lags larger than necessary may significantly decrease the robustness of the process monitoring measures, since the extra dimensionality captures additional noise, which may be difficult to characterize with limited data. The procedure used for this book follows Larimore's suggestion of selecting the number of lags h as that minimizing the small sample AIC criterion using an ARX model (see Section 7.5). This ensures that the number of lags is large enough to capture all the statistically significant information in the data.

As described in Part III, the selection of the reduction order is critical to developing efficient measures for process monitoring. The order selection methods described in Part III will be used. The parallel analysis method (see Section 4.3) is applied to select a in PCA and DPCA. The information criterion (5.12) is used to determine a for FDA and DFDA. The small sample AIC (7.37) is applied to CVA to determine the state order k.

Although it is popularly referred to in the literature, the cross-validation method is not used here for any of the process monitoring methods. Cross-validation is computationally expensive when dealing with several large data sets. More importantly, there can be problems with cross-validation when serial correlations in the data exist [183].

9.6 Fault Detection

The proficiencies of PCA, DPCA, and CVA for detecting faults were investigated on the TEP. The measures applied for each method, the corresponding equation numbers, and the distributions used to determine the thresholds for the measures are listed in Table 9.2. For instance, the first row indicates that PCA is used to generate the T^2 statistic according to (4.12) and the threshold is calculated according to (4.14). The distribution listed as "TR" means that the threshold is set to be the tenth highest value for Fault 0 of the *testing* set, in which the number of observations $n = 960$. The threshold corresponds to a level of significance $\alpha = 0.01$ by considering the probability distribution of the statistics for Fault 0. A thorough discussion of the measures is available in the respective chapters, and more information related to applying these measures to the TEP is contained in Section 10.6.

Table 9.2. The measures employed for fault detection

Method	Basis	Equation	Distribution
PCA	T^2	4.12	4.14
PCA	Q	4.21	4.22
DPCA	T^2	4.12†	4.14†
DPCA	Q	4.21†	4.22†
CVA	T_s^2	7.41	7.42
CVA	T_r^2	7.43	7.41
CVA	Q	7.46	TR††

† Applied to the data matrix with lags

††TR - Threshold set based on testing data for Fault 0

There exist techniques to increase the sensitivity and robustness of the PCA and DPCA process monitoring measures as described in Section 4.8, for example, through the use of the CUSUM or EWMA version of the measures. However, these techniques compromise the response time of the measures. Although such techniques can be highly useful in practice, the process monitoring methods applied in Chapter 10 do not employ them because it would complicate the comparison of the process monitoring methods. The measures investigated for each process monitoring method are designed to detect and diagnose the faults with the smallest delay. Applying the CUSUM and EWMA versions of PCA and DPCA is left as a homework problem.

9.7 Fault Identification

The proficiencies of PCA, DPCA, and CVA for identifying faults were investigated on the TEP. The measures applied for each method and the corresponding equation numbers are presented in Table 9.3. A discussion on how to apply the measures based on PCA, DPCA, and CVA can be found in Sections 4.5, 4.7, and 7.8, respectively. A thorough discussion of the measures is available in the respective chapters, and more information related to applying these measures to the TEP is contained in Section 10.7.

Table 9.3. The measures employed for fault identification

Method	Basis	Equation
PCA	*CONT*	4.25
PCA	*RES*	4.32
DPCA	*CONT*	4.25 with 4.44
DPCA	*RES*	4.32 with 4.44
CVA	*CONT*	4.25 with 7.22
CVA	*RES*	4.32 with 7.45

9.8 Fault Diagnosis

The proficiencies of the fault diagnosis methods described in Part III were investigated on the TEP. Fault diagnosis measures based on discriminant analysis that use no dimensionality reduction are given in (3.7). When this multivariate statistic (MS) is applied to data with no lags, it will be referred to as the T_0^2 statistic. When the multivariate statistic is applied to data with 1 lag, it will be referred to as the T_1^2 statistic. These are considered in Chapter 10 to serve as a benchmark for the other measures, as the dimensionality should only be reduced if it decreases the misclassification rate for a testing set. The fault diagnosis measures and the corresponding equation or section numbers are presented in Table 9.4. The statistic(s) upon which each measure is based is also listed in the table. A thorough discussion of the measures is available in the respective chapters, and more information related to applying these measures to the TEP is contained in Section 10.8.

Table 9.4. The measures employed for fault diagnosis

Method	Basis	Equation/Section
PCAm	T^2	Equation 4.35†
PCA1	T^2	Equation 4.33†
PCAm	Q	Equation 4.37
PCAm	T^2 & Q	Equation 4.38††
DPCAm	T^2	Equations 4.35† and 4.44
DPCAm	Q	Equations 4.37 and 4.44
DPCAm	T^2 & Q	Equations 4.38†† and 4.44
FDA	T^2	Equation 5.16†
FDA/PCA1	T^2	Equations 5.17† and 5.16
FDA/PCA2	T^2	Equations 5.17† and 5.16
DFDA/DPCA1	T^2	Equations 5.17†, 5.16, and 4.44
CVA	T_s^2	Equations 4.35† and 7.41
CVA	T_r^2	Equations 4.35 and 7.43
CVA	Q	Equations 4.37 and 7.46
PLS1	–	Section 6.3
PLS2	–	Section 6.3
PLS1_{adj}	–	Section 6.4
PLS2_{adj}	–	Section 6.4
MS	T_0^2	Equation 3.7
MS	T_1^2	Equation 3.7

† Applied to the score space only
†† $c_i = 0.5$ and $\alpha = 0.01$

10. Results and Discussion

10.1 Introduction

In this chapter, the process monitoring methods in Part III are compared and contrasted through application to the **Tennessee Eastman process** (TEP). The proficiencies of the process monitoring statistics listed in Tables 9.2-9.4 are investigated for fault detection, identification, and diagnosis. The evaluation and comparison of the statistics are based on criteria that quantify the process monitoring performance. To illustrate the strengths and weaknesses of each statistic, Faults 1, 4, 5, and 11 are selected as specific case studies in Sections 10.2, 10.3, 10.4, and 10.5, respectively. Sections 10.6, 10.7, and 10.8 present and apply the quantitative criteria for evaluating the fault detection, identification, and diagnosis statistics, respectively. The *overall* results of the statistics are evaluated and compared. Results corresponding to the case studies are highlighted in boldface in Tables 10.6 to 10.20.

10.2 Case Study on Fault 1

In the normal operating condition (Fault 0), Stream 4 in Figure 8.1 contains 0.485, 0.005, and 0.510 mole fraction of A, B, and C, respectively [72]. When Fault 1 occurs, a step change is induced in the A/C feed ratio in Stream 4, which results in an increase in the C feed and a decrease in the A feed in Stream 4. This results in a decrease in the A feed in the recycle Stream 5 and a control loop reacts to increase the A feed in Stream 1 (see Figure 10.1). These two effects counteract each other over time, which results in a constant A feed composition in Stream 6 after enough time (see Figure 10.2).

The variations in the flow rates and compositions of Stream 6 to the reactor causes variations in the reactor level (see Figure 8.1), which affects the flow rate in Stream 4 through a cascade control loop (see Figure 10.3). The flow rate of Stream 4 eventually settles to a steady-state value lower than its value at the normal operating conditions.

Since the ratio of the reactants A and C changes, the distribution of the variables associated with material balances (*i.e.*, level, pressure, composition) changes correspondingly. Since more than half of the variables monitored deviate significantly from their normal operating behavior, this fault is expected to be easily detected. Process monitoring statistics that show poor performance on Fault 1 are likely to perform poorly on other faults as well.

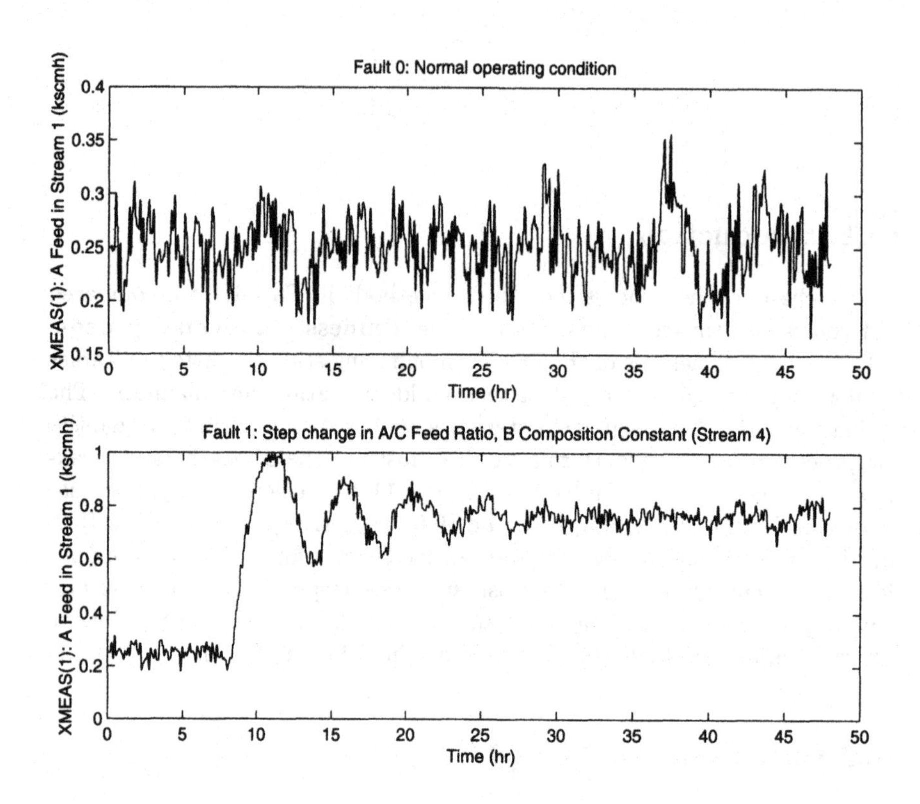

Fig. 10.1. Comparison of XMEAS(1) for Faults 0 and 1

The (D)PCA-based and CVA-based statistics for fault detection are shown in Figures 10.4 and 10.5, respectively. The dotted line in each figure is the threshold for the statistic, the statistic above its threshold indicates that a fault is detected (the statistic is shown as a solid line). The first eight hours were operated under normal operating conditions. Thus, all statistics are expected to fall below the thresholds for the first eight hours, which they did. The quantitative fault detection results are shown in Table 10.1. All of the statistics produced nearly zero missed detection rates. For a fault that significantly changes the distribution of the variables monitored, all fault detection statistics perform very well.

Assuming that process data collected during a fault are represented by a previous fault class, the objective of the fault diagnosis statistics in Table 9.4 is to classify the data to the *correct* fault class. That is, a highly proficient fault diagnosis statistic produces small misclassification rates when applied

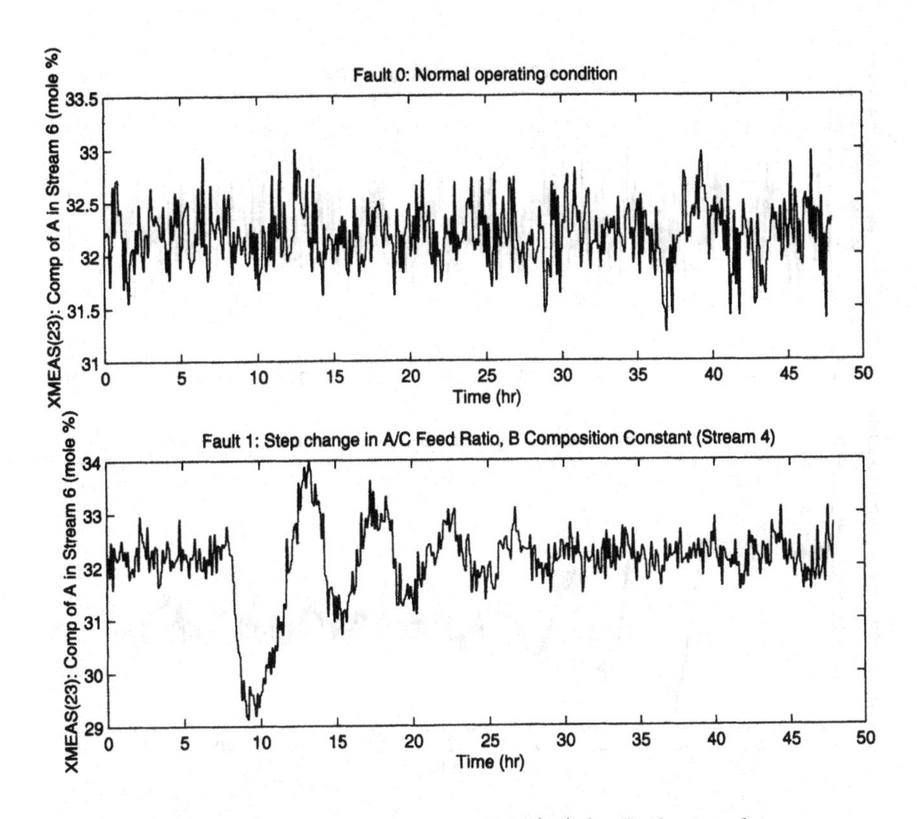

Fig. 10.2. Comparison of XMEAS(23) for Faults 0 and 1

Table 10.1. Missed detection rates for Faults 1, 4, 5, and 11

Method	Fault Basis	1	4	5	11
PCA	T^2	0.008	0.956	0.775	0.794
PCA	Q	0.003	0.038	0.746	0.356
DPCA	T^2	0.006	0.939	0.756	0.801
DPCA	Q	0.005	0	0.748	0.193
CVA	T_s^2	0.001	0.688	0	0.515
CVA	T_r^2	0	0	0	0.195
CVA	Q	0.003	0.975	0	0.669

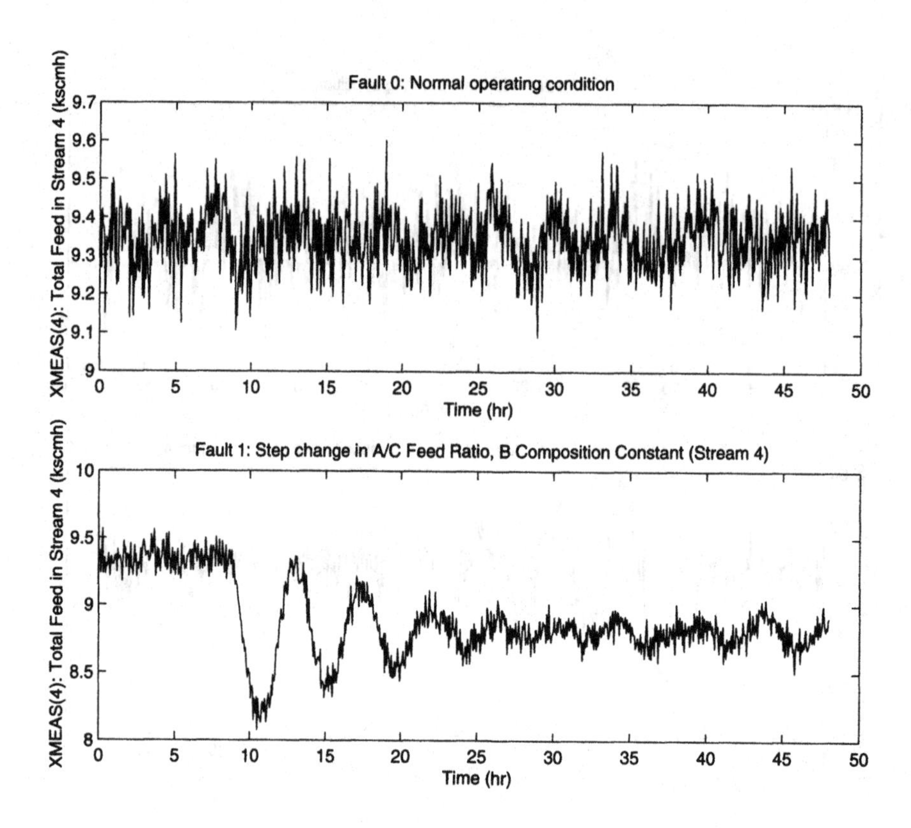

Fig. 10.3. Comparison of XMEAS(4) for Faults 0 and 1

to data independent of the training set. As shown in Table 10.2, most of the fault diagnosis statistics performed very well (Fault 1 being correctly diagnosed > 96% of the time).

10.3 Case Study on Fault 4

Fault 4 involves a step change in the reactor cooling water inlet temperature (see Figure 8.1). A significant effect of Fault 4 is to induce a step change in the reactor cooling water flow rate (see Figure 10.6). When the fault occurs, there is a sudden temperature increase in the reactor (see Figure 10.7 at time = 8 hr), which is compensated by the control loops. The other 50 measurement and manipulated variables remain steady after the fault occurs; the mean and standard deviation of each variable differ less than 2% between Fault 4 and the normal operating condition. This makes the fault detection and diagnosis tasks more challenging than for Fault 1.

Table 10.2. The overall misclassification rates for Faults 1, 4, 5, and 11

Method	Fault Basis	1	4	5	11
PCAm	T^2	0.680	0.810	0.956	0.989
PCA1	T^2	0.024	0.163	0.021	0.234
PCAm	Q	0.028	0.951	0.913	0.859
PCAm	T^2&Q	0.041	1.000	0.973	0.968
DPCAm	T^2	0.880	0.720	0.874	0.948
DPCAm	Q	0.035	0.964	0.856	0.843
DPCAm	T^2&Q	0.038	1.000	1.000	0.983
PLS1	–	0.013	0.170	0.006	0.989
PLS2	–	0.013	0.119	0.008	0.979
PLS1$_{adj}$	–	0.019	0.364	0.044	0.859
PLS2$_{adj}$	–	0.019	0.320	0.043	0.886
CVA	T_s^2	0.028	0.981	0.061	0.904
CVA	T_r^2	0.026	0.358	0.040	0.139
CVA	Q	0.245	0.890	0.174	0.901
FDA	T^2	0.025	0.176	0.020	0.245
FDA/PCA1	T^2	0.024	0.163	0.020	0.244
FDA/PCA2	T^2	0.025	0.176	0.020	0.245
DFDA/DPCA1	T^2	0.026	0.159	0.023	0.118
MS	T_0^2	0.025	0.178	0.020	0.245
MS	T_1^2	0.035	0.427	0.040	0.121

The extent to which the (D)PCA-based and CVA-based statistics are sensitive to Fault 4 can be examined in Figure 10.8 and Figure 10.9 respectively. The quantitative fault detection results are shown in Table 10.1. The variation in the residual space was captured by T_r^2, but not by the CVA-based Q statistic. The potential advantage of applying T_r^2 to capture variation in the residual space is clearly shown. It is interesting to see that the PCA and DPCA-based Q statistics were able to detect Fault 4, but the CVA-based Q statistic did not. The CVA-based T_s^2 statistic passes the threshold much of time after the fault occurs, but does not have the persistence of the CVA-based T_r^2 statistic (see Figure 10.9). Although the PCA and DPCA-based Q statistics both are able to detect the fault, the DPCA-based Q statistic outperformed the PCA-based statistic in terms of exceeding the threshold by a greater degree. This indicates the potential advantage of taking serial correlation into account when developing fault detection procedures.

For this fault the PCA and DPCA-based Q statistics were more sensitive than the PCA and DPCA-based T^2 statistics, and the CVA-based T_r^2

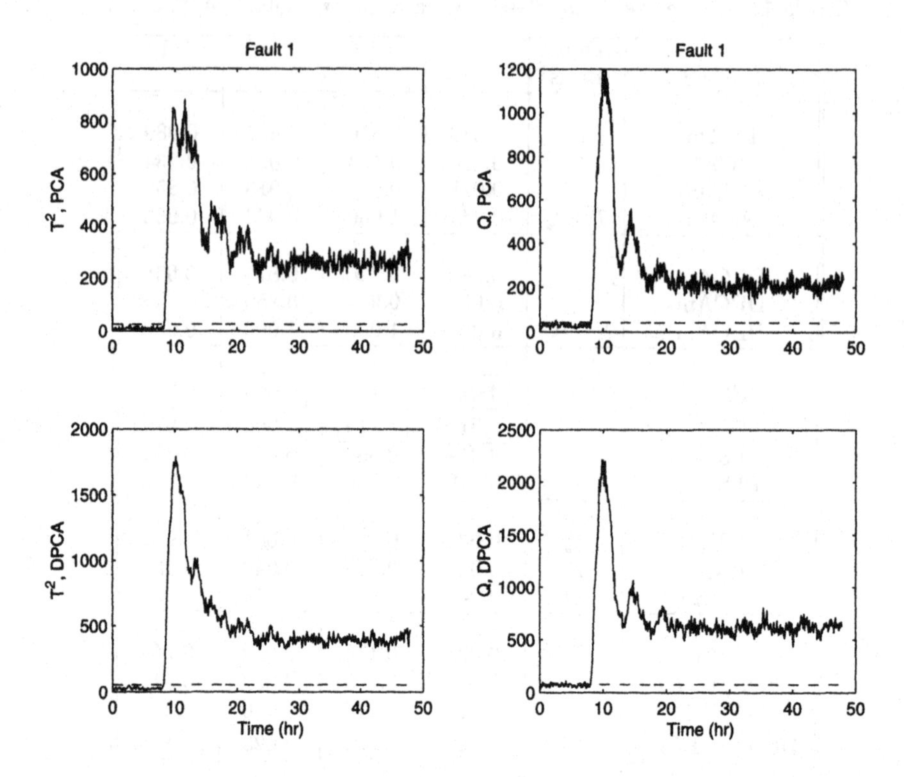

Fig. 10.4. The (D)PCA multivariate statistics for fault detection for Fault 1

statistic was more sensitive than the CVA-based T_s^2 statistic (see Table 10.1). These statistics quantifying variations in the residual space were overall more sensitive to Fault 4 than the statistics quantifying the variations in the score or state space. In other words, the fault created new states in the process rather than magnifying the states based on in-control operations. Although this conclusion does not hold for all faults, it certainly is true for a large portion of them.

Recall that Fault 4 is associated with a step change in the reactor cooling water inlet temperature (see Table 8.4), which is unmeasured. Engineering judgment and an examination of Figure 8.1 and Tables 8.1-8.3 indicate that the most closely related observation variable is the reactor cooling water flow rate. The fault identification statistics in Table 9.3 provide a rank ordering of the observation variables from most relevant to least relevant in terms of being associated with the fault. For Fault 4, the third column of Table 10.3 lists where the reactor cooling water flow rate was ranked by the various fault identification methods. All of the methods correctly ranked the reactor

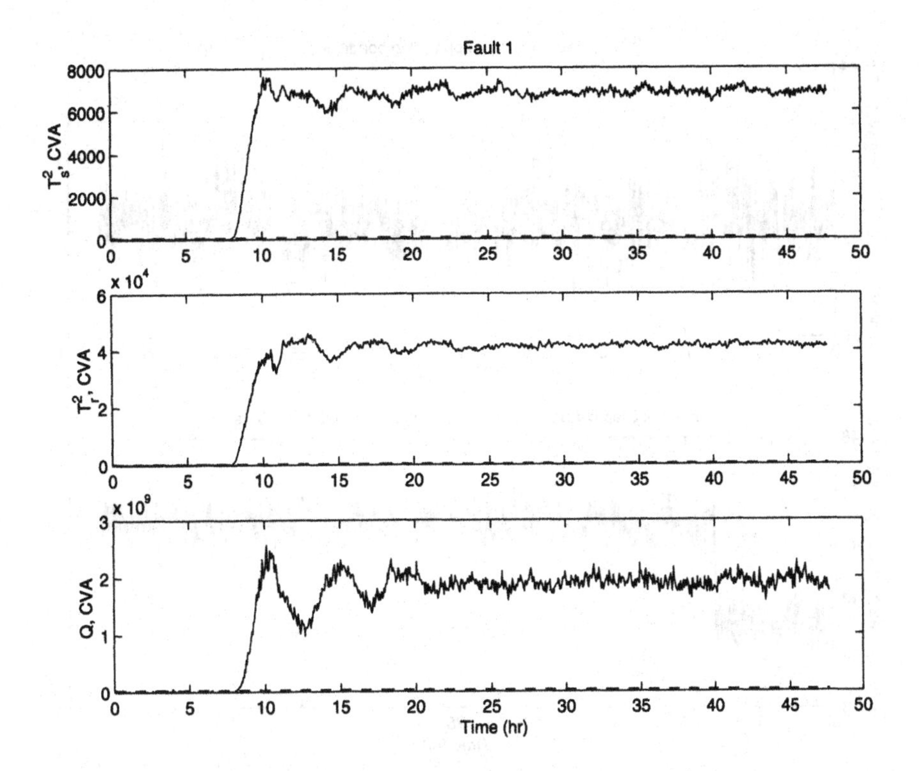

Fig. 10.5. The CVA multivariate statistics for fault detection for Fault 1

cooling water flow rate as most closely related to Fault 4 except for the CVA-based *CONT* statistic.

Table 10.3. The overall rankings for Faults 4 and 11

Method	Fault Basis	4	11
PCA	*CONT*	1	1
PCA	*RES*	1	1
DPCA	*CONT*	1	1
DPCA	*RES*	1	1
CVA	*CONT*	11	13
CVA	*RES*	1	1

The CVA-based *CONT* statistic did not perform well because the inverse of the matrix $\hat{\Sigma}_{pp}$ in (7.40) allowed certain observation variables to dominate the statistic. In particular, the maximum values of the J_k matrix corresponding to the observation variables x_{12}, x_{15}, x_{17}, x_{48}, x_{49}, and x_{52} are above 50

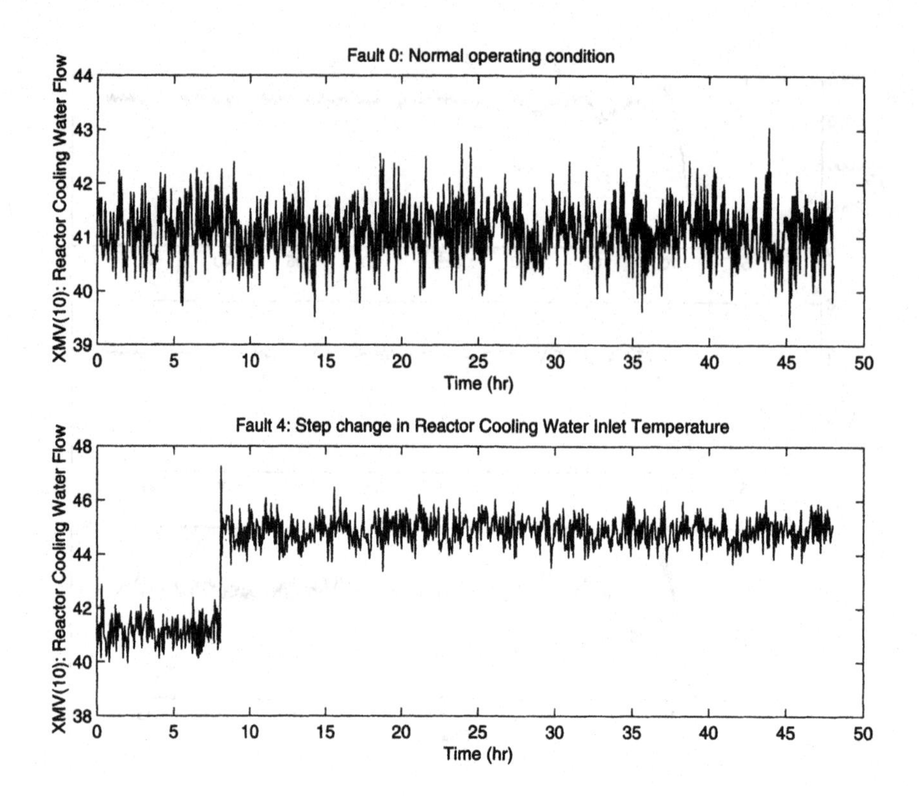

Fig. 10.6. Comparison of XMV(10) for Faults 0 and 4

while the elements of J_k corresponding to all the other variables are less than 3 (see Figure 10.10). The dominance of the observation variables x_{12}, x_{15}, x_{17}, x_{48}, x_{49}, and x_{52} in J_k was observed for all of the other faults investigated as well.

For fault diagnosis, many of the statistics performed poorly for Fault 4 (see Table 10.2). PLS2 gave the lowest misclassification rates. This indicates that discriminant PLS can outperform FDA for some faults although it would be expected theoretically that FDA should be better in most cases. PLS1 had a similar misclassification rate as all the FDA-based statistics, PCA1, and MS T_0^2. PLS1 and PLS2 gave significantly lower misclassification rates than PLS1_{adj} and PLS2_{adj}. This makes the point that the adjustment procedure described in Section 6.4 does not always improve fault diagnosis.

DFDA/DPCA1 produced similar misclassification rates as the static FDA methods. However, including lagged variables actually degraded the performance of the MS statistic.

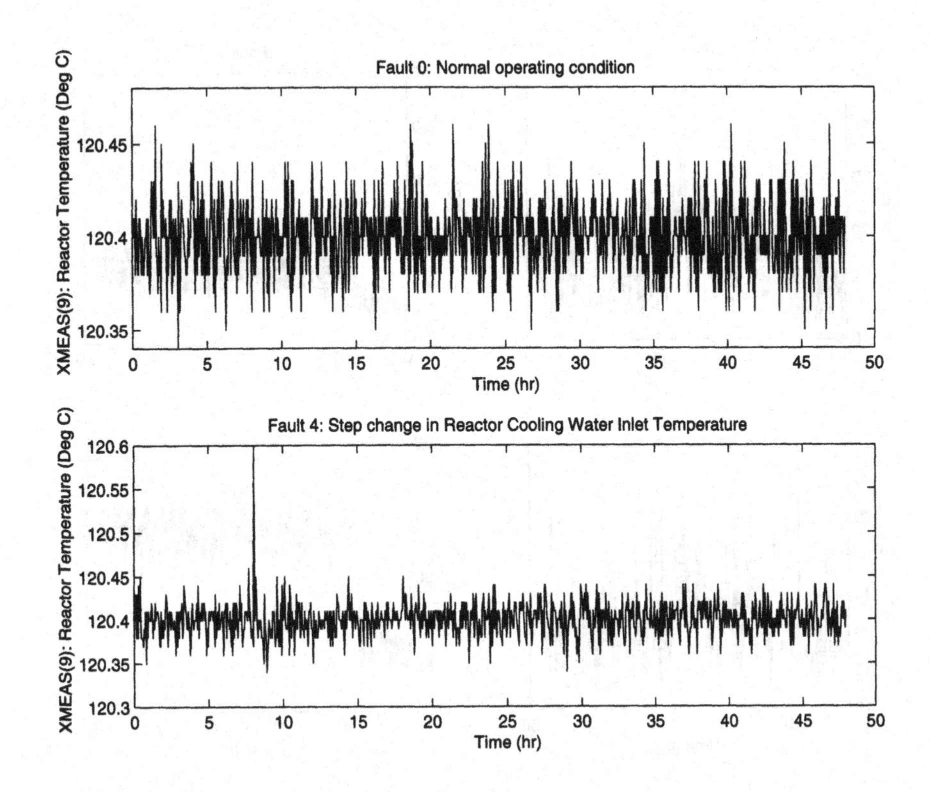

Fig. 10.7. Comparison of XMEAS(9) for Faults 0 and 4

10.4 Case Study on Fault 5

Fault 5 involves a step change in the condenser cooling water inlet temperature (see Figure 8.1). The significant effect of the fault is to induce a step change in the condenser cooling water flow rate (see Figure 10.11). When the fault occurs, the flow rate of the outlet stream from the condenser to the vapor/liquid separator also increases, which results in an increase in temperature in the vapor/liquid separator, and thus the separator cooling water outlet temperature (see Figure 10.12). Similar to Fault 4, the control loops are able to compensate for the change and the temperature in the separator returns to its set-point. The time it takes to reach the steady state is about 10 hours. For the rest of the 50 variables that are being monitored, 32 variables have similar transients that settle in about 10 hours. Detecting and diagnosing such a fault should not be a challenging task.

The (D)PCA-based and CVA-based statistics for fault detection are shown in Figures 10.13 and 10.14, respectively. The quantitative fault detection results are shown in Table 10.1, where it is seen that the (D)PCA-based

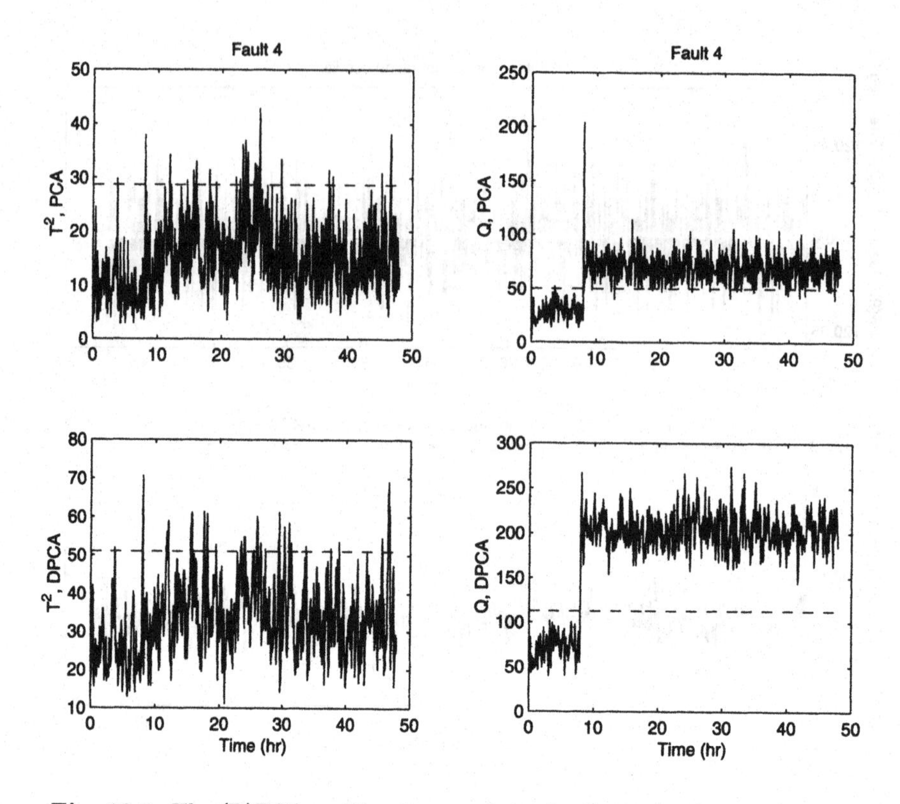

Fig. 10.8. The (D)PCA multivariate statistics for fault detection for Fault 4

statistics had a high missed detection rate, and all the CVA statistics had a zero missed detection rate. The reason for the apparent poor behavior of the (D)PCA-based statistics is clear from plotting the observation variables over time. Most variables behaved similarly to Figure 10.12—they returned to their set-points 10 hours after the fault occurred. The (D)PCA-based statistics fail to indicate a fault 10 hours after the fault occurs (see Figure 10.13). On the other hand, all the CVA statistics stayed above their thresholds (see Figure 10.14).

The persistence of a fault detection statistic (the CVA statistic in this case) is important in practice. At any given time a plant operator has several simultaneous tasks to perform and typically does not focus on all tasks with the same degree of attentiveness. Also, it usually takes a certain amount of time to track down the cause of abnormal process operation. When the time to locate the source of a fault is longer than the persistence of the fault detection statistic, a plant operator may conclude that the fault has "corrected itself" and assume that the process is again operating in normal operating conditions. In contrast, a persistent fault detection statistic will

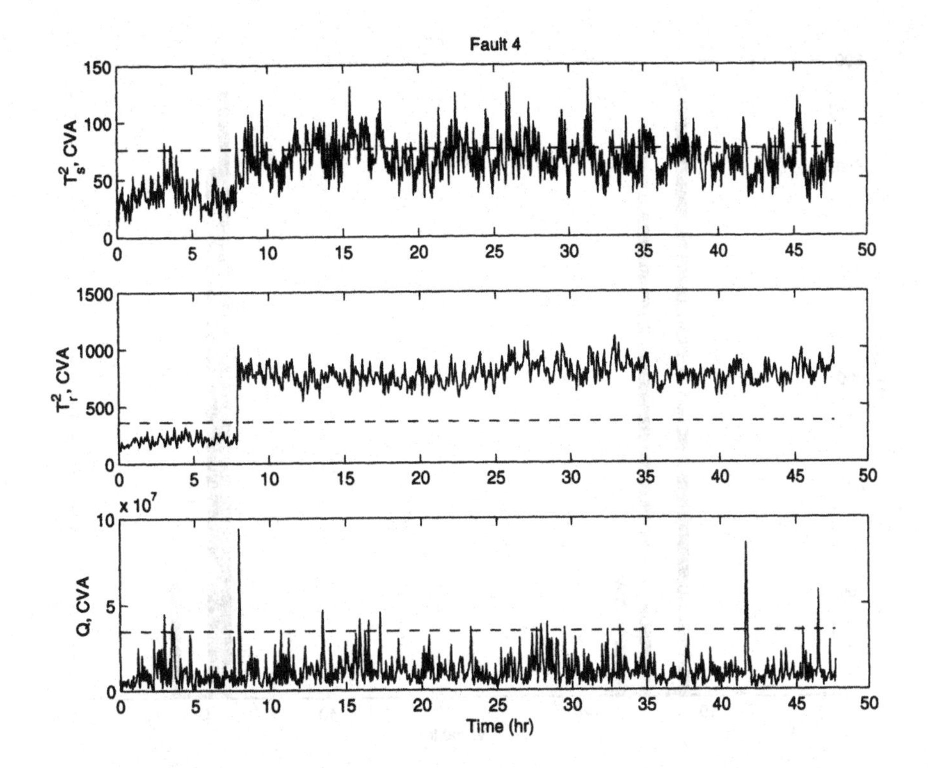

Fig. 10.9. The CVA multivariate statistics for fault detection for Fault 4

continue to inform the operator of a process abnormality although all the process variables will appear to have returned to their normal values.

It is somewhat interesting that examination of the canonical variables ($J\mathbf{p_t}$) for Fault 5 reveals that the canonical variable corresponding to the 99th generalized singular value is solely responsible for the out-of-control T_r^2 values between 10-40 hours after the fault occurred.

10.5 Case Study on Fault 11

Similar to Fault 4, Fault 11 induces a fault in the reactor cooling water inlet temperature. The fault in this case is a random variation. As seen in Figure 10.15, the fault induces large oscillations in the reactor cooling water flow rate, which results in a fluctuation of reactor temperature (see Figure 10.16). The other 50 variables are able to remain around the set-points and behave similarly as in the normal operating conditions.

The extent to which the (D)PCA-based and CVA-based statistics are sensitive to Fault 11 can be examined in Figure 10.17 and Figure 10.18,

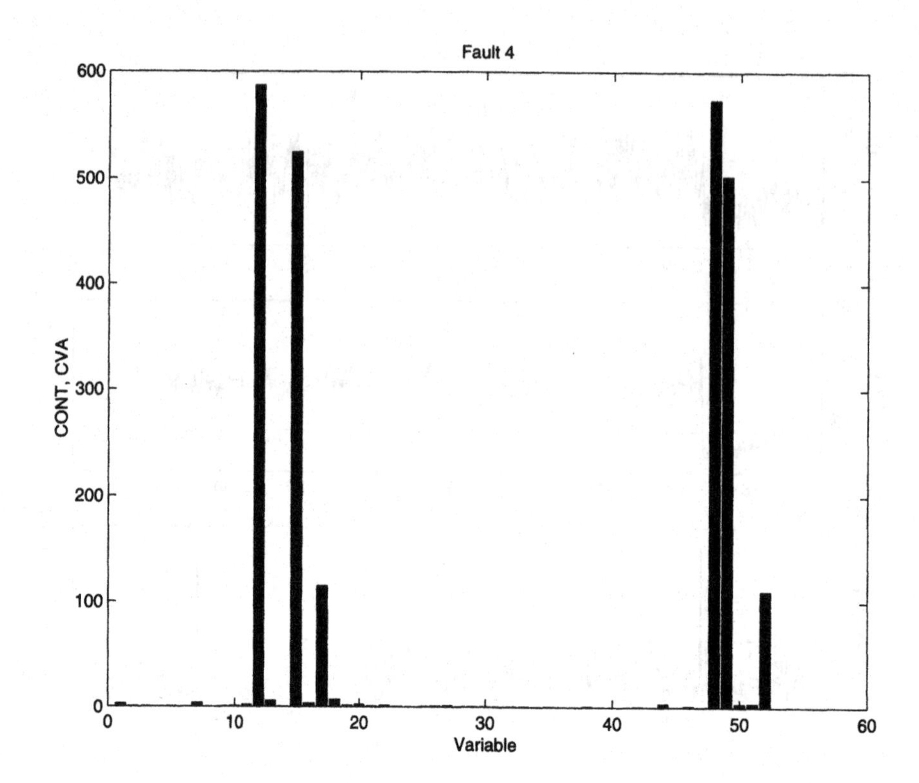

Fig. 10.10. The average contribution plot for Fault 4 for the CVA-based *CONT*

respectively. The quantitative fault detection results are shown in Table 10.1. The (D)PCA-based Q statistics performed better than the (D)PCA-based T^2 statistics. Similarly to Fault 4, the variation in residual space was captured better by T_r^2 than the CVA-based Q statistic. Overall, the DPCA-based Q statistic gave the lowest missed detection rate (see Table 10.1).

As Fault 11 and Fault 4 affect the same process variable, the fault was expected to influence the reactor cooling water flow the most. Similarly to Fault 4, the CVA-based *RES* and the (D)PCA-based statistics gave superior results, in terms of correctly identifying the reactor cooling water flow as the variable responsible for this fault (see Table 10.3). The improper dominance of the observation variables x_{12}, x_{15}, x_{17}, x_{48}, x_{49}, and x_{52} was again responsible for the poor performance of the CVA-based *CONT* (see Figure 10.19).

Some fault diagnosis techniques more easily diagnosed Fault 4 while others did better diagnosing Fault 11 (see Table 10.2). The lowest misclassification rates were provided by the MS T_1^2, DFDA/DPCA1 T^2, and CVA T_r^2 statistics, all of which take serial correlation into account. It is interesting that 'dynamic' versions of PCA which are designed to take serial correlation into account did

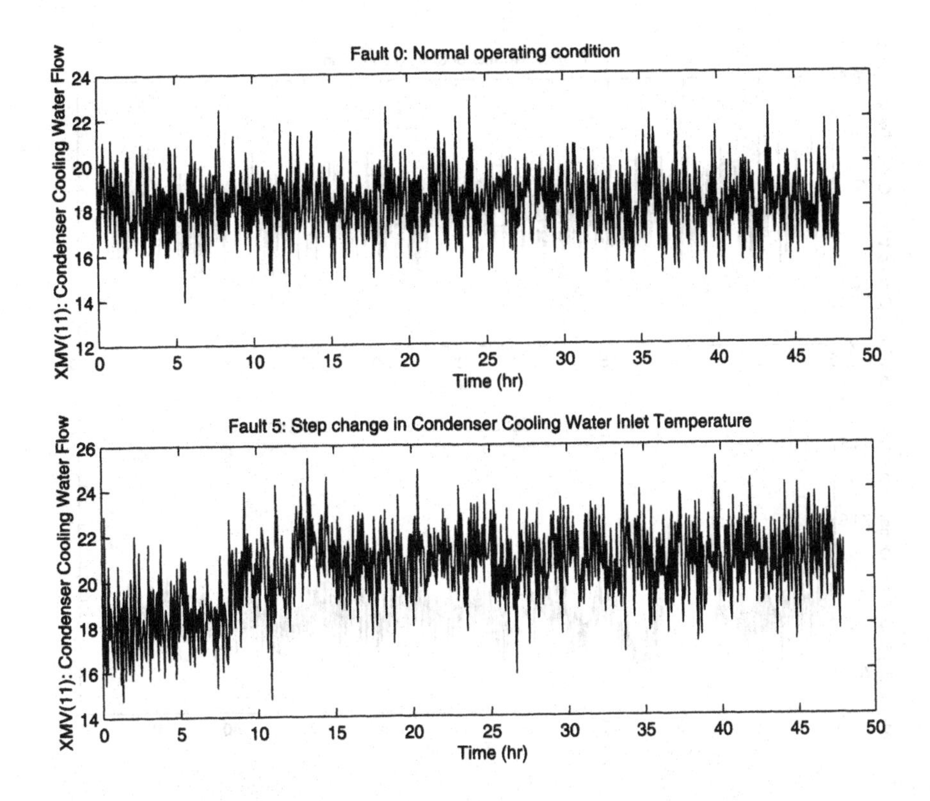

Fig. 10.11. Comparison of XMV(11) for Faults 0 and 5

not provide significantly improved fault diagnosis over their static versions for Fault 11.

10.6 Fault Detection

The objectives of a fault detection statistic are to be *robust* to data independent of the training set, *sensitive* to all the possible faults of the process, and *prompt* to the detection of the faults. The robustness of each statistic in Table 9.2 is determined by calculating the false alarm rate for the normal operating condition of the testing set and comparing it against the level of significance upon which the threshold is based. The sensitivity of the statistics is quantified by calculating the missed detection rates for Faults 1-21 of the testing set. The promptness of the statistics is based on the detection delays for Faults 1-21 of the testing set.

Prior to applying each of the statistics to the testing set, the parameter values associated with each statistic need to be specified. The orders deter-

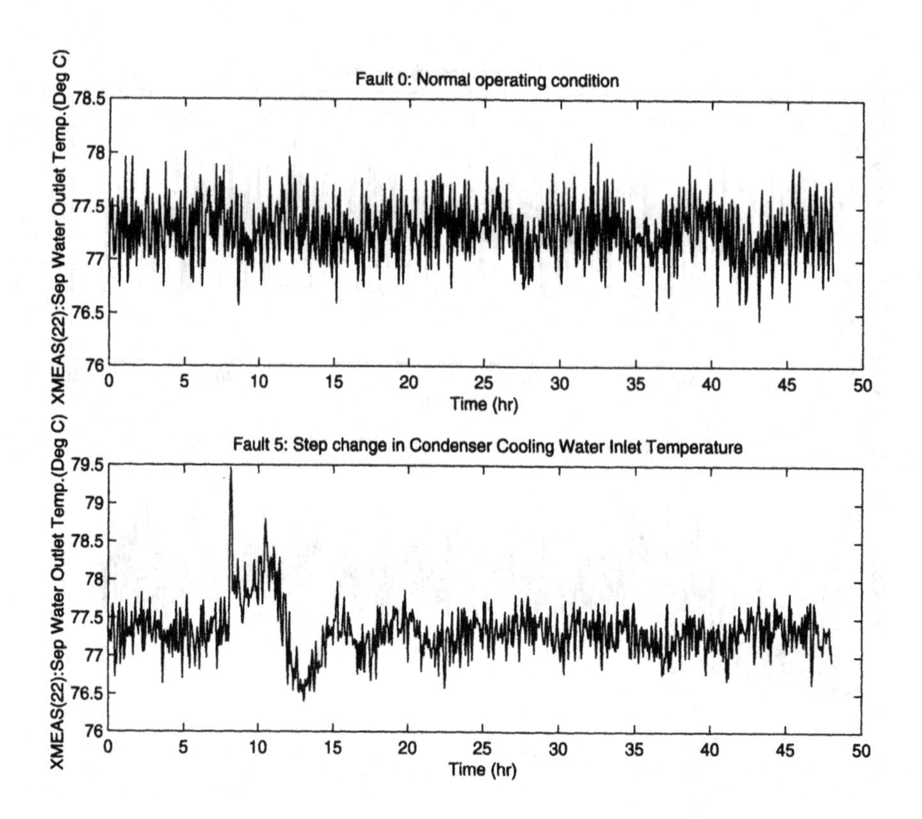

Fig. 10.12. Comparison of XMEAS(22) for Faults 0 and 5

mined for PCA, DPCA, PLS, and CVA and the number of lags h determined for DPCA and CVA are listed in Table 10.4. The orders and the number of lags were determined by applying the procedures described in Section 9.5 to the pretreated data for the normal operating condition of the training set.

The probability distributions used to determine the threshold for each statistic are listed in Table 9.2. Using a level of significance $\alpha = 0.01$, the false alarm rates of the training and testing sets were computed and tabulated in Table 10.5. The false alarm rates for the PCA and DPCA-based T^2 statistics are comparable in magnitude to $\alpha = 0.01$. The CVA-based statistics and the DPCA-based Q statistic resulted in relatively high false alarm rates for the testing set compared to the other multivariate statistics. The lack of robustness for T_s^2 and T_r^2 can be explained by the inversion of $\hat{\Sigma}_{pp}$ in (7.40). The high false alarm rate for the DPCA-based Q statistic may be due to a violation of the assumptions used to derive the threshold (4.22) (see Homework Problem 12 for a further exploration of this issue).

It would not be fair to directly compare the fault detection statistics in terms of missed detection rates when they have such widely varying false

Table 10.4. The lags and orders for the various models. The PLS models are all based on discriminant PLS.

Fault	ARX	PCAm	PCA1	DPCAm	PLS1	PLS2	$PLS1_{adj}$	$PLS2_{adj}$	FDA	FDA/ PCA1	FDA/ PCA2	DFDA/ DPCA1	CVA
	h	a	a	a	a	a	a	a	a	a	a	a	k
0	3	11	47	29	13	45	16	41	52	50	52	51	29
1	2	8	47	13	13	45	16	41	52	50	52	51	25
2	2	8	47	19	13	45	16	41	52	50	52	51	28
3	2	12	47	27	13	45	16	41	52	50	52	51	26
4	2	12	47	26	13	45	16	41	52	50	52	51	26
5	2	8	47	15	13	45	16	41	52	50	52	51	27
6	6	6	47	11	13	45	16	41	52	50	52	51	3
7	2	6	47	10	13	45	16	41	52	50	52	51	27
8	2	8	47	13	13	45	16	41	52	50	52	51	24
9	3	11	47	26	13	45	16	41	52	50	52	51	30
10	3	10	47	24	13	45	16	41	52	50	52	51	28
11	3	9	47	26	13	45	16	41	52	50	52	51	28
12	2	6	47	7	13	45	16	41	52	50	52	51	25
13	2	8	47	12	13	45	16	41	52	50	52	51	24
14	3	13	47	27	13	45	16	41	52	50	52	51	25
15	3	11	47	27	13	45	16	41	52	50	52	51	30
16	3	12	47	27	13	45	16	41	52	50	52	51	26
17	2	9	47	24	13	45	16	41	52	50	52	51	27
18	2	4	47	4	13	45	16	41	52	50	52	51	27
19	3	16	47	32	13	45	16	41	52	50	52	51	29
20	3	11	47	25	13	45	16	41	52	50	52	51	32
21	2	14	47	25	13	45	16	41	52	50	52	51	26

The ARX model is used to determine the lag orders for CVA, DPCA, and FDA as described in Section 7.5

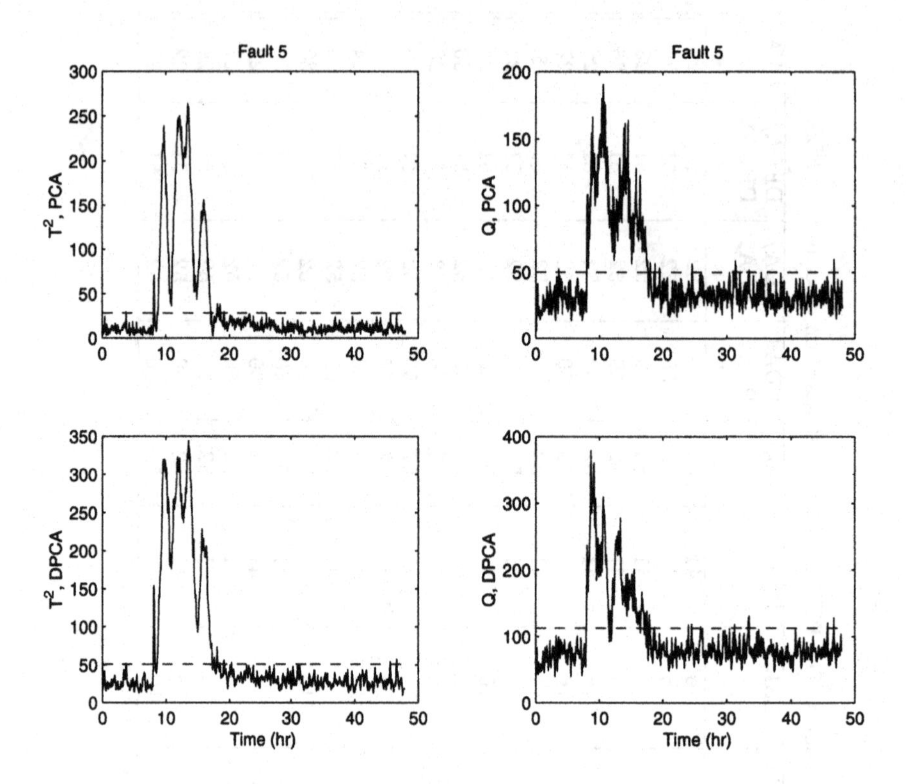

Fig. 10.13. The (D)PCA multivariate statistics for fault detection for Fault 5

Table 10.5. False alarm rates for the training and testing sets

Method	Measures	Training Set	Testing Set
PCA	T^2	0.002	0.014
PCA	Q	0.004	0.016
DPCA	T^2	0.002	0.006
DPCA	Q	0.004	0.281
CVA	T_s^2	0.027	0.083
CVA	T_r^2	0	0.126
CVA	Q	0.009	0.087

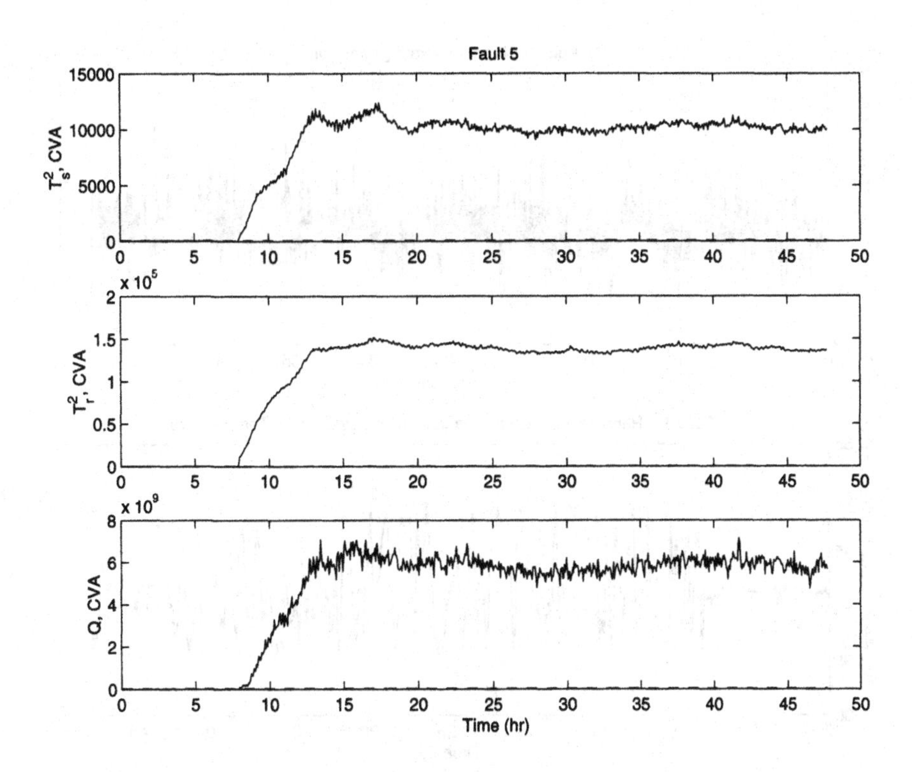

Fig. 10.14. The CVA multivariate statistics for fault detection for Fault 5

alarm rates. In computing the missed detection rates for Faults 1-21 of the testing set, the threshold for each statistic was adjusted to the tenth highest value for the normal operating condition of the *testing* set. The adjusted thresholds correspond to a level of significance $\alpha = 0.01$ by considering the probability distributions of the statistics for the normal operating condition. For statistics which showed low false alarm rates, the adjustment only shifted the thresholds slightly. For each statistic which showed a high false alarm rate, the adjustment increased the threshold by approximately 50%. Numerous simulation runs for the normal operating conditions confirmed that the adjusted thresholds indeed corresponded to a level of significance $\alpha = 0.01$. It was felt that this adjustment of thresholds provides a fairer basis for the comparison of the sensitivities of the statistics. For each statistic, the missed detection rates for all 21 faults were computed and tabulated in Table 10.6.

The missed detection rates for Faults 3, 9, and 15 are very high for all the fault detection statistics. No observable change in the mean or the variance can be detected by visually comparing the plots of each observation variable associated with Faults 3, 9, and 15 to the plots associated with the normal operating condition (Fault 0). It is conjectured that any statistic will result

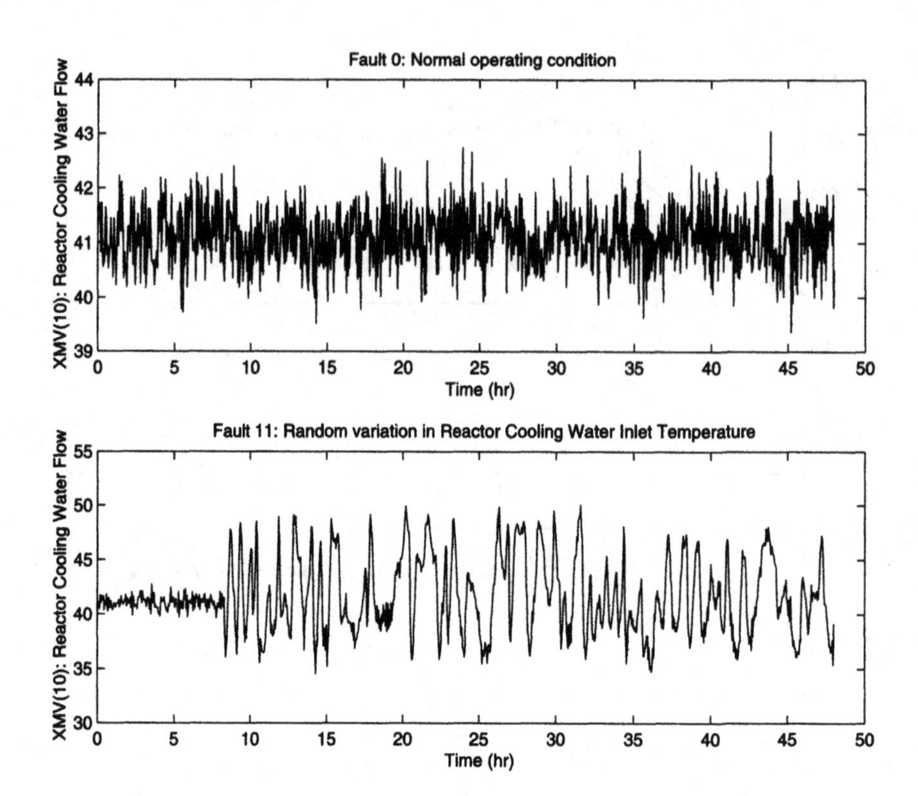

Fig. 10.15. Comparison of XMV(10) for Faults 0 and 11

in high missed detection rates for those faults, in other words, Faults 3, 9, and 15 are *unobservable* from the data. Including the missed detection rates for these faults would skew the comparison of the statistics, and therefore these faults are not analyzed when comparing the overall performance of the statistics.

The minimum missed detection rate achieved for each fault except Faults 3, 9, and 15 is contained in a box in Table 10.6. The T_r^2 statistic with the threshold rescaled as described above had the lowest missed detection rate except for the unobservable Faults 3 and 9. The conclusion that the T_r^2 statistic with a scaled threshold will *always* give lower missed detection rates than the other statistics would be *incorrect*, since another method may be better for a different amount of data or a different process. In particular, a fault that does not affect the states in the T_r^2 statistic will be invisible to this statistic. Since many of the statistics have comparable missed detection rates for many of the faults, it seems to have an advantage to incorporate the T_r^2 statistics with other statistics for fault detection.

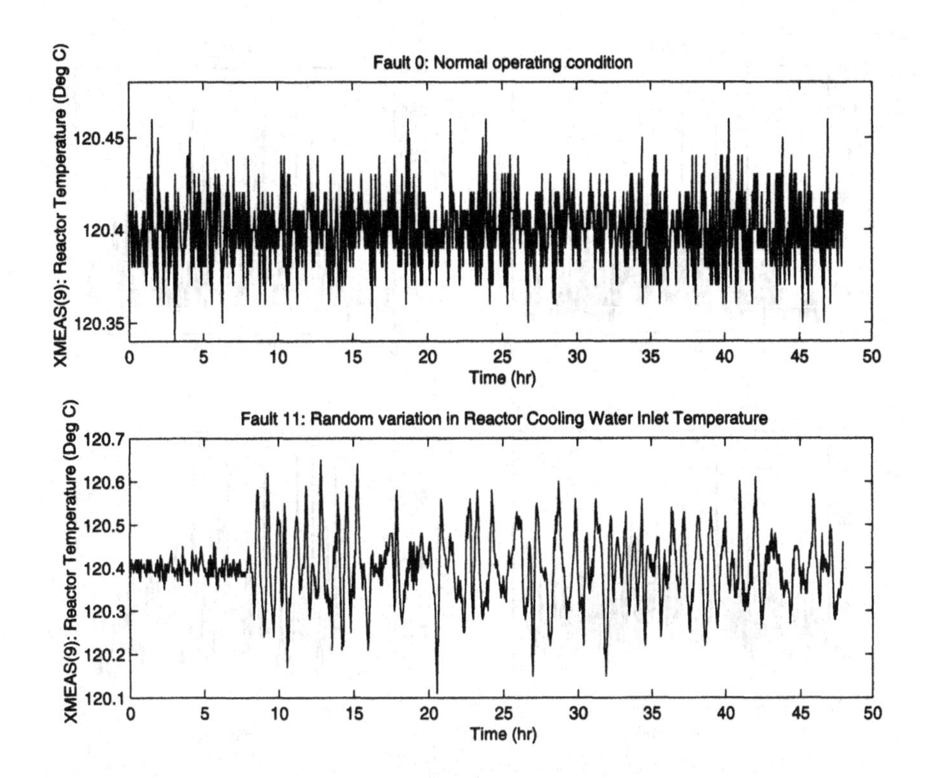

Fig. 10.16. Comparison of XMEAS(9) for Faults 0 and 11

The CVA-based Q statistic gave missed detection rates similar to those from the T_r^2 statistic for some faults, but performed more poorly for others. Other results, not shown here for brevity, showed that a slight shift in the lag order h or state order k can result in a large variation of the CVA-based Q statistic. Tweaking these parameters may improve the CVA-based Q statistic enough to give fault detection performance more similar to the T_r^2 statistic.

The number of minimums achieved with the residual-based statistics is far more than the number of minimums achieved with state- or score-based statistics. Residual-based multivariate statistics tended to be more sensitive to the faults of the TEP than the state or score-based statistics. The better performance of residual-based statistics supports the claims in the literature, based on either theoretical analysis [345] or case studies [183], that residual-based statistics tend to be more sensitive to faults. A comparison of *all* the fault detection statistics revealed that the residual-based T_r^2 statistic was overall the most sensitive to the faults of the TEP. However, the T_r^2 statistic was found not to be very robust compared to most of the other statistics, due to the inversion of the matrix $\hat{\Sigma}_{pp}$ in (7.40). Also, recall that the threshold

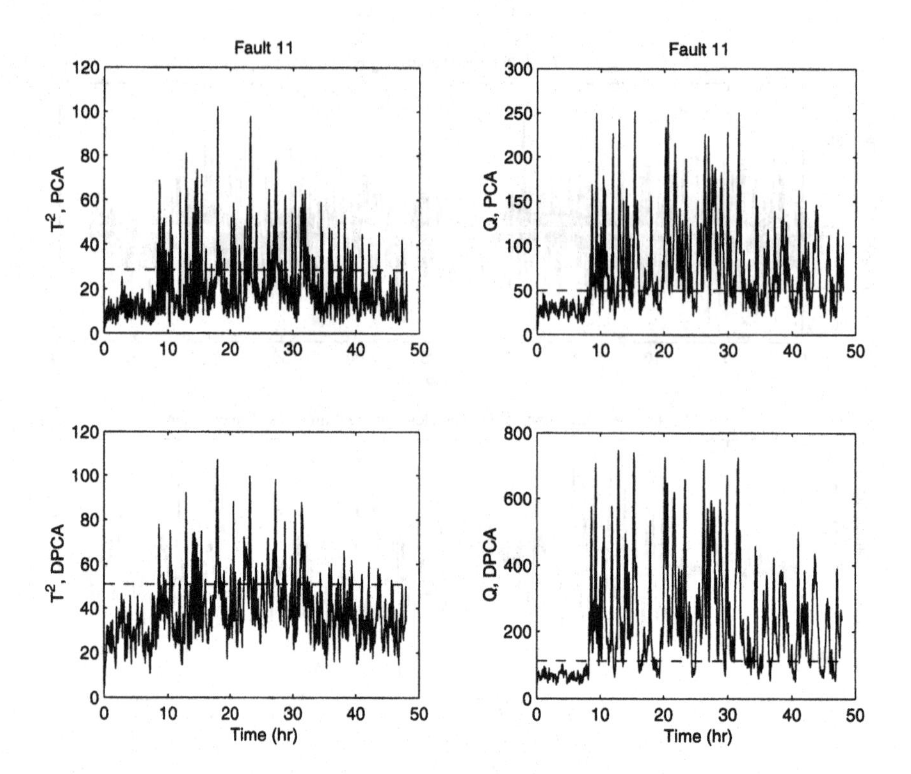

Fig. 10.17. The (D)PCA multivariate statistics for fault detection for Fault 11

used here was rescaled based on the testing set to give a false alarm rate of 0.01, as described in Section 10.6. The behavior of the T_r^2 statistic with the threshold (7.44) can give large false alarm rates, as was discussed earlier.

On average, the DPCA-based statistics were somewhat more sensitive to the faults than the PCA-based statistics, although the overall difference was not very large. The high false alarm rates found for the DPCA-based Q statistic (see Table 10.5) indicate that the threshold (4.22) may need to be rescaled based on an additional set of data as was done here.

Most statistics performed well for the faults that affect a significant number of observation variables (Faults 1, 2, 6, 7, 8, 14, and 18). In these cases, most variables deviated significantly from their distribution in the normal operating conditions. The other faults had a limited number of the observation variables deviate from their distribution in the normal operating conditions. Detecting such faults is relatively more challenging.

Since false alarms are inevitable, it is often difficult to determine whether the out-of-control value of a statistic is the result of a fault or of a false alarm. In order to decrease the rate of false alarms, it is common to show an

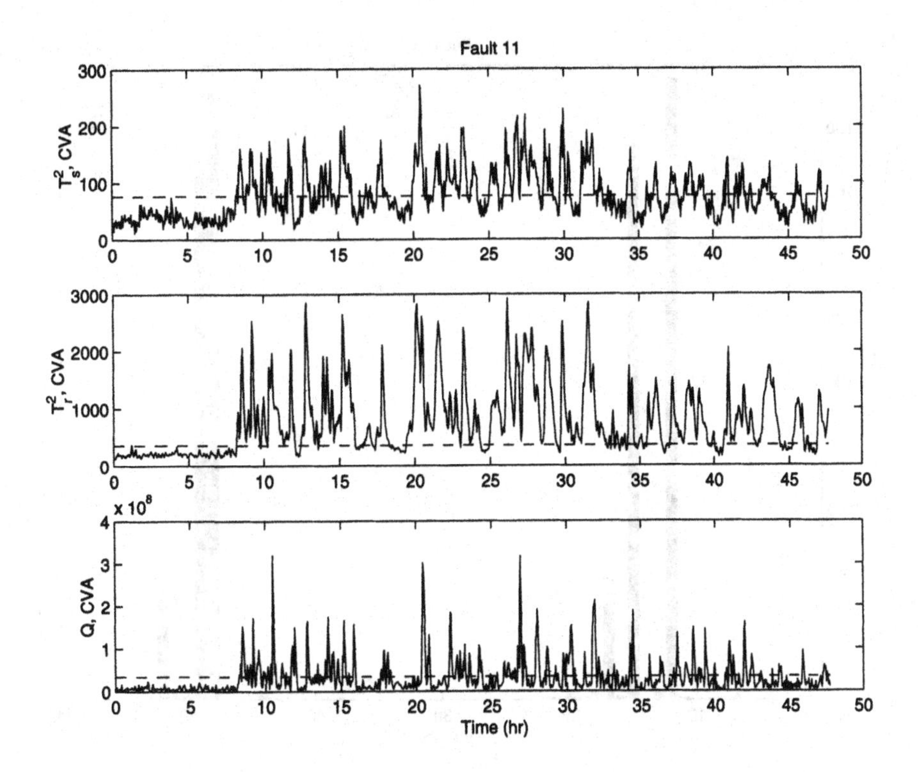

Fig. 10.18. The CVA multivariate statistics for fault detection for Fault 11

alarm only when several consecutive values of a statistic have exceeded the threshold. In computing the detection delays for the statistics in Table 10.7, a fault is indicated only when six consecutive measure values have exceeded the threshold, and the detection delay is recorded as the first time instant in which the threshold was exceeded. Assuming independent observations and $\alpha = 0.01$, this corresponds to a false alarm rate of $0.01^6 = 1 \times 10^{-12}$. The detection delays for all 21 faults listed in Table 10.7 were obtained by applying the same thresholds as used to determine the missed detection rates.

For the multivariate statistics, a close examination of Tables 10.6 and 10.7 reveals that the statistics exhibiting small detection delays tend to exhibit small missed detection rates and *vice versa*. Since the detection delay results correlate well with the missed detection rate results, all of the conclusions for missed detection rates apply here.

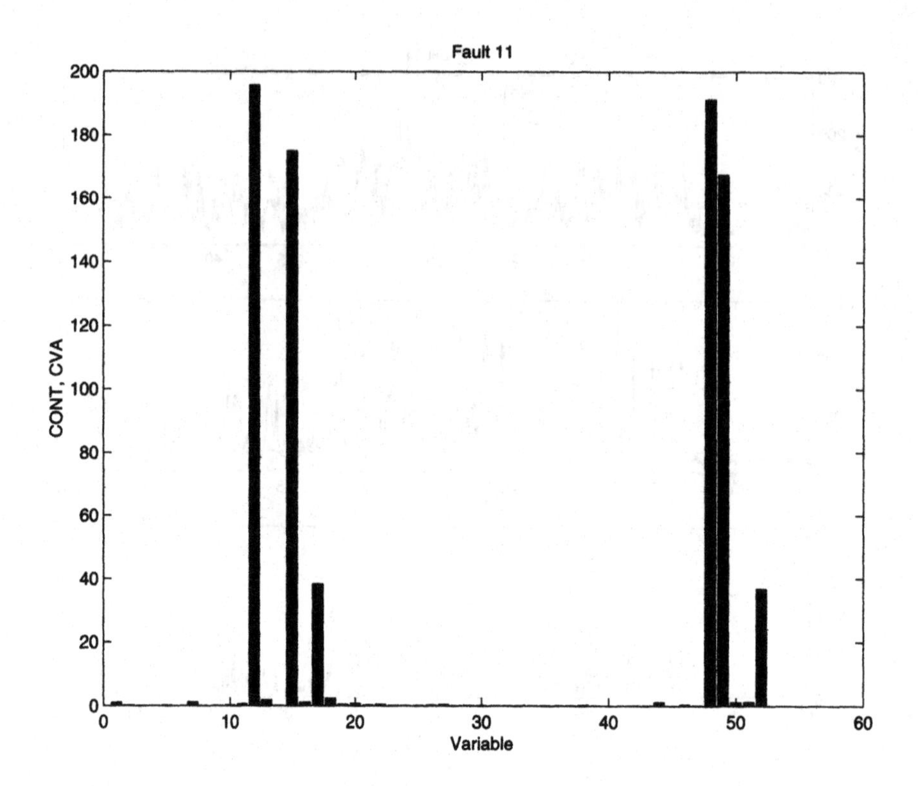

Fig. 10.19. The average contribution plot for Fault 11 for the CVA-based *CONT*

10.7 Fault Identification

The objective of a fault identification statistic is to identify the observation variable(s) most closely related to the fault. The challenge in developing a good criterion for comparing the different statistics is choosing which observation variable(s) is most relevant to diagnosing the fault. This, of course, depends on the knowledge and expertise of the plant operators and engineers. The only faults investigated here for fault identification are those in which a direct and clear link between the fault and an observation variable could be determined. The faults investigated in this section for fault identification and the observation variables directly related to each fault are listed in Table 10.8. The ranking of these observation variables for each fault is the criterion used to compare the different statistics listed in Table 9.3.

The statistics investigated in this section are listed in Table 9.3, and the parameter values associated with the statistics are listed in Table 10.4. The rankings of the observation variables listed in Table 10.8 for each statistic and fault are contained in Tables 10.9, 10.10, and 10.11. These tables list the

Table 10.6. Missed detection rates for the testing set

Fault	PCA T^2	PCA Q	DPCA T^2	DPCA Q	CVA T_s^2	CVA T_r^2	CVA Q
1	**0.008**	**0.003**	**0.006**	**0.005**	**0.001**	0	**0.003**
2	0.020	0.014	0.019	0.015	0.011	0.010	0.026
3	0.998	0.991	0.991	0.990	0.981	0.986	0.985
4	**0.956**	**0.038**	**0.939**	0	**0.688**	0	**0.975**
5	**0.775**	**0.746**	**0.758**	**0.748**	0	0	0
6	0.011	0	0.013	0	0	0	0
7	0.085	0	0.159	0	0.386	0	0.486
8	0.034	0.024	0.028	0.025	0.021	0.016	0.486
9	0.994	0.981	0.995	0.994	0.986	0.993	0.993
10	0.666	0.659	0.580	0.665	0.166	0.099	0.599
11	**0.794**	**0.356**	**0.801**	**0.193**	**0.515**	**0.195**	**0.669**
12	0.029	0.025	0.010	0.024	0	0	0.021
13	0.060	0.045	0.049	0.049	0.047	0.040	0.055
14	0.158	0	0.061	0	0	0	0.122
15	0.988	0.973	0.964	0.976	0.928	0.903	0.979
16	0.834	0.755	0.783	0.708	0.166	0.084	0.429
17	0.259	0.108	0.240	0.053	0.104	0.024	0.138
18	0.113	0.101	0.111	0.100	0.094	0.092	0.102
19	0.996	0.873	0.993	0.735	0.849	0.019	0.923
20	0.701	0.550	0.644	0.490	0.248	0.087	0.354
21	0.736	0.570	0.644	0.558	0.440	0.342	0.547

rankings for the average statistic values over the time periods 0-5 hours, 5-24 hours, and 24-40 hours, after the fault occurred. A ranking of 1 in the tables indicates that the observation variable listed in Table 10.8 had the largest average statistic value, and a ranking of 52 indicates that the observation variable listed in Table 10.8 had the smallest average statistic value. The best ranking for each fault is contained in a box. The results are divided into three tables because it is useful to analyze how the proficiencies of the statistics change with time. It is best to identify the fault properly as soon as it occurs, and therefore the results during the time period 0-5 hours after the fault are tabulated separately. The results for the time period between 5-24 and 24-40 hours after the fault occurred were tabulated separately, because this is useful in determining the robustness of the statistics.

As shown in Tables 10.9-10.11, the (D)PCA-based $CONT$ performed well. The better performance of the (D)PCA-based $CONT$ to the (D)PCA-based RES suggests that the abstraction of structure provided by PCA was even more critical to fault identification than fault detection. For the faults where fault propagation occurred, the performance of the data-driven statistics de-

Table 10.7. Detection delays (minutes) for the testing set

Fault	PCA T^2	PCA Q	DPCA T^2	DPCA Q	CVA T_s^2	CVA T_r^2	CVA Q
1	**21**	**9**	**18**	**15**	**6**	**9**	**6**
2	51	36	48	39	39	45	75
3	–	–	–	–	–	–	–
4	–	**9**	**453**	**3**	**1386**	**3**	–
5	**48**	**3**	**6**	**6**	**3**	**3**	**0**
6	30	3	33	3	3	3	0
7	3	3	3	3	3	3	0
8	69	60	69	63	60	60	63
9	–	–	–	–	–	–	–
10	288	147	303	150	75	69	132
11	**912**	**33**	**585**	**21**	**876**	**33**	**81**
12	66	24	9	24	6	6	0
13	147	111	135	120	126	117	129
14	12	3	18	3	6	3	3
15	–	2220	–	–	2031	–	–
16	936	591	597	588	42	27	33
17	87	75	84	72	81	60	69
18	279	252	279	252	249	237	252
19	–	–	–	246	–	33	–
20	261	261	267	252	246	198	216
21	1689	855	1566	858	819	1533	906

Table 10.8. The variables assumed to be most closely related to each disturbance

Fault	Process Variable	Data Variable	Variable Description
2	XMV(6)	x_{47}	Purge Valve (stream 9)
4	XMV(10)	x_{51}	Reactor Cooling Water Flow
5	XMEAS(22)	x_{22}	Sep. Cooling Water Outlet Temp
6	XMV(3)	x_{44}	A Feed Flow (stream 1)
11	XMV(10)	x_{51}	Reactor Cooling Water Flow
12	XMEAS(22)	x_{22}	Sep. Cooling Water Outlet Temp
14	XMV(10)	x_{51}	Reactor Cooling Water Flow
21	XMV(4)	x_{45}	A, B, and C Feed Flow (stream 4)

Table 10.9. The rankings for the time period 0-5 hours after the fault occurred

Fault	PCA *CONT*	PCA *RES*	DPCA *CONT*	DPCA *RES*	CVA *CONT*	CVA *RES*
2	2	4	2	5	10	2
4	1	1	1	1	**10**	1
5	**12**	**21**	**11**	8	**15**	**17**
6	1	6	3	2	6	6
11	1	1	1	1	**10**	1
12	1	6	1	3	10	14
14	2	2	1	2	11	1
21	52	40	48	48	52	52

Table 10.10. The rankings for the time period 5-24 hours after the fault occurred

Fault	PCA *CONT*	PCA *RES*	DPCA *CONT*	DPCA *RES*	CVA *CONT*	CVA *RES*
2	2	5	2	7	10	3
4	1	1	1	1	**12**	1
5	**31**	**34**	**30**	**31**	**18**	**14**
6	5	52	8	45	8	3
11	1	1	1	1	**13**	1
12	1	12	1	3	13	24
14	2	2	1	2	10	1
21	52	46	51	51	52	52

Table 10.11. The rankings for the time period 24-40 hours after the fault occurred

Fault	PCA *CONT*	PCA *RES*	DPCA *CONT*	DPCA *RES*	CVA *CONT*	CVA *RES*
2	2	5	3	12	10	4
4	1	1	1	1	**11**	1
5	9	**35**	**14**	**30**	**16**	**16**
6	7	51	11	45	1	3
11	1	1	1	1	**13**	1
12	10	21	4	36	17	26
14	2	2	1	2	11	1
21	52	48	52	52	52	50

teriorated as the effect of the fault evolved. Robustness may be achieved by applying model-based fault identification statistics that are able to take into account the propagation of the fault (see Chapter 11).

All fault identification statistics performed poorly for Fault 21 (see Tables 10.9-10.11). The A/B/C feed flow valve for Stream 4 was fixed at the steady state position (see Figure 8.1). The valve was stuck, indicating that the signals from this valve were constant, which corresponds to zero variance. The RES and $CONT$-based statistics had great difficulty identifying the A/B/C feed flow as the variable associated with the fault because these statistics are designed to detect *positive shift in variance* only. This illustrates the importance in such cases of implementing statistics such as Equation 4.29 which can detect a *negative shift in variance.* This type of statistic implemented in the appropriate manner would have detected Fault 21 rather easily. In general it is suggested that such a statistic should be applied to each process variable, with the α level set to keep the false alarm rate low.

The performance of a fault identification statistic can significantly deteriorate over time for faults whose effects on the process variables change over time. For instance, the effect of Fault 12 propagates over the interval 5 to 40 hours after the fault occurred. As a result, there is only one statistic producing a ranking below 10 in Table 10.11 while all but one statistic produced a ranking at or above 10 in Table 10.9. For Fault 6, the performance of the (D)PCA-based fault identification statistics substantially degraded over time, while the performance of the CVA-based statistics actually improved.

10.8 Fault Diagnosis

Assuming that process data collected during a fault are represented by a previous fault class, the objective of the fault diagnosis statistics in Table 9.4 is to classify the data to the *correct* fault class. That is, a highly proficient fault diagnosis statistic produces small misclassification rates when applied to data independent of the training set. Such a statistic usually has an accurate representation of each class, more importantly such a statistic separates each class from the others very well. Recall that all the methods listed in Table 9.4 are based on supervised classification. For the discriminant PLS, PCA1, MS, and FDA methods, one model is built for all fault classes. For the other methods listed in Table 9.4, a separate model is built for each fault class. The proficiencies of the statistics in Table 9.4 are investigated in this section based on the misclassification rates for Faults 1-21 of the testing set. The parameters for each statistic were determined from Faults 1-21 of the training set. The lags and orders associated with the statistics are listed in Table 10.4.

The overall misclassification rate for each statistic when applied to Faults 1-21 of the testing set is listed in Table 10.12. For each statistic, the misclassification rates for all 21 faults were computed and tabulated in Tables

10.13-10.20. The minimum misclassification rate achieved for each fault except Faults 3, 9, and 15 is contained in a box.

Table 10.12. The overall misclassification rates

Method	Basis	Misclassification Rate
PCAm	T^2	0.742
PCA1	T^2	0.212
PCAm	Q	0.609
PCAm	T^2 & Q	0.667
DPCAm	T^2	0.724
DPCAm	Q	0.583
DPCAm	T^2 & Q	0.662
PLS1	–	0.565
PLS2	–	0.567
PLS1_{adj}	–	0.576
PLS2_{adj}	–	0.574
CVA	T_s^2	0.501
CVA	T_r^2	0.213
CVA	Q	0.621
FDA	T^2	0.195
FDA/PCA1	T^2	0.206
FDA/PCA2	T^2	0.195
DFDA/DPCA1	T^2	0.192
MS	T_0^2	0.214
MS	T_1^2	0.208

When applying the fault diagnosis statistics, it was assumed that the *a priori* probability for each class i was equal to $P(\omega_i) = 1/p$ where $p = 21$ is the number of fault classes. DFDA/DPCA1 produced the lowest overall misclassification rate (0.192), followed by the rest of the FDA-based methods, as shown in Table 10.12. The CVA-based T_r^2, PCA1, and MS statistics produced comparable overall misclassification rates.

To compare the FDA/PCA1 and FDA/PCA2 methods for diagnosing faults, the overall misclassification rates for the training and testing sets and the information criterion (5.12) are plotted for various orders using FDA, FDA/PCA1, and FDA/PCA2 (see Figures 10.20, 10.21, and 10.22), respectively. The overall misclassification rates for the testing set using FDA/PCA1 and FDA/PCA2 was lower than that of the FDA for most orders $a \geq p$. The performance of FDA/PCA1 and FDA/PCA2 was very similar, indicating that

Table 10.13. The misclassification rates for 0-40 hours after the Faults 1-21 occurred

Fault	PCAm T^2	PCA1 T^2	PCAm Q	PCAm T^2&Q	DPCAm T^2	DPCAm Q	DPCAm T^2&Q	CVA T_s^2	CVA T_r^2	CVA Q
1	**0.680**	**0.024**	**0.028**	**0.041**	**0.880**	**0.035**	**0.038**	**0.028**	**0.026**	**0.245**
2	0.410	0.018	0.024	0.035	0.441	0.060	0.034	0.010	0.090	0.155
3	0.939	0.783	0.991	1.000	0.701	0.995	1.000	0.940	0.821	0.978
4	**0.810**	**0.163**	**0.951**	**1.000**	**0.720**	**0.964**	**1.000**	**0.981**	**0.358**	**0.890**
5	**0.956**	**0.021**	**0.913**	**0.973**	**0.874**	**0.856**	**1.000**	**0.061**	**0.040**	**0.174**
6	0.100	0	0.050	0.076	0.049	0.063	0.089	0.001	0.001	0.014
7	0.978	0	0.405	0.496	0.868	0.336	0.633	0.638	0.001	0.578
8	0.998	0.030	0.270	0.409	1.000	0.170	0.398	0.518	0.055	0.670
9	0.993	0.779	0.995	1.000	0.988	0.998	1.000	0.969	0.848	0.969
10	0.849	0.126	0.988	1.000	0.743	0.995	1.000	0.745	0.098	0.816
11	**0.989**	**0.234**	**0.859**	**0.968**	**0.948**	**0.843**	**0.983**	**0.904**	**0.139**	**0.901**
12	0.850	0.021	0.216	0.204	0.700	0.203	0.215	0.009	0.020	0.294
13	1.000	0.235	0.501	0.754	1.000	0.441	0.721	0.495	0.328	0.591
14	0.244	0.036	0.273	0.438	0.564	0.110	0.153	0.203	0.001	0.450
15	0.963	0.768	0.994	1.000	0.964	0.996	1.000	0.964	0.666	0.984
16	0.841	0.200	0.984	1.000	0.801	0.989	1.000	0.568	0.145	0.859
17	0.563	0.193	0.415	0.413	0.648	0.320	0.403	0.218	0.638	0.217
18	0.360	0.410	0.393	0.324	0.294	0.395	0.298	0.540	0.134	0.829
19	0.401	0.124	0.651	0.876	0.789	0.564	0.956	0.470	0.005	0.929
20	0.761	0.143	0.916	1.000	0.708	0.948	1.000	0.306	0.090	0.588
21	0.899	0.138	0.979	1.000	0.529	0.953	1.000	0.948	0.611	0.924
overall	0.742	0.212	0.609	0.667	0.724	0.583	0.662	0.501	0.213	0.621

Table 10.14. The misclassification rates for 0-40 hours after the Faults 1-21 occurred

Fault	PLS1	PLS2	$PLS1_{adj}$	$PLS2_{adj}$	FEA	FEA/PCA1	FEA/PCA2	DFEA/DPCA1	MS	MS
	–	–	–	–	T^2	T^2	T^2	T^2	T_0^2	T_1^2
1	**0.013**	**0.013**	**0.019**	**0.019**	**0.025**	**0.024**	**0.025**	**0.026**	**0.025**	**0.035**
2	0.014	0.024	0.024	0.024	0.019	0.019	0.019	0.019	0.019	0.033
3	0.961	0.970	0.869	0.876	0.780	0.734	0.780	0.735	0.780	0.886
4	**0.170**	**0.119**	**0.364**	**0.320**	**0.176**	**0.163**	**0.176**	**0.159**	**0.176**	**0.427**
5	**0.006**	**0.008**	**0.044**	**0.043**	**0.020**	**0.020**	**0.020**	**0.023**	**0.020**	**0.040**
6	0.435	0.778	0.834	0.831	0	0	0	0	0	0
7	0	0	0	0.001	0	0	0	0	0	0
8	0.851	0.789	0.848	0.850	0.003	0.004	0.003	0.026	0.030	0.019
9	0.981	0.981	0.899	0.915	0.773	0.780	0.773	0.801	0.773	0.872
10	0.661	0.591	0.586	0.569	0.131	0.158	0.131	0.101	0.131	0.098
11	**0.989**	**0.979**	**0.859**	**0.886**	**0.245**	**0.244**	**0.245**	**0.118**	**0.245**	**0.121**
12	0.988	0.953	0.869	0.866	0.018	0.016	0.018	0.030	0.018	0.005
13	0.646	0.625	0.751	0.738	0.239	0.246	0.239	0.229	0.239	0.208
14	0.995	0.998	0.931	0.930	0.013	0.013	0.013	0.004	0.013	0.001
15	0.988	0.981	0.926	0.925	0.764	0.780	0.764	0.784	0.764	0.725
16	0.894	0.660	0.658	0.558	0.193	0.184	0.193	0.218	0.193	0.255
17	0.146	0.164	0.388	0.378	0.150	0.145	0.150	0.043	0.150	0.038
18	0.775	0.843	0.839	0.796	0.315	0.399	0.315	0.154	0.750	0.431
19	0.913	0.945	0.800	0.778	0.039	0.055	0.039	0.142	0.039	0.003
20	0.334	0.274	0.509	0.525	0.126	0.125	0.126	0.176	0.126	0.158
21	0.098	0.096	0.068	0.066	0.044	0.198	0.030	0.261	0.004	0.003
overall	0.565	0.568	0.576	0.574	0.195	0.206	0.195	0.192	0.214	0.208

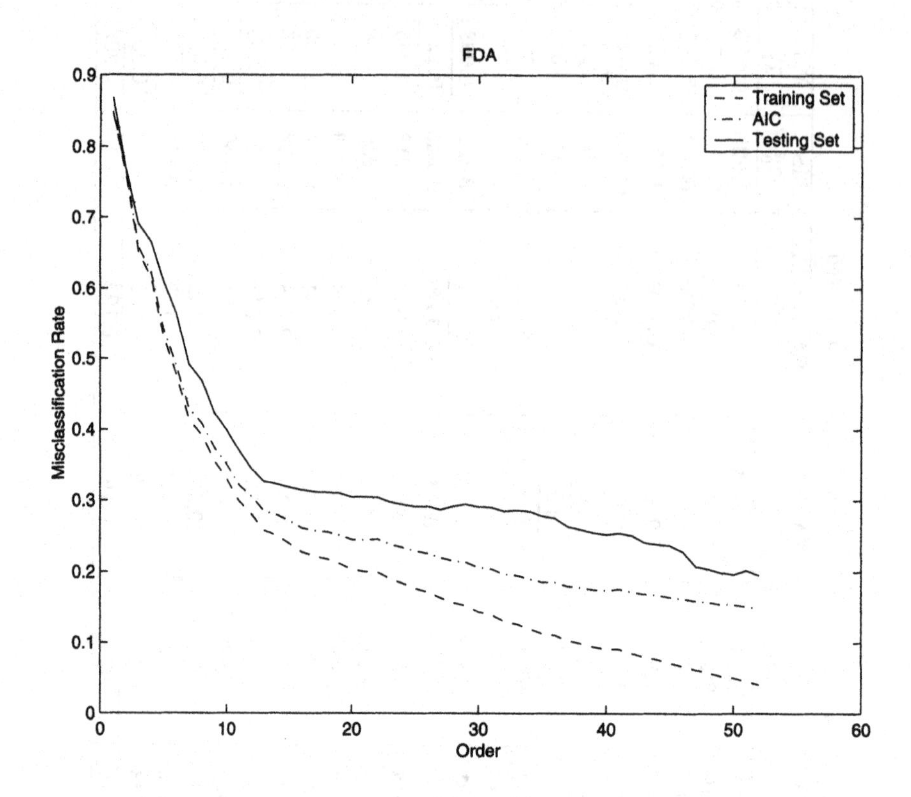

Fig. 10.20. The overall misclassification rates for the training and testing sets and the information criterion (AIC) for various orders using FDA

using PCA1 to rank the $m - p + 1$ eigenvectors corresponding to the zero eigenvalues in FDA is a reasonable approach. A close comparison of Figures 10.21 and 10.22 indicates that for $20 \leq a \leq 48$, the overall misclassification rate for the testing set using FDA/PCA1 is lower than FDA/PCA2. Because of this advantage of using FDA/PCA1 over FDA for this problem, lag variables will be included only on the data for FDA/PCA1 when investigating the proficiency of the methods for removing serial correlations of the data.

To evaluate the potential advantage of including lagged variables in FDA/PCA1 to capture correlations, the overall misclassification rates for the training and testing sets and the information criterion (5.12) are plotted for various orders using FDA/PCA1 and DFDA/DPCA1 (see Figures 10.21 and 10.23), respectively. FDA/PCA1 and DFDA/DPCA1 select excellent vectors for projecting to a lower-dimensional space for small a. Figures 10.21 and 10.23 show that most of the separation between the fault classes occurs in the space provided by the first 13 generalized eigenvectors. The misclassification rate with $a = 13$ for FDA/PCA1 is 0.33 and DFDA/DPCA1 is 0.34.

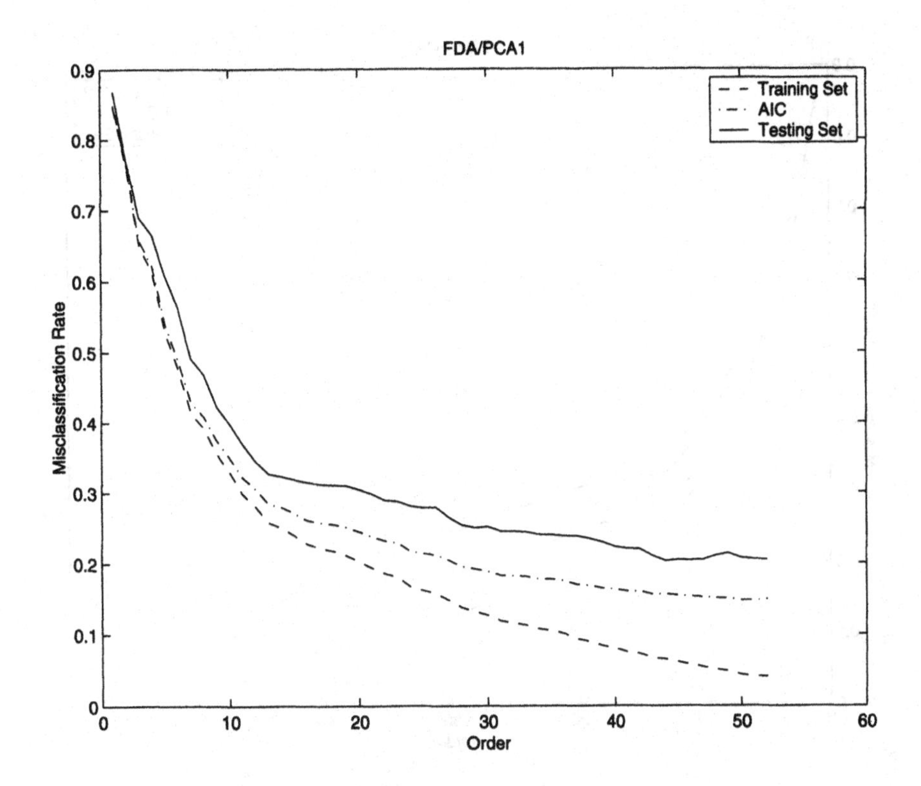

Fig. 10.21. The overall misclassification rates for the training and testing sets and the information criterion (AIC) for various orders using FDA/PCA1

The FDA/PCA and DFDA/DPCA1-based statistics were able to separate the fault classes well for the space spanned by the first $p-1$ generalized eigenvectors. The proficiency was slightly increased as the dimensionality was increased further for FDA/PCA1 and DFDA/DPCA1. DFDA/DPCA1 produced the lowest overall misclassification rate among all of the fault diagnosis methods investigated in this chapter. Including lagged variables in FDA/PCA1 can give better fault diagnosis performance. The advantage becomes especially clear when DFDA/DPCA1 is applied to a system with a short sampling time (see Homework Problem 11).

The information criterion performed relatively well, as the slope of the misclassification rate of the testing set is fairly equivalent to the slope of the information criterion for $a = 15$ to 50 in Figures 10.20-10.23. The AIC captures the shape and slope of the misclassification rate curve for the testing data. The AIC weighs the prediction error term and the model complexity term fairly. If one desires to have a lower-dimensional FDA model for diag-

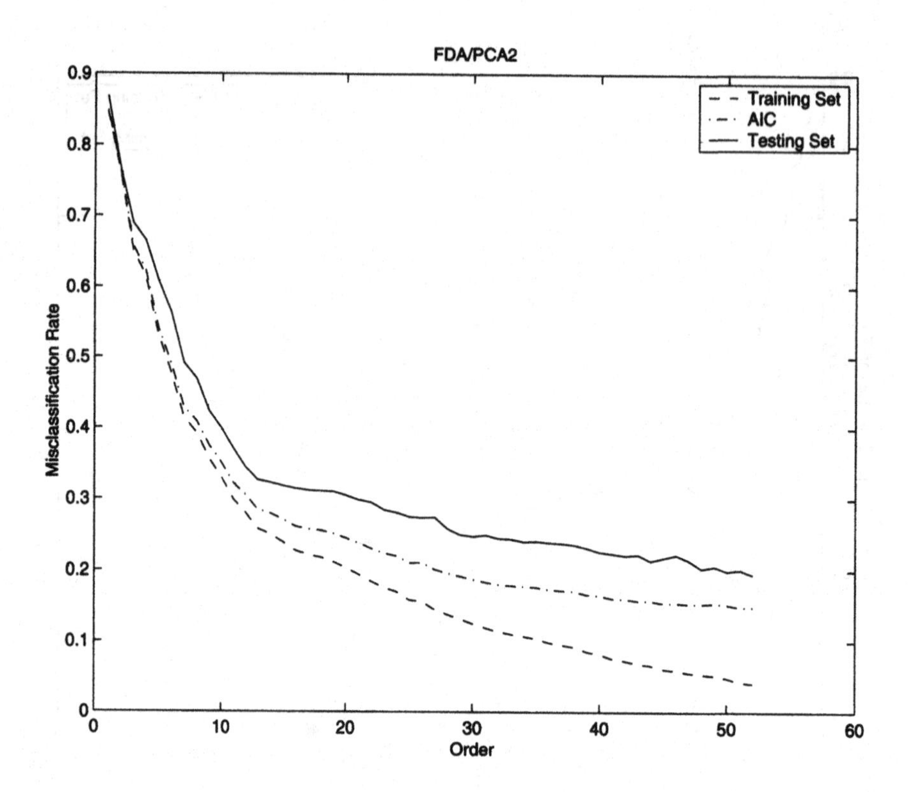

Fig. 10.22. The overall misclassification rates for the training and testing sets and the information criterion (AIC) for various orders using FDA/PCA2

nosing faults, the model complexity term can be weighed more heavily (see Homework Problem 5).

Figure 10.24 plots the overall misclassification rates for the training and testing sets and the information criterion (5.12) for various orders using PLS1 and PLS2. The reduction order c is the point at which the information criterion is minimized. The reduction order for each class in PLS1 is $c_1 = 13$ and the reduction order for PLS2 $c_2 = 45$. In general, the overall misclassification rate of PLS1 is lower than that of PLS2 for a fixed order, especially when $a < c_1$. Also, the performance of PLS1 is less sensitive to order selection than PLS2. The misclassification rate on average is the same for the best reduction orders for PLS1 and PLS2, as shown in Table 10.12.

Figure 10.25 plots the overall misclassification rates for the training and testing sets and the information criterion (5.12) for various orders using PLS1_{adj} and PLS2_{adj}. Figures 10.24 and 10.25 show similar trends. Regardless of order selected, PLS1_{adj} performs better than PLS2_{adj} in terms of lower overall misclassification rates. The reduction orders that minimize the AIC

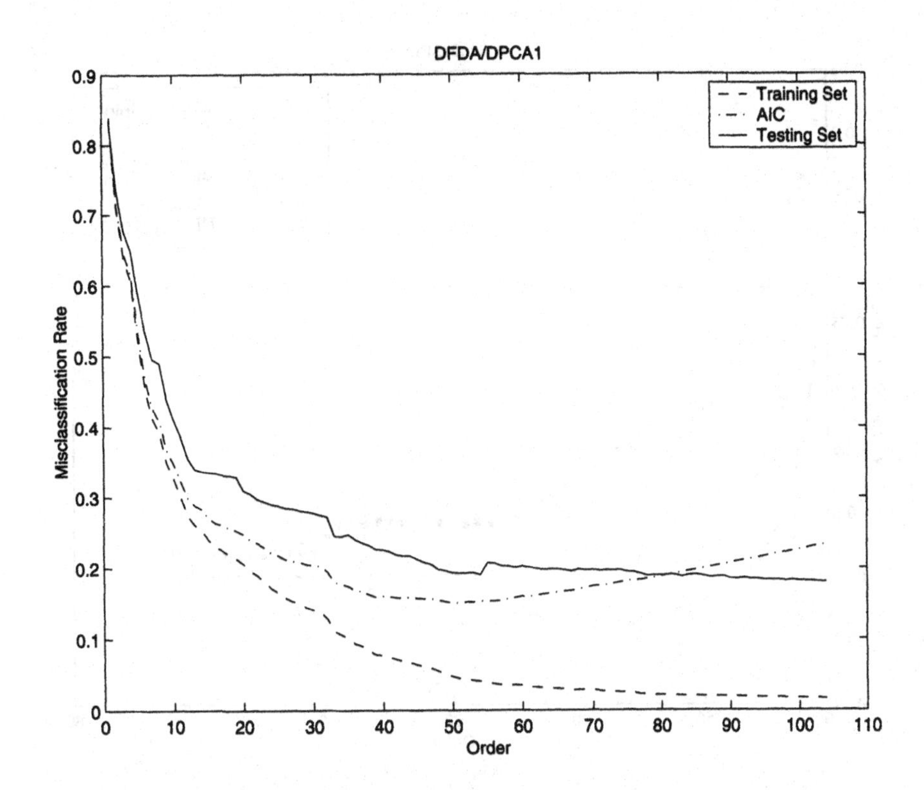

Fig. 10.23. The overall misclassification rates for the training and testing sets and the information criterion (AIC) for various orders using DFDA/DPCA1

(5.12) for PLS1_{adj} and PLS2_{adj} are 16 and 41, respectively, which are close to the orders for PLS1 and PLS2 (c_1 and c_2), respectively. In terms of overall misclassification rates, PLS1_{adj} and PLS2_{adj} have similar performance to PLS1 and PLS2, respectively. For a fixed model order, the PLS1 methods almost always gave better fault diagnosis than the PLS2 methods. The performance of the PLS1 methods was also less sensitive to order selection than the PLS2 methods, and with the AIC resulting in lower model orders (see Table 10.4).

The information criterion worked fairly well for all discriminant PLS methods. The overall misclassification rate for the testing set with the reduction order using the information criterion for PLS1_{adj} is 0.58 while that for the other three PLS methods is 0.57. The minimum overall misclassification rate for the testing set is 0.56 for PLS1_{adj} and PLS2_{adj} and 0.55 for PLS1 and PLS2. The AIC curves (see Figures 10.24 and 10.25) nearly overlap the misclassification rate curves for PLS2 and adjusted PLS2, which indicates that the AIC will give similar model orders as cross-validation in these cases.

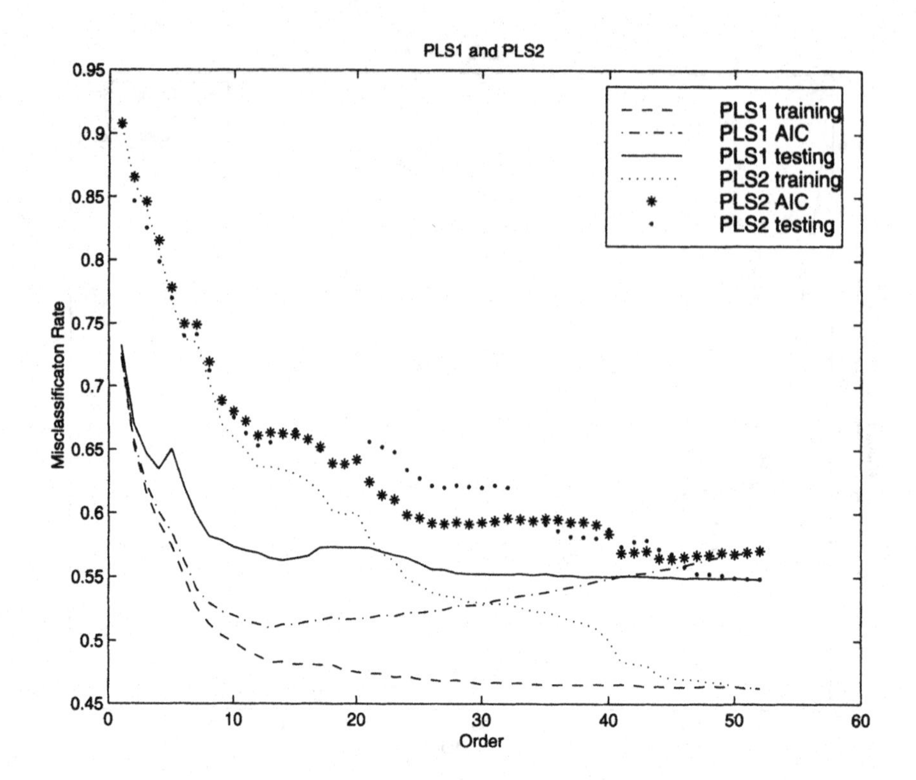

Fig. 10.24. The overall misclassification rates for the training and testing sets and the information criterion (AIC) for various orders using PLS1 and PLS2

For PLS1 and adjusted PLS1, the AIC does not overlap with the classification rate curves, but does have a minimum at approximately the same order as where the misclassification rate curves for the testing data flatten out. This indicates that the AIC provided good model orders for the PLS1 methods.

Figure 10.26 plots the overall standard deviation of misclassification rates for the testing sets for various orders using PLS1, PLS2, PLS1_{adj}, and PLS2_{adj}. The standard deviations for PLS1_{adj} and PLS2_{adj} were 10-25% lower than that of PLS1 and PLS2 (respectively) for most orders. This indicates that PLS1_{adj} and PLS2_{adj} provided a more consistent prediction quality than PLS1 and PLS2. For example, 7 of 21 classes had misclassification rates between 0.90 to 1.00 using PLS1 and PLS2, respectively (see Table 10.14). However, only 2 of 21 classes were between 0.90 and 1.00 using PLS1_{adj} and PLS2_{adj} and the highest misclassification rate was 0.93. This also means that when PLS1 and PLS2 produced low misclassification rates, PLS1_{adj} and PLS2_{adj} tended to produce higher misclassification rates. There was an

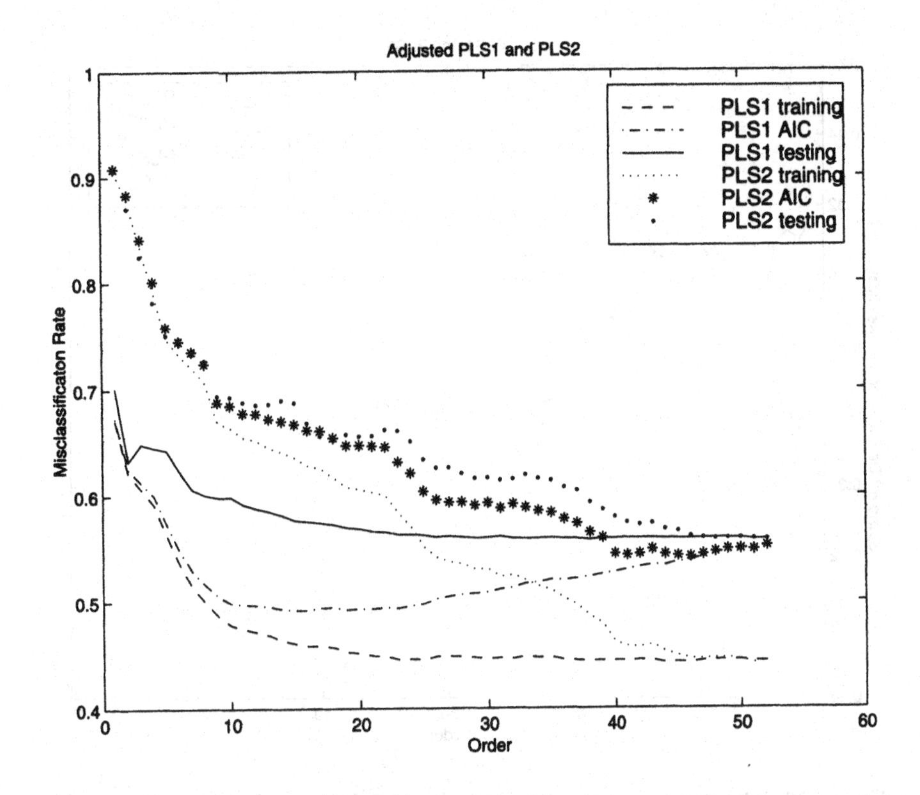

Fig. 10.25. The overall misclassification rates for the training and testing sets and the information criterion (AIC) for various orders using PLS1_{adj} and PLS2_{adj}

advantage to apply PLS1_{adj} and PLS2_{adj} when PLS1 and PLS2 performed poorly.

Although PLS1 was able to capture a large amount of variance using only a few factors, it does require more computation time. Recall that in the calibration steps, PLS1 needs to run the NIPALS p times whereas PLS2 only needs to run the NIPALS one time, and that NIPALS runs from (6.10) to (6.20) for each PLS component. Since iteration from (6.10) to (6.13) is needed for PLS2, NIPALS requires a longer computation time in PLS2. Assume that it takes t_1 computation time to run from (6.22) to (6.27) for PLS1, and that it takes PLS2 $t_1 + \epsilon$ computation time. The total computation time t_{train} in the calibration steps is equal to pat_1 and $a(t_1 + \epsilon)$ for PLS1 and PLS2, respectively, where $a = \min(m, n)$. In the prediction steps, assume it takes t_2 computation time unit to run from (6.30) to (6.32), and that the total computation time t_{test} in the prediction step is equal to pc_1t_2 and c_2t_2 for PLS1 and PLS2, respectively. The ratio r_t of the total computation time between PLS1 and PLS2 is

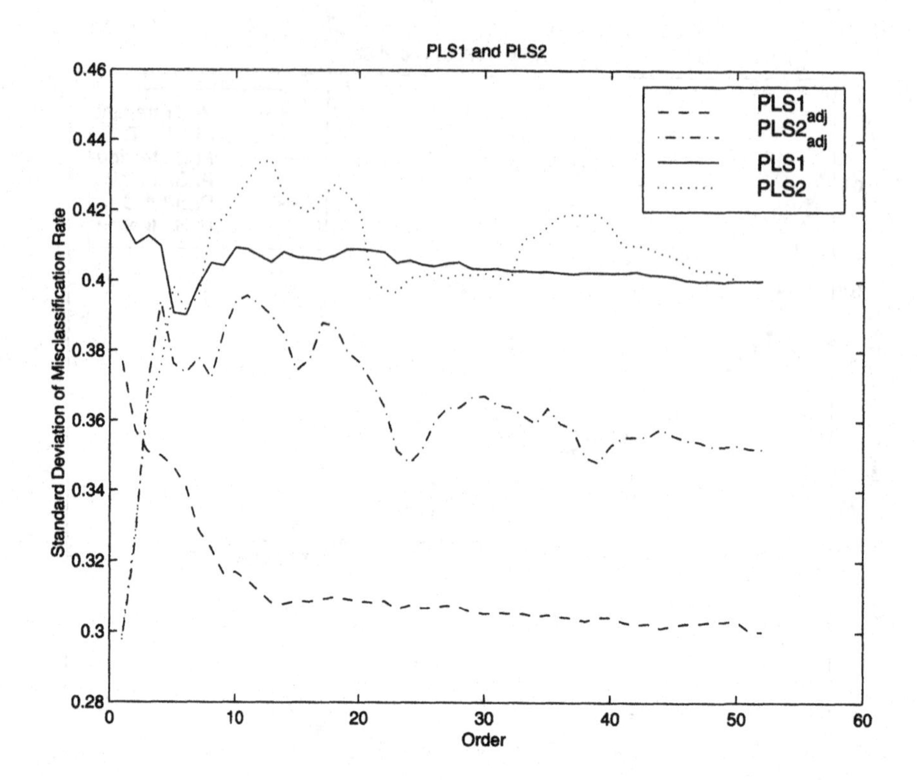

Fig. 10.26. The standard deviation of misclassification rates for the testing set for various orders using PLS1, PLS2, PLS1_{adj}, and PLS2_{adj}

$$r_t = \frac{pat_1 + pc_1t_2}{a(t_1 + \epsilon) + c_2t_2} \tag{10.1}$$

This ratio is much greater than 1 when p is large.

The overall misclassification rates for the training and testing sets and the information criterion (5.12) for various orders using PCA1 are plotted in Figure 10.27. At $a = 52$, the overall misclassification rates for the T^2 statistics based on PCA1 and MS were the same (0.214). This verifies the discussion in Section 4.6 that PCA1 reduces to MS when $a = m$. Regardless of order selected, all FDA methods always gave a lower overall misclassification rate than PCA1 (see Figure 10.20, 10.21, and 10.27). This suggests that FDA model has an advantage over PCA model for diagnosing faults.

It is interesting to see that when all of the factors are included in the FDA methods, the overall misclassification rates were about 0.20, which were different from the overall misclassification rate produced by MS. This is because, when $a = m$, the matrices W_a in (5.16) and $W_{mix,a}$ in (5.17) are not necessarily orthogonal, and so may not project the data into an orthogonal space.

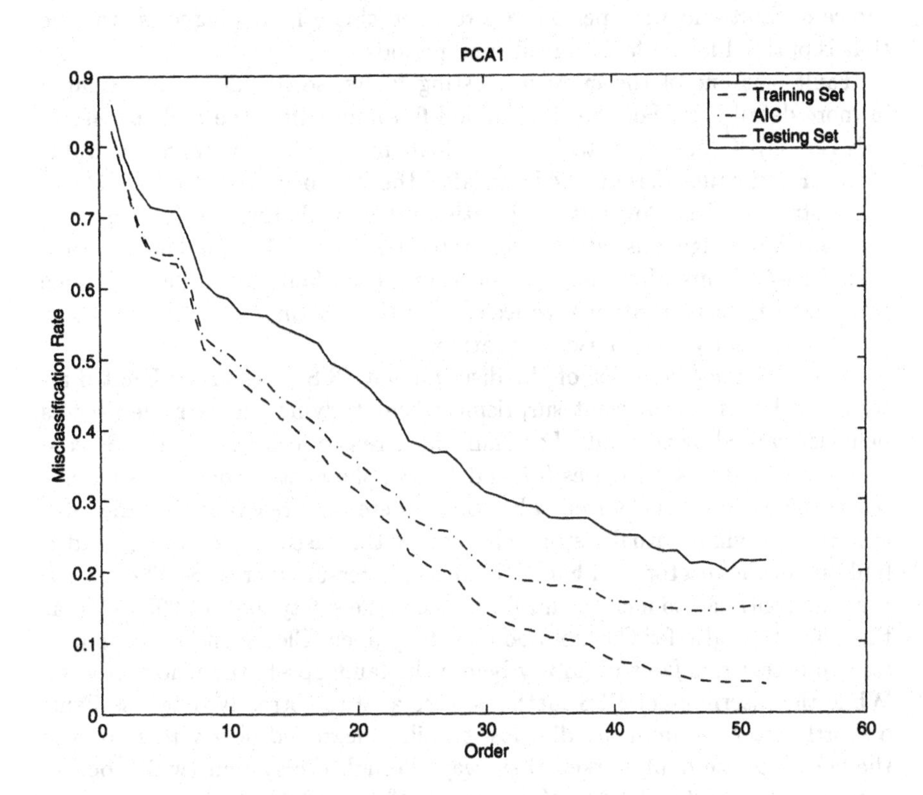

Fig. 10.27. The overall misclassification rates for the training and testing sets and the information criterion (AIC) for various orders using PCA1

The PCAm-based and DPCAm-based statistics produced high overall misclassification rates (see Table 10.12). A weakness of the PCAm-based statistics is that PCAm reduces the dimensionality of each class by using the information in only one class but not the information from all the classes. As shown in Table 10.13, the T^2 statistic based on PCA1 gave a much lower misclassification rate than the statistic based on PCAm for almost all faults.

Now let us consider the PCA, DPCA, and CVA fault diagnosis statistics, all of which separate the dimensionality into a state or score space, and a residual space. For some faults the state or score space version of the statistic gave lower misclassification rates; in other cases the residual space statistics gave lower misclassification rates. Hence, a complete fault diagnosis approach should contain score/state space and residual statistics.

The misclassification rates for the 21 faults were separated into three time periods after the occurrence of the fault (0-5, 5-24, and 24-40 hours), and have been tabulated in Tables 10.15 to 10.20. These tables indicate that each fault diagnosis statistic gives the lowest misclassification rate for some choice of fault and time period. There is no single fault diagnosis statistic that is optimal for all faults or all time periods.

Fault 6 is one of the more interesting faults, so it will be investigated in more detail here. For the time period 0-5 hours after the fault occurred, only the (D)PCAm-based statistics had high misclassification rates (see Table 10.15). For the time period 5-24 hours after the fault occurred, the (D)PCAm-based statistics have low misclassification rates, while the discriminant PLS methods have high misclassification rates (see Table 10.17). For the time period 24-40 hours after the fault occurred, each fault diagnosis technique has a zero misclassification rate except for the discriminant PLS methods, which have nearly 100% misclassification.

The very poor behavior of the discriminant PLS method for Fault 6 after $t = 5$ hours is somewhat surprising when studying the extreme process behavior caused by the fault. For Fault 6, there is a feed loss of A in Stream 1 at $t = 8$ hours (see Figures 8.1 and 10.28), the control loop on Stream 1 reacts to fully open the A feed valve. Since there is no reactant A in the feed, the reaction will eventually stop. This causes the gaseous reactants D and E build up in the reactor, and hence the reactor pressure increases. The reactor pressure continues to increase until it reaches the safety limit of 2950 kPa, at this point the valve for Control Loop 6 is fully open. Clearly, it is very important to detect this fault promptly before the fault upsets the whole process. While the discriminant PLS methods were able to correctly diagnose Fault 6 shortly after the fault, its diagnostic ability degraded nearly to zero once the effects of the fault worked their way through the system (which occurs approximately at $t = 8 + 5 = 13$ hours, see Figure 10.28).

For these data sets it was found that the FDA-based methods gave the lowest misclassification rates averaged over all fault classes (see Table 10.12), and that the MS, PCA1, and CVA T_r^2 statistics gave comparable overall misclassification rates as the FDA methods. Based only on this information, one might hypothesize that dimensionality reduction techniques are not useful for fault diagnosis as their performance is very similar to MS. However, this conclusion would be *incorrect*, even for this particular application. For particular faults and particular time periods, substantially lower misclassification rates were provided by the statistics that used dimensionality reduction (see Tables 10.15 to 10.20). For example, 24-40 hours after Fault 18 occurred, two dimensionality reduction statistics resulted in a zero misclassification rate while one MS statistic had a 70% misclassification rate and the other had a 100% misclassification rate (see Table 10.20).

There are several general reasons that fault diagnosis statistics based on dimensionality reduction are useful in practice. First, there are inherent lim-

Table 10.15. The misclassification rates for 0-5 hours after the Faults 1-11 occurred

Method	Fault Basis	**1**	2	3	**4**	**5**	6	7	8	9	10	**11**
PCAm	T^2	**1**	1	0.980	**0.830**	**0.910**	0.720	1	1	1	0.870	**1**
PCA1	T^2	**0.190**	0.140	0.790	**0.110**	**0.170**	0	0	0.160	0.880	0.240	**0.360**
PCAm	Q	**0.210**	0.160	1	**0.890**	**0.900**	0.400	0.480	0.250	1	0.980	**0.860**
PCAm	T^2&Q	**0.330**	0.280	1	**1**	**0.990**	0.610	0.870	0.400	1	1	**0.970**
DPCAm	T^2	**1**	0.960	0.690	**0.740**	**0.470**	0.380	0.800	1	0.990	0.860	**0.970**
DPCAm	Q	**0.240**	0.340	1	**0.950**	**0.930**	0.500	0.370	0.240	1	1	**0.870**
DPCAm	T^2&Q	**0.300**	0.270	1	**1**	**1**	0.710	0.840	0.440	1	1	**0.980**
PLS1	–	**0.090**	0.090	0.940	**0.160**	**0**	0	0	0.840	0.950	0.810	**0.990**
PLS2	–	**0.090**	0.100	0.950	**0.120**	**0.010**	0	0	0.850	0.940	0.790	**0.990**
PLS1$_{adj}$	–	**0.140**	0.180	0.830	**0.330**	**0.110**	0.100	0	0.840	0.810	0.790	**0.860**
PLS2$_{adj}$	–	**0.140**	0.180	0.870	**0.310**	**0.110**	0.070	0	0.840	0.850	0.770	**0.870**
CVA	T_s^2	**0.200**	0.080	0.950	**0.990**	**0.430**	0.010	0.530	0.270	0.980	0.740	**0.910**
CVA	T_r^2	**0.210**	0.140	0.900	**0.330**	**0.280**	0.010	0.010	0.210	0.950	0.210	**0.200**
CVA	Q	**0.570**	0.460	0.970	**0.910**	**0.480**	0.110	0.330	0.620	0.950	0.830	**0.890**
FDA	T^2	**0.200**	0.150	0.800	**0.140**	**0.160**	0	0	0.160	0.910	0.240	**0.370**
FDA/PCA1	T^2	**0.190**	0.150	0.750	**0.110**	**0.160**	0	0	0.160	0.910	0.280	**0.340**
FDA/PCA2	T^2	**0.200**	0.150	0.800	**0.140**	**0.160**	0	0	0.160	0.910	0.240	**0.370**
DFDA/DPCA1	T^2	**0.210**	0.150	0.740	**0.130**	**0.180**	0	0	0.160	0.930	0.210	**0.150**
MS	T_0^2	**0.200**	0.150	0.800	**0.140**	**0.160**	0	0	0.160	0.910	0.240	**0.370**
MS	T_1^2	**0.280**	0.260	0.950	**0.350**	**0.320**	0	0	0.130	0.970	0.220	**0.190**

Table 10.16. The misclassification rates for 0-5 hours after the Faults 12-21 occurred

Method	Fault Basis	12	13	14	15	16	17	18	19	20	21	Avg.
PCAm	T^2	0.810	1	0.200	0.970	0.870	0.670	0.990	0.500	0.790	0.620	0.844
PCA1	T^2	0.010	0.360	0.040	0.780	0.190	0.320	0.840	0.040	0.740	0.240	0.314
PCAm	Q	0.190	0.440	0.330	0.990	0.980	0.510	0.750	0.570	0.930	0.990	0.658
PCAm	T^2&Q	0.003	0.700	0.500	1	1	0.470	0.820	0.890	1	1	0.754
DPCAm	T^2	0.640	1	0.480	0.950	0.870	0.650	1	0.900	0.850	0.190	0.780
DPCAm	Q	0.240	0.440	0.130	1	0.990	0.390	0.800	0.410	0.960	0.970	0.656
DPCAm	T^2&Q	0.050	0.650	0.160	1	1	0.490	[0.710]	0.940	1	1	0.740
PLS1	–	0.970	0.840	0.990	0.970	0.940	0.270	0.770	0.940	0.620	0.550	0.606
PLS2	–	0.970	0.830	0.990	0.990	0.870	0.270	0.800	0.970	0.600	0.880	0.663
$PLS1_{adj}$	–	0.950	0.940	0.920	0.910	0.740	0.490	0.840	0.820	0.770	0.530	0.614
$PLS2_{adj}$	–	1.000	0.950	0.930	0.890	0.710	0.470	0.850	0.810	0.790	0.520	0.616
CVA	T_s^2	[0]	0.610	0.160	0.980	0.620	0.280	0.930	0.210	0.730	1	0.553
CVA	T_r^2	0.010	0.390	[0]	0.560	[0.160]	0.200	0.830	[0]	[0.540]	0.990	0.340
CVA	Q	0.240	0.690	0.450	0.980	0.960	0.320	0.730	0.880	0.820	0.950	0.673
FDA	T^2	0.010	0.360	0.020	0.760	0.180	0.280	0.840	0.020	0.640	0.100	0.302
FDA/PCA1	T^2	0.010	0.370	0.020	0.840	0.200	0.320	0.830	0.040	0.630	0.230	0.311
FDA/PCA2	T^2	0.010	0.360	0.020	0.760	0.180	0.280	0.840	0.020	0.640	0.099	0.301
DFDA/DPCA1	T^2	0.020	0.360	[0]	0.800	0.300	0.200	0.790	0.090	0.610	0.480	0.310
MS	T_0^2	0.010	0.360	0.020	0.760	0.180	0.280	0.840	0.020	0.640	[0.020]	0.298
MS	T_1^2	[0]	[0.320]	[0]	0.640	0.220	[0.170]	0.770	[0]	0.610	0.030	[0.306]

Table 10.17. The misclassification rates for 5-24 hours after the Faults 1-11 occurred

Method	Fault Basis	1	2	3	4	5	6	7	8	9	10	11
PCAm	T^2	**0.645**	0.345	0.939	**0.832**	**0.934**	0.021	0.953	0.995	0.995	0.845	**0.984**
PCA1	T^2	**0**	0	0.721	**0.189**	**0**	0	0	0.021	0.782	0.121	**0.226**
PCAm	Q	**0.003**	0.008	0.992	**0.963**	**0.903**	0	0.416	0.321	0.997	0.989	**0.863**
PCAm	T^2&Q	**0**	0	1	**1**	**0.979**	0	0.511	0.466	1	1	**0.974**
DPCAm	T^2	**0.834**	0.352	0.732	**0.737**	**0.876**	0.003	0.792	1	0.984	0.771	**0.916**
DPCAm	Q	**0.011**	0.037	0.997	**0.966**	**0.853**	0	0.318	0.208	0.997	0.995	**0.850**
DPCAm	T^2&Q	**0**	0	1	**1**	**1**	0	0.634	0.466	1	1	**0.979**
PLS1	–	**0**	0	0.955	**0.158**	**0.008**	0.076	0	0.840	0.950	0.810	**0.990**
PLS2	–	**0**	0	0.974	**0.095**	**0.008**	0.797	0	0.803	0.995	0.518	**0.984**
PLS1$_{adj}$	–	**0**	0	0.840	**0.355**	**0.034**	0.887	0	0.879	0.940	0.505	**0.868**
PLS2$_{adj}$	–	**0**	0	0.850	**0.324**	**0.034**	0.890	0	0.879	0.961	0.492	**0.890**
CVA	T_s^2	**0.005**	0	0.929	**0.984**	**0.016**	0	0.553	0.582	0.963	0.742	**0.887**
CVA	T_r^2	**0**	0.003	0.789	**0.400**	**0.008**	0	0	0.050	0.842	0.079	**0.100**
CVA	Q	**0.213**	0.150	0.979	**0.890**	**0.121**	0	0.582	0.620	0.950	0.830	**0.890**
FDA	T^2	**0**	0	0.716	**0.213**	**0**	0	0	0.021	0.771	0.118	**0.221**
FDA/PCA1	T^2	**0**	0	0.716	**0.184**	**0**	0	0	0.029	0.787	0.142	**0.226**
FDA/PCA2	T^2	**0**	0	0.716	**0.213**	**0**	0	0	0.021	0.771	0.118	**0.221**
DFDA/DPCA1	T^2	**0**	0	0.726	**0.174**	**0**	0	0	0.013	0.824	0.092	**0.097**
MS	T_0^2	**0**	0	0.716	**0.213**	**0**	0	0	0.021	0.771	0.118	**0.221**
MS	T_1^2	**0**	0	0.866	**0.468**	**0**	0	0	0.005	0.874	0.063	**0.095**

Table 10.18. The misclassification rates for 5-24 hours after the Faults 12-21 occurred

Method	Fault Basis	12	13	14	15	16	17	18	19	20	21	Avg.
PCAm	T^2	0.813	1	0.247	0.958	0.871	0.461	0.018	0.397	0.758	0.900	0.710
PCA1	T^2	0.005	0.211	0.018	0.700	0.187	0.145	0.124	0.084	0.061	0.095	0.176
PCAm	Q	0.200	0.621	0.242	0.995	0.992	0.416	0.124	0.587	0.900	0.989	0.596
PCAm	T^2&Q	0.126	0.850	0.397	1	1	0.405	0.063	0.839	1	1	0.648
DPCAm	T^2	0.697	1	0.555	0.971	0.868	0.616	0.103	0.808	0.658	0.403	0.699
DPCAm	Q	0.161	0.558	0.100	0.995	0.995	0.308	0.116	0.489	0.934	0.976	0.565
DPCAm	T^2&Q	0.147	0.818	0.126	1	1	0.387	0.058	0.942	1	1	0.646
PLS1	–	0.979	0.942	0.995	0.982	0.874	0.097	0.587	0.916	0.266	0.050	0.541
PLS2	–	0.947	0.942	1.000	0.979	0.882	0.129	0.561	0.945	0.200	0.053	0.562
$PLS1_{adj}$	–	0.866	0.982	0.929	0.913	0.634	0.358	0.711	0.795	0.458	0	0.569
$PLS2_{adj}$	–	0.882	0.976	0.934	0.921	0.640	0.345	0.700	0.761	0.468	0	0.569
CVA	T_s^2	0	0.516	0.218	0.953	0.526	0.205	0.155	0.421	0.253	0.982	0.471
CVA	T_r^2	0.003	0.324	0.003	0.650	0.132	0.045	0.063	0.003	0.024	0.966	0.214
CVA	Q	0.229	0.571	0.413	0.992	0.816	0.184	0.711	0.929	0.542	0.940	0.602
FDA	T^2	0.005	0.213	0.003	0.716	0.184	0.116	0.132	0.024	0.063	0.013	0.168
FDA/PCA1	T^2	0.003	0.221	0.003	0.732	0.163	0.090	0.129	0.032	0.068	0.192	0.177
FDA/PCA2	T^2	0.005	0.213	0.003	0.716	0.184	0.116	0.132	0.024	0.063	0.005	0.168
DFDA/DPCA1	T^2	0.003	0.195	0.003	0.737	0.174	0.013	0.116	0.100	0.092	0.416	0.180
MS	T_0^2	0.005	0.213	0.003	0.716	0.184	0.116	0.516	0.024	0.063	0	0.186
MS	T_1^2	0	0.168	0.003	0.661	0.240	0.018	0.108	0.003	0.097	0	0.175

Table 10.19. The misclassification rates for 24-40 hours after the Faults 1-11 occurred

Method	Fault Basis	1	2	3	4	5	6	7	8	9	10	11
PCAm	T^2	**0.622**	0.303	0.925	**0.779**	**1**	0	1	1	0.988	0.847	**0.991**
PCA1	T^2	**0**	0	0.853	**0.147**	**0**	0	0	0	0.744	0.097	**0.203**
PCAm	Q	**0**	0	0.988	**0.956**	**0.928**	0	0.369	0.216	0.991	0.988	**0.853**
PCAm	$T^2\&Q$	**0**	0	1	**1**	**0.959**	0	0.363	0.344	1	1	**0.959**
DPCAm	T^2	**0.897**	0.384	0.669	**0.694**	**0.997**	0	0.978	1	0.991	0.672	**0.978**
DPCAm	Q	**0**	0	0.991	**0.966**	**0.837**	0	0.347	0.103	0.997	0.994	**0.825**
DPCAm	$T^2\&Q$	**0**	0	1	**1**	**1**	0	0.566	0.303	1	1	**0.988**
PLS1	–	**0.003**	0.003	0.972	**0.188**	**0.006**	0.997	0.003	0.806	0.969	0.684	**0.984**
PLS2	–	**0.003**	0.003	0.969	**0.147**	**0.006**	0.997	0.003	0.753	0.975	0.619	**0.969**
$PLS1_{adj}$	–	**0.003**	0.003	0.916	**0.384**	**0.034**	0.997	0.003	0.879	0.940	0.505	**0.868**
$PLS2_{adj}$	–	**0.003**	0.003	0.909	**0.319**	**0.031**	0.997	0.003	0.816	0.881	0.597	**0.891**
CVA	T_s^2	**0**	0	0.950	**0.975**	**0**	0	0.772	0.519	0.972	0.750	**0.922**
CVA	T_r^2	**0**	0	0.834	**0.316**	**0.003**	0	0	0.013	0.822	0.084	**0.166**
CVA	Q	**0.181**	0.066	0.978	**0.884**	**0.141**	0	0.650	0.650	0.972	0.834	**0.900**
FDA	T^2	**0**	0	0.850	**0.144**	**0**	0	0	0	0.731	0.113	**0.234**
FDA/PCA1	T^2	**0**	0	0.800	**0.153**	**0**	0	0	0	0.731	0.138	**0.234**
FDA/PCA2	T^2	**0**	0	0.850	**0.144**	**0**	0	0	0	0.731	0.113	**0.234**
DFDA/DPCA1	T^2	**0**	0	0.743	**0.151**	**0**	0	0	0	0.734	0.078	**0.132**
MS	T_0^2	**0**	0	0.850	**0.144**	**0**	0	0	0	0.731	0.113	**0.234**
MS	T_1^2	**0**	0	0.890	**0.401**	**0**	0	0	0	0.841	0.100	**0.132**

Table 10.20. The misclassification rates for 24-40 hours after the Faults 12-21 occurred

Method	Fault Basis	12	13	14	15	16	17	18	19	20	21	Avg.
PCAm	T^2	0.906	1	0.253	0.966	0.797	0.403	0.569	0.375	0.756	0.984	0.736
PCA1	T^2	0.044	0.225	0.056	0.844	0.219	0.209	0.616	0.197	0.053	0.156	0.222
PCAm	Q	0.244	0.378	0.291	0.994	0.975	0.384	0.600	0.753	0.931	0.962	0.610
PCAm	T^2&Q	0.350	0.656	0.466	1	1	0.403	0.478	0.916	1	1	0.662
DPCAm	T^2	0.722	1	0.600	0.959	0.700	0.684	0.300	0.731	0.722	0.784	0.736
DPCAm	Q	0.241	0.303	0.116	0.997	0.981	0.313	0.600	0.700	0.959	0.919	0.580
DPCAm	T^2&Q	0.347	0.597	0.181	1	1	0.394	0.600	0.978	1	1	0.664
PLS1	–	0.997	0.244	0.997	0.991	0.894	0.163	1.000	0.903	0.316	0.006	0.577
PLS2	–	0.953	0.181	0.997	0.988	0.903	0.172	1.000	0.947	0.253	0.006	0.564
$PLS1_{adj}$	–	0.847	0.419	0.938	0.944	0.669	0.391	0.997	0.803	0.488	0.003	0.571
$PLS2_{adj}$	–	0.856	0.388	0.925	0.938	0.666	0.388	0.997	0.790	0.509	0.003	0.567
CVA	T_s^2	0.022	0.434	0.197	0.972	0.600	0.213	0.875	0.609	0.237	0.890	0.519
CVA	T_r^2	0.044	0.313	0	0.719	0.156	0.044	0	0.009	0.028	0.072	0.173
CVA	Q	0.388	0.584	0.494	0.975	0.878	0.225	1.000	0.944	0.569	0.897	0.629
FDA	T^2	0.034	0.231	0.022	0.822	0.206	0.150	0.369	0.063	0.041	0.063	0.194
FDA/PCA1	T^2	0.034	0.238	0.022	0.819	0.203	0.156	0.584	0.088	0.034	0.194	0.211
FDA/PCA2	T^2	0.034	0.231	0.022	0.822	0.206	0.150	0.369	0.063	0.041	0.047	0.193
DFDA/DPCA1	T^2	0.066	0.229	0.006	0.834	0.245	0.028	0	0.207	0.141	0.009	0.172
MS	T_0^2	0.034	0.231	0.022	0.822	0.206	0.150	1.000	0.063	0.041	0.031	0.222
MS	T_1^2	0.013	0.219	0	0.828	0.285	0.019	0.709	0.003	0.088	0	0.216

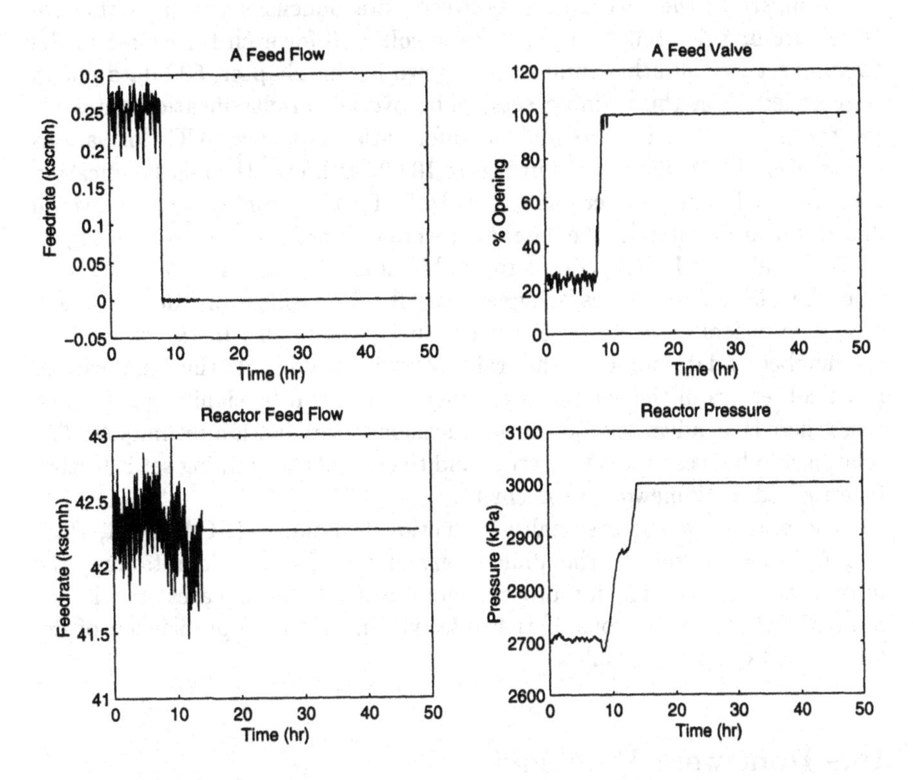

Fig. 10.28. Closed loop simulation for a step change of A feed loss in Stream 1 (Fault 6)

itations due to round-off errors that usually prevent the construction of full-dimensional models for large-scale systems such as industrial plants. Second, there can be limitations on the size of the models used by process monitoring methods that can be implemented in real time on the computer hardware connected to a particular process. While this limitation is becoming less of an issue over time, the authors are aware of industrial control systems still using older control computers.

The main reason for dimensionality reduction is based on the amount of data usually available in practice that has been sufficiently characterized for use in process monitoring. This data, for example, should be cleaned of all outliers caused by computer or database programming errors [255]. For the application of fault diagnosis methods it is required to label each observation as being associated with normal operating conditions or with a particular fault class. These requirements can limit the available training data, especially for the purposes of computing fault diagnosis statistics, to less than what was used in this chapter.

To illustrate the relationship between data dimensionality and the size of the training set, 100 data points were collected for each fault class in the training set (for all other simulations shown in this chapter, 500 data points were collected in the training sets). The overall misclassification rates for the training and testing sets and the information criterion (AIC) for various orders using PCA1 are plotted in Figure 10.29. Although the misclassification rates reduced nearly to zero as a goes to 52 for the training set, the overall misclassification rates for the testing set were very high as compared to Figure 10.27. Recall that PCA1 reduces to the MS statistic when $a = 52$, this shows that the MS statistic gives a higher overall misclassification rate for many reduction orders ($a = 20$ to 45, as seen in Figure 10.29). In the case where the number of data points in the training set is insufficient (the usual case in practice), errors in the sample covariance matrix will be significant. In such cases there is an advantage to using dimensionality reduction techniques. The relationship between reduction order and the size of the training set is further investigated in Homework Problem 11.

The purpose of dimensionality reduction techniques (PCA, FDA, PLS, and CVA) is to reduce the dimensions of the data while retaining the most useful information for process monitoring. In most cases, the lower-dimensional representations of the data will improve the proficiency of detecting and diagnosing faults.

10.9 Homework Problems

1. A co-worker at a major company suggested that false alarms were not an issue with fault identification and that it may be useful to apply all the scores (not just the first a scores) for the PCA, DPCA, and CVA-based $CONT$ as shown in Section 4.5. Evaluate the merits of the proposal. Apply this idea to the data collected from the Tennessee Eastman plant simulator (`http://brahms.scs.uiuc.edu`). What are your conclusions?
2. Apply the similarity index (4.41) and mean overlap (4.42) to the data collected from the Tennessee Eastman plant simulator. Relate your results with these two measures with the misclassification rates of the fault diagnosis statistics as reported in this chapter. Do the similarity index and mean overlap assess the likelihood of successful diagnosis? Explain in detail why one measure performs better than the other.
3. As discussed in Chapter 5, (D)FDA only ranks the eigenvectors associated with the non-zero eigenvalues. Propose a method other than PCA1 to rank the eigenvectors associated with the zero eigenvalues. Evaluate your proposal using the data collected from the Tennessee Eastman plant simulator.
4. In addition to the original 21 faults for the TEP, simulate 39 additional multiple faults (combination of two faults) of your choice. Apply FDA,

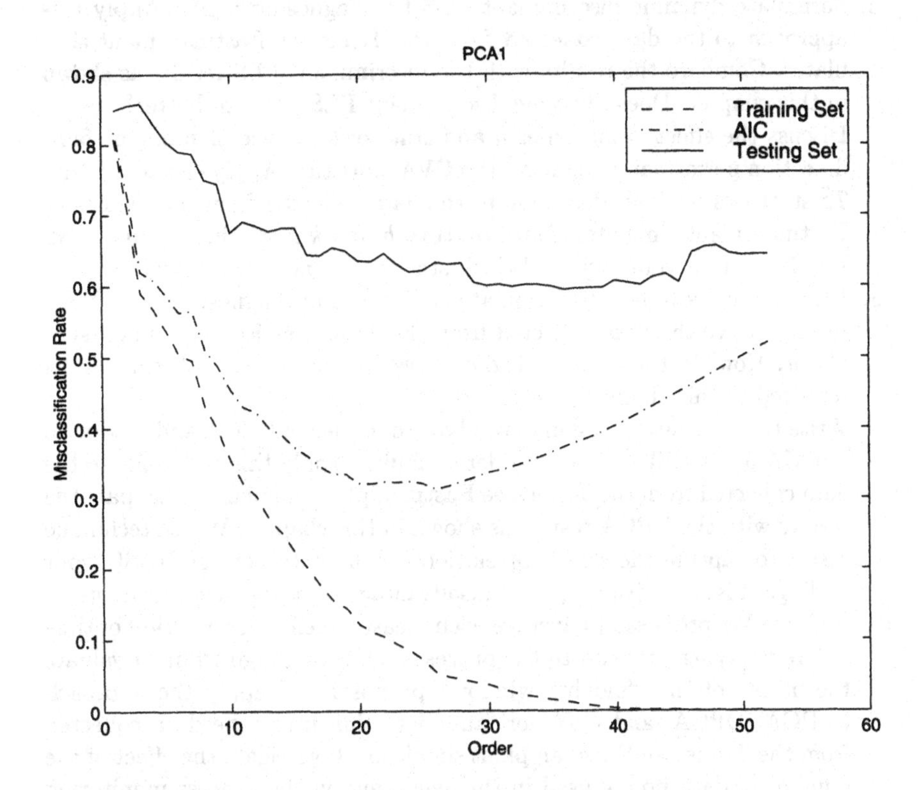

Fig. 10.29. The overall misclassification rates for the training and testing sets and the information criterion (AIC) for various orders using PCA1 with 100 data points in the training set

FDA/PCA1, FDA/PCA2, and their corresponding dynamic version to diagnose these 60 faults and comment on your findings.

5. A co-worker at a major company proposed to modify the model complexity term in the information criterion (5.12) to $1.5a/\tilde{n}$. Based only on the performance as given by Figure 10.23 which was obtained by an application of the original information criterion (5.12) to a simulated industrial plant, evaluate the relative merits of the co-worker's proposal. Another co-worker suggested to modify the model complexity term in the information criterion (5.12) to a/n. Evaluate the relative merits of the second proposal. Based on Figure 10.23, propose a modification of the model complexity term which will give the best results for the simulated industrial plant. How well does your modified model complexity term perform? [Note that designing the best information criterion for one specific process application does not necessarily give the best possible information criterion for other process applications.]

6. Formulate dynamic discriminant PLS for diagnosing faults. Apply this approach to the data collected from the Tennessee Eastman plant simulator. Compare the results with the discriminant PLS results as shown in this chapter. Does dynamic discriminant PLS perform better?
7. Discuss the effect of lag order h and state order k selection on the fault detection performance using all the CVA statistics. Apply the Q, T_s^2, and T_r^2 statistics for fault detection to the data collected from the Tennessee Eastman plant simulator. Now, perturb h and k from their optimal values. Report on your results. Which statistic deviates the most? Why?
8. Describe in detail how to formulate CVA for fault diagnosis. Apply these techniques to the data collected from the Tennessee Eastman plant simulator. How do these fault diagnosis results compared with the results reported in this chapter?
9. Write a report describing in detail how to implement PCA and PLS with EWMA and CUSUM charts to detect faults. Apply this technique to the data collected from the Tennessee Eastman plant simulator. Compare the results with the DPCA results as shown in this chapter. Which technique seems to capture the serial correlations of the data better? Justify your findings. List an advantage and disadvantage of using each technique.
10. A co-worker proposed to average each measurement over a period of time before applying the data to the process monitoring algorithms. Evaluate the merits of this "moving window" proposal and apply the approach to PCA, DPCA, and CVA for fault detection using the data collected from the Tennessee Eastman plant simulator. Investigate the effect of the number of data points used in the averaging on the process monitoring performance. Was it possible to improve on DPCA and CVA using this approach? Justify your answers.
11. Evaluate the effects of the size of training set and the sampling interval on the reduction order and process monitoring performance. Construct training and testing data sets for the TEP using (i) 150 points with a sampling interval of 10 minutes, (ii) 1500 points with a sampling interval of 1 minute, and (iii) 1500 points with a sampling interval of 10 minutes. Implement all process monitoring statistics described in this book. How is the relative performance of each process monitoring statistic affected? Why? How is the reduction order affected? Compare the techniques in terms of the sensitivity of their performance to changes in the size of the training set and the sampling interval.
12. While the threshold for the Q statistic (Equation 4.22) is widely used in practice, its derivation relies on certain assumptions that are not always true (as mentioned in Section 10.6). Write a report on the exact distribution for Q and how to compute the exact threshold for the Q statistic. Under what conditions is Equation 4.22 a valid approximation? Would these conditions be expected to hold for most applications to process data collected from large-scale industrial plants? (Hint: Several papers

that describe the exact distribution for Q are cited at the end of the paper by Jackson and Mudholkar [145].)

Part V

Analytical and Knowledge-based Methods

11. Analytical Methods

11.1 Introduction

As discussed in Section 1.2, process monitoring measures can be characterized as being data-driven, analytical, or knowledge-based. Part III focused mostly on the data-driven methods, which include control charts (Shewhart, CUSUM, and EWMA charts) and dimensionality reduction techniques (PCA, PLS, FDA, and CVA). A well-trained engineer should also have some familiarity with the analytical and knowledge-based approaches since they have advantages for some process monitoring problems. Also, many measures can be associated with more than one approach. For example, the CVA method, while being entirely data driven, can also be characterized as being an analytical method since a state-space model can be constructed from the Kalman states (see Chapter 7). Other measures at the intersection of more than one approach are discussed in Chapter 12.

Based on the measured input $\mathbf{u}$ and output $\mathbf{y}$, the analytical methods generate features using detailed mathematical models. Commonly used features include residuals $\mathbf{r}$, parameter estimates $\hat{\mathbf{p}}$, and state estimates $\hat{\mathbf{x}}$. Faults are detected or diagnosed by comparing the observed features with the features associated with normal operating conditions either directly or after some transformation.

Analytical methods that use residuals as features are commonly referred to as **analytical redundancy** methods. The residuals are the outcomes of consistency checks between the plant observations and a mathematical model. The residuals will be non-zero due to faults, disturbances, noise, and/or modeling errors. As we will see, part of the challenge in designing a process monitoring system based on analytical redundancy is distinguishing between residuals caused by faults, and residuals caused by the other variations. In the preferred situation, the residuals or transformations of the residuals will be relatively large when faults are present, and small in the presence of disturbances, noise, and/or modeling errors. In this case the presence of faults can be detected by defining appropriate thresholds. In any case, an analytical redundancy method will arrive at a diagnostic decision based on the residuals [87, 101, 221].

The three main ways to generate residuals are **parameter estimation**, **observers**, and **parity relations** [94].

1. **Parameter estimation.** For parameter estimation, the residuals are the difference between the nominal model parameters and the estimated model parameters. Deviations in the model parameters serve as the basis for detecting and isolating faults [20, 135, 136, 163].
2. **Observers.** The observer-based method reconstructs the output of the system from the measurements or a subset of the measurements with the aid of observers. The difference between the measured outputs and the estimated outputs is used as the vector of residuals [54, 68, 86].
3. **Parity relations.** This method checks the consistency of the mathematical equations of the system with the measurements. The parity relations are subjected to a linear dynamic transformation, with the transformed residuals used for detecting and isolating faults [63, 101, 226, 227].

When an accurate first-principles or other mathematical model is available, the analytical approach can provide improved process monitoring compared to data-driven or knowledge-based approaches. Analytical approaches can also incorporate process flowsheet information in a straightforward way.

As mentioned in Section 1.1, process monitoring terminology varies across disciplines. The definition of fault detection is fairly consistent, while a variety of overlapping definitions is used for fault identification and fault diagnosis. A term not defined in Section 1.1 is **fault isolation**, which is commonly defined as determining the exact location of the fault or faulty component, that is, to determine which component is faulty [101]. Fault isolation provides more information than a fault identification procedure as defined in Section 1.1, in which only the observation variables associated with the fault are determined. Fault isolation does not provide as much information as a fault diagnosis procedure as defined in Section 1.1, in which the type, magnitude, and time of the fault are determined. More specifically, a single component may have a variety of different types of faults associated with it (e.g., a valve may be stuck closed, or may just have occasional sticking). A fault isolation procedure may locate the component (e.g., the valve), but a fault diagnosis procedure would be needed to determine the type of fault associated with the component (e.g., "stuck closed" versus "occasional sticking"). A commonly used term in the literature is the **FDI system**, which is a process monitoring method that contains both fault detection and isolation stages.

Most of the analytical methods described in this chapter can be characterized as being FDI systems. Enough background is provided on each method so that the reader can determine which approach is likely to be most promising in a particular application. Plenty of references are given for the reader to learn more about implementation. The chapter begins in Section 11.2 by defining additive and multiplicative faults, and describing how these faults affect the process dynamics. Analytical approaches based on parameter estimation, state estimators/observers, and parity relations are discussed in Sections 11.3, 11.4, and 11.5, respectively.

11.2 Fault Descriptions

For a plant with input $\mathbf{u} \in \mathcal{R}^{m_u}$ and output $\mathbf{y} \in \mathcal{R}^{m_y}$, the discrete-time linear state-space model (without faults, disturbance, and noise) is

$$\mathbf{x}(t+1) = A\mathbf{x}(t) + B\mathbf{u}(t) \tag{11.1}$$

$$\mathbf{y}(t) = C\mathbf{x}(t) + D\mathbf{u}(t) \tag{11.2}$$

where $\mathbf{x} \in \mathcal{R}^{a}$ is the state vector, t is the discrete-time index, and the state-space matrices A, B, C, and D specify the state-space model.

Faults that can be modeled as unknown changes in signals in the system are called **additive faults**. Additive faults include

- actuator faults $\Delta\mathbf{u}(t)$,
- sensor faults $\Delta\mathbf{y}(t)$,
- some plant faults (such as, leaks) $\Delta\mathbf{u_p}(t)$.

An example of an actuator fault is a sticking valve or a burnt-out motor. A sensor fault is a corroded thermocouple, or a leak in the pressure line to a differential pressure gauge. An example of a plant fault that acts as an additive fault is a leak in a pipe containing process fluid.

Now consider the effect of additive faults on the observed values of the inputs and outputs. As shown in Figure 11.1, the observed values of the input $\mathbf{u}(t)$ and output $\mathbf{y}(t)$ are related to the true values (those acting on or arising from the plant) $\mathbf{u}^{\circ}(t)$ and $\mathbf{y}^{\circ}(t)$ as

$$\mathbf{u}^{\circ}(t) = \mathbf{u}(t) + \Delta\mathbf{u}(t), \tag{11.3}$$

and

$$\mathbf{y}^{\circ}(t) = \mathbf{y}(t) + \Delta\mathbf{y}(t). \tag{11.4}$$

The plant faults affect both the true output and the observed output.

Now the above equations are augmented to include additive noise and disturbances. Consider additive plant disturbance $\mathbf{d}(t)$ and the following noise signals:

- actuator noise $\delta\mathbf{u}(t)$,
- sensor noise $\delta\mathbf{y}(t)$,
- plant noise $\delta\mathbf{u_p}(t)$.

The *observed* and *true* values for the plant input $\mathbf{u}$ and output $\mathbf{y}$ are related to the additive faults and noise signals by

$$\mathbf{u}^{\circ}(t) = \mathbf{u}(t) + \Delta\mathbf{u}(t) + \delta\mathbf{u}(t), \tag{11.5}$$

and

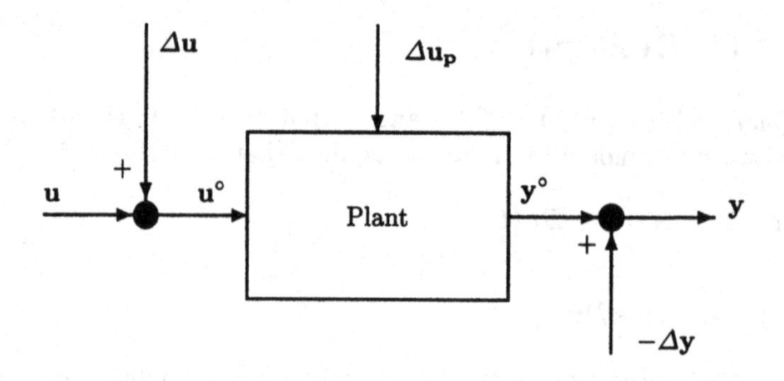

Fig. 11.1. Relationship between the additive faults and the plant variables

$$\mathbf{y}^{\circ}(t) = \mathbf{y}(t) + \Delta\mathbf{y}(t) + \delta\mathbf{y}(t). \tag{11.6}$$

Define the combined vector of additive faults as

$$\mathbf{f}(t) = \begin{bmatrix} \Delta\mathbf{u}(t) \\ \Delta\mathbf{u_p}(t) \\ \Delta\mathbf{y}(t) \end{bmatrix} \tag{11.7}$$

and the combined vector of additive noise as

$$\mathbf{n}(t) = \begin{bmatrix} \delta\mathbf{u}(t) \\ \delta\mathbf{u_p}(t) \\ \delta\mathbf{y}(t) \end{bmatrix}. \tag{11.8}$$

Extending the state equations (11.1) and (11.2) to include the additive noise, disturbances, and faults gives

$$\mathbf{x}(t+1) = A\mathbf{x}(t) + B\mathbf{u}(t) + B_f\mathbf{f}(t) + B_d\mathbf{d}(t) + B_n\mathbf{n}(t) \tag{11.9}$$

$$\mathbf{y}(t) = C\mathbf{x}(t) + D\mathbf{u}(t) + D_f\mathbf{f}(t) + D_d\mathbf{d}(t) + D_n\mathbf{n}(t) \tag{11.10}$$

where the subscript f is for matrices associated with faults, d is for matrices associated with disturbances, and n is for matrices associated with noise. The state-space matrices are usually highly structured, especially for the matrices associated with the faults (B_f, D_f) and disturbances (B_d, D_d) in which entire rows or columns of zeros are common. For example, the column of B_f associated with a sensor fault in $\mathbf{f}$ is commonly equal to zero, since a sensor fault may affect the output equation without affecting the states.

By introducing the shift operator [19]

$$q\mathbf{x}(t) = \mathbf{x}(t+1), \tag{11.11}$$

the state equations (11.9) and (11.10) can be rewritten in terms of transfer functions:

$$\mathbf{y}(t) = P(q)\mathbf{u}(t) + P_f(q)\mathbf{f}(t) + P_d(q)\mathbf{d}(t) + P_n(q)\mathbf{n}(t) \tag{11.12}$$

with the transfer functions being described by

$$\begin{aligned} P(q) &= C(qI - A)^{-1}B + D, \\ P_f(q) &= C(qI - A)^{-1}B_f + D_f, \\ P_d(q) &= C(qI - A)^{-1}B_d + D_d, \\ P_n(q) &= C(qI - A)^{-1}B_n + D_n. \end{aligned} \tag{11.13}$$

Equation 11.12 describes the effects of additive faults, disturbances, and noise on the plant output. Each effect enters the output equation only as changes in signals, not as changes in the transfer functions (the state-space matrices are assumed fixed).

Alternatively, some faults are best modeled as being **multiplicative faults**, which are written in state-space form as

$$\mathbf{x}(t+1) = (A + \Delta A)\mathbf{x}(t) + (B + \Delta B)\mathbf{u}(t) \tag{11.14}$$

$$\mathbf{y}(t) = (C + \Delta C)\mathbf{x}(t) + (D + \Delta D)\mathbf{u}(t). \tag{11.15}$$

Using the shift operator, the state equations can be written in transfer function form

$$\mathbf{y}(t) = P^{\circ}(q)\mathbf{u}(t) \tag{11.16}$$

where

$$P^{\circ}(q) = (C + \Delta C)(qI - A - \Delta A)^{-1}(B + \Delta B) + D + \Delta D \tag{11.17}$$

where $P^{\circ}(q)$ is the true transfer function for the physical system.

The discrepancy $\Delta P(q)$ between the model and the true system is defined by

$$P^{\circ}(q) = P(q) + \Delta P(q). \tag{11.18}$$

Introducing the expression for the process model $P(q)$ from (11.13) and rearranging gives

$$\begin{aligned} \Delta P(q) &= P^{\circ}(q) - P(q) \\ &= (C + \Delta C)(qI - A - \Delta A)^{-1}(B + \Delta B) + D + \Delta D \\ &\quad - (C(qI - A)^{-1}B + D). \end{aligned} \tag{11.19}$$

The discrepancy may be due to parametric faults, where the plant has deviated from its earlier normal behavior, which was properly represented by the model. Instead, the discrepancy could be due to modeling error, which may be present since the implementation of the algorithm. The modeling error may be due to inaccuracy in some of the physical parameters, or due to unmodeled dynamics caused by simplifying a higher-order model with a lower-order model. Another common source of modeling error is from approximating a nonlinear model with a linear model, or by making simplifying assumptions in the derivation of a first-principles model for the plant.

In the absence of the additive faults, disturbances, or noise, the plant output would be

$$\mathbf{y}(t) = (P(q) + \Delta P(q))\mathbf{u}(t) = P(q)\mathbf{u}(t) + \Delta P(q)\mathbf{u}(t). \tag{11.20}$$

This equation shows why the discrepancy in (11.18) is said to be multiplicative rather than additive. By comparing (11.20) with (11.12), we see that multiplicative faults and additive faults affect the plant output in a different manner. Additive faults and disturbances are signals that are related to the output through time-invariant transfer functions. On the other hand, parametric faults and model errors cause a discrepancy in the input-output transfer function. This discrepancy is multiplied by the plant input.

Let us consider a specific case, where the plant input $\mathbf{u}$ is doubled in size. For an additive fault (11.12), this doubling would not affect the mapping between the faults and the plant output. For a multiplicative fault (11.20), doubling the magnitude of the plant input $\mathbf{u}$ doubles the magnitude of the effect of the discrepancy on the plant output. This example is useful to keep in mind when classifying a particular type of fault as being additive or multiplicative.

In the above presentation, the state-space model (11.9) and (11.10) was written in discrete-time form and the transfer function form of the input-output relationship (11.12) was derived using the shift operator. An alternative approach is to use a continuous-time state-space model:

$$\frac{d\mathbf{x}(t)}{dt} = \bar{A}\mathbf{x}(t) + \bar{B}\mathbf{u}(t) + \bar{B}_f\mathbf{f}(t) + \bar{B}_d\mathbf{d}(t) + \bar{B}_n\mathbf{n}(t) \tag{11.21}$$

$$\mathbf{y}(t) = \bar{C}\mathbf{x}(t) + \bar{D}\mathbf{u}(t) + \bar{D}_f\mathbf{f}(t) + \bar{D}_d\mathbf{d}(t) + \bar{D}_n\mathbf{n}(t). \tag{11.22}$$

Applying the Laplace transform on (11.21) and (11.22) and rearranging results in the transfer function form for the input-output relationship:

$$\mathbf{y}(s) = P(s)\mathbf{u}(s) + P_f(s)\mathbf{f}(s) + P_d(s)\mathbf{d}(s) + P_n(s)\mathbf{n}(s) \tag{11.23}$$

where

$$\begin{aligned}
P(s) &= \bar{C}(sI - \bar{A})^{-1}\bar{B} + \bar{D}, \\
P_f(s) &= \bar{C}(sI - \bar{A})^{-1}\bar{B}_f + \bar{D}_f, \\
P_d(s) &= \bar{C}(sI - \bar{A})^{-1}\bar{B}_d + \bar{D}_d, \\
P_n(s) &= \bar{C}(sI - \bar{A})^{-1}\bar{B}_n + \bar{D}_n.
\end{aligned} \tag{11.24}$$

Similarly, multiplicative faults can be written as

$$\mathbf{y}(s) = P^{\circ}(s)\mathbf{u}(s) \tag{11.25}$$

where

$$P^{\circ}(s) = P(s) + \Delta P(s) \tag{11.26}$$

and the model discrepancy $\Delta P(s)$ is defined by

$$\begin{aligned} \Delta P(s) &= P^{\circ}(s) - P(s) \\ &= (\bar{C} + \Delta\bar{C})(sI - \bar{A} - \Delta\bar{A})^{-1}(\bar{B} + \Delta\bar{B}) + \bar{D} + \Delta\bar{D} \\ &\quad - (\bar{C}(sI - \bar{A})^{-1}\bar{B} + \bar{D}) \end{aligned} \tag{11.27}$$

and $\Delta\bar{A}$, $\Delta\bar{B}$, $\Delta\bar{C}$, and $\Delta\bar{D}$ are the perturbations in the state-space matrices for the continuous-time system.

The next three sections describe how additive and multiplicative faults can be detected and isolated using parameter estimation, observers, and parity relations. As we will see, parameter estimation is especially suited for handling multiplicative faults, whereas additive faults are more naturally addressed using observers or parity relations.

11.3 Parameter Estimation

The parameter estimation method is appropriate if the process faults are associated with changes in model parameters (*i.e.*, multiplicative faults), and appropriate mathematical models are available. The model parameters are generally unmeasured, but can be estimated using standard parameter estimation techniques [25, 199], which can be implemented recursively to reduce computational requirements. Constructing the models from first-principles facilitates relating the model parameters directly to parameters that have physical meaning in the process. Thresholds can be placed on the individual differences between the nominal model parameters and the parameter estimates, or on some combination of these differences. Many papers based on the parameter estimation method are available [58, 135, 263].

The parameter estimation method consists of the following steps:

1. Write the process equations for the measurable input variables $\mathbf{u}(t)$ and output variables $\mathbf{y}(t)$ using conservation equations and phenomenological relationships (e.g., phase equilibria, fluid constitutive equations). The process equations relate the input variables $\mathbf{u}(t)$ and the physical model parameters p_j to the output variables $\mathbf{y}(t)$.

2. If necessary, make simplifying assumptions or lump the physical model parameters p_j together so the parameter estimation problem for the new parameters θ_j is observable, that is, so that the new parameters can be uniquely determined. During this step, it is also useful to re-define variables so that the new variables θ_j enter linearly in the process equations, as this will simplify the parameter estimation problem.
3. Estimate the model parameters θ_j from the current and recent past measurements of the input variables $\mathbf{u}(t)$ and output variables $\mathbf{y}(t)$ [22, 25, 199]. If the θ_j appear linearly in the process equations, then it is possible to stack the equations so that

$$\mathbf{z} = \Psi\theta + \mathbf{e} \tag{11.28}$$

where $\mathbf{z}$ is a vector the elements of which are known functions of the measured variables, Ψ is a matrix of measured variables, θ is the vector of parameters to be estimated, and $\mathbf{e}$ is the vector of the equation errors. If the measurement noise is relatively small, then the vector of estimated parameters $\hat{\theta}$ can be obtained by minimizing the sum-of-squared-errors function $\mathbf{e}^T\mathbf{e}$ by least squares:

$$\hat{\theta} = (\Psi^T\Psi)^{-1}\Psi^T\mathbf{z}. \tag{11.29}$$

These parameter estimates will be biased if there is significant measurement noise. If there is significant measurement noise or the θ_j appear nonlinearly, then more sophisticated parameter estimation algorithms should be used [25, 199, 355].
4. Calculate estimates of the physical parameters $\hat{p}_j$ from the estimated model parameters $\hat{\theta}_j$. If lumping was used, then in some cases only combinations of the physical parameters $\hat{p}_j$ can be determined.
5. Faults are indicated if changes in the physical parameters are larger than those observed in training data. Isolate faults by comparing changes in the physical parameters with observations stored in historical databases.

In the parameter estimation method, it is required that the signals have sufficiently high persistent excitation. This motivates keeping the number of independent parameters as small as possible, by simplifying the model or by lumping several parameters together (Step 2).

For fault detection and isolation, Step 5 compares the parameter estimates to their nominal values by computing the differences

$$\Delta p_j = p_j - \hat{p}_j \tag{11.30}$$

where p_j is the nominal value for the physical parameter. Even if no faults are occurring in the plant, the Δp_j will not be equal to zero due to process disturbances and noise. In other words, the Δp_j will be stochastic variables, and a threshold must be used to indicate whether a fault has occurred. A fault is detected when a single Δp_j is larger than some threshold, or some combination of Δp_j is greater than some threshold. The parameters associated with the threshold violation are those associated with the fault. Thresholds can be defined using the T^2 statistics with training data as discussed in Parts II and III, or by more sophisticated statistics [141]. The process monitoring procedure can be made more sensitive to slow drifts by applying exponential moving averages or cumulative sums on the parameter differences (11.30), in a way similar to that in univariate or multivariate control charts (see Chapter 2).

The procedure of detecting faults using the parameter estimation method is illustrated using a gravity flow tank (see Figure 11.2) [245]. The single-input-single-output system is governed by the material balance equation:

$$A_c \frac{dh}{dt} = F_i - ch \tag{11.31}$$

where A_c is the cross-sectional area of the tank, h is the liquid level, c is a constant which depends on the valve, and F_i is the measured inlet flow rate. The outlet flow rate F_o is measured, and is nominally equal to ch. Equation 11.31 can be written in terms of the state-space equations

$$\frac{d\mathbf{x}}{dt} = A\mathbf{x}(t) + B\mathbf{u}(t) \tag{11.32}$$

$$\mathbf{y}(t) = C\mathbf{x}(t) \tag{11.33}$$

where $\mathbf{u} = F_i$, $\mathbf{y} = F_o$, $\mathbf{x} = h$, $A = -c/A_c$, $B = 1/A_c$, and $C = c$. All measured signals are assumed to have additive normally distributed noise with zero mean and variance with magnitude of 10^{-4}.

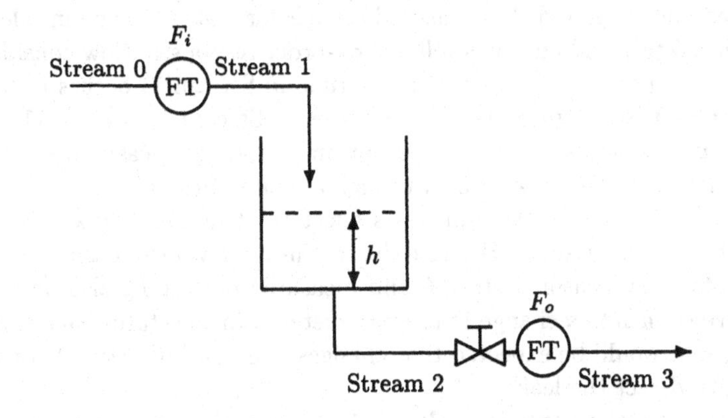

Fig. 11.2. A gravity flow tank system with one measured input F_i, one measured output F_o, and one measured state h. The FT is standard nomenclature for a flow transmitter [134].

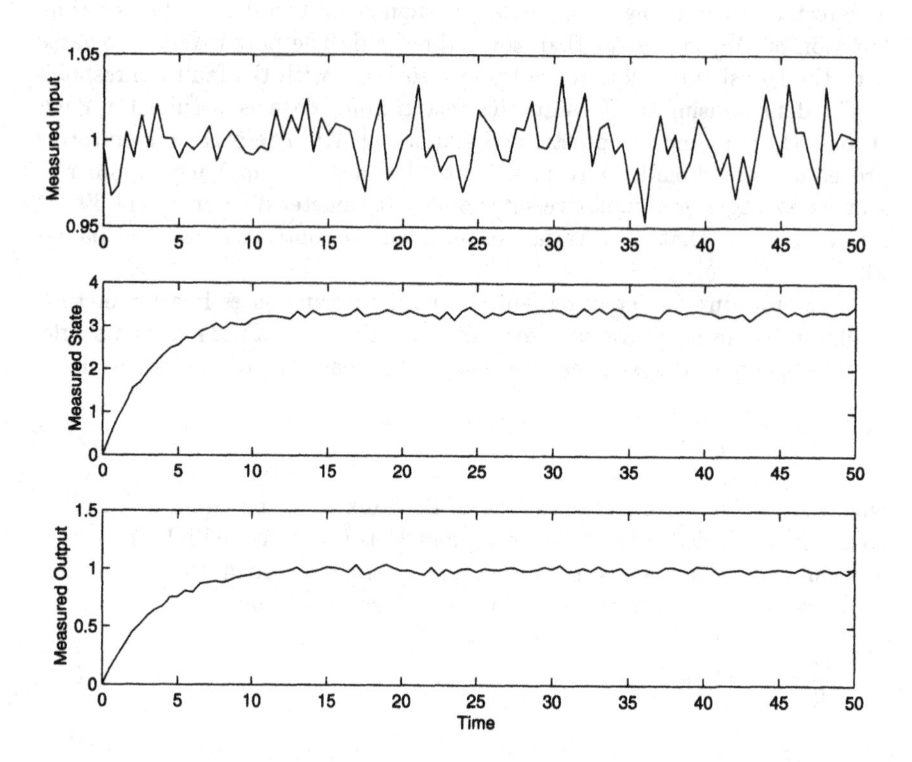

Fig. 11.3. The measured input (inlet flow rate), the measured state (liquid height), and the measured output (outlet flow rate) of the gravity tank during normal operating conditions

Assuming $c = 0.3$ and $A_c = 1$, Figure 11.3 plots the measured input u, the measured output y, and the measured state x for a step change in inlet flow rate. The state x and output y follow first-order responses. Now consider the case where a leak (a 30% drop of magnitude in Stream 1) occurs in Stream 1 for $t \geq 14.5$ (see Figure 11.2 for definition of Stream 1). Figure 11.4 plots the measured input u, the measured output y, and the measured state x for a step change in inlet flow rate. Although a leak in Stream 1 does not affect the measured input F_i, the fault does affect the true input flow rate to the tank, which is unmeasured. Because the true input flow rate to the tank drops at $t = 14.5$, the measured state h and measured output F_o also drop. This fault corresponds to a change in the parameter B in the state-space equation (11.32), so it would be expected that changes in an on-line estimate of B can be used to detect the leak.

To estimate the parameter B in (11.32), the process model equation is written in the form:

$$z(t) = \psi^T(t)\theta + e(t) \tag{11.34}$$

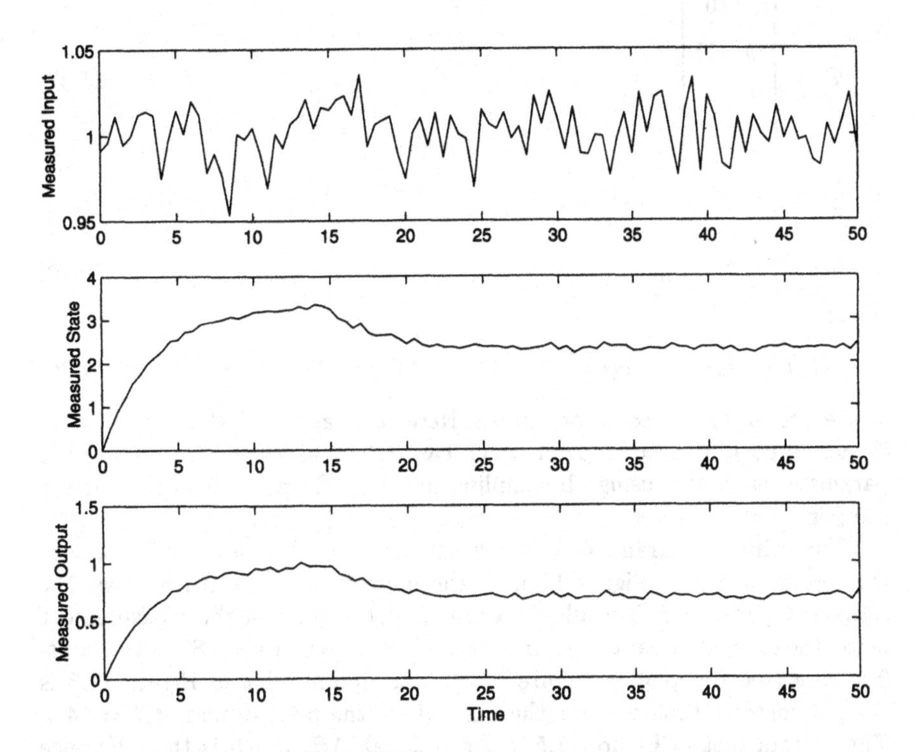

Fig. 11.4. The measured input (inlet flow rate), the measured state (liquid height), and the measured output (outlet flow rate) of the gravity tank when a leak in Stream 1 occurs at $t = 14.5$

where $z(t) = dx/dt$, $\psi^T(t) = [x, u]$, $\theta = [A, B]^T$, and $e(t)$ is the equation error. The derivative of the state can be numerically approximated by backward difference applied to the measured state

$$\frac{dx(t)}{dt} \approx \frac{x(t) - x(t - T_0)}{T_0} \tag{11.35}$$

where $T_0 = 0.5$ was the sampling interval.

Better parameter estimates are obtained by using consecutive time intervals. To put this system into the standard form for parameter estimation, stack the elements of $z(t)$ and $\psi^T(t)$ into vectors

$$\mathbf{z} = [z(0), z(1), \cdots, z(n)]^T \tag{11.36}$$

and

$$\Psi_j = \begin{bmatrix} \psi^T(0) \\ \psi^T(1) \\ \vdots \\ \psi^T(n) \end{bmatrix}. \tag{11.37}$$

Then

$$\mathbf{z} = \Psi\theta + \mathbf{e} \tag{11.38}$$

where

$$e = [e(0), e(1), \cdots, e(n)]^T \tag{11.39}$$

is the vector of the equation errors. Here it is assumed that the process is monitored during startup, in which case it is reasonable to compute the parameter estimates using all sampling instances from $t = 0$ to the current sampling instance $t = n$.

The estimated parameter vector $\hat{\theta}$ was determined by least squares, with the results shown in Figure 11.5. In the noise-free and fault-free case, the estimated parameter $\hat{B}$ would be equal to 1. Because of the measurement noise, the estimated parameter $\hat{B}$ is actually approximately 0.88 in the fault-free case (see top plot in Figure 11.5). The middle plot in Figure 11.5 is the parameter estimate $\hat{B}$ for the case where the fault occurs at $t = 14.5$. The bottom plot in Figure 11.5 is the residual ΔB, which is the difference between the estimated parameter $\hat{B}$ in the normal operating conditions (0.88) and the parameter in the case where the fault occurs at $t = 14.5$. The residual significantly deviates from zero at $t = 21$, indicating that a fault is detected. The detection delay is 13 sampling intervals. The fact that the estimated model parameter $\hat{B}$ is decreasing with time suggests that the fault is due to a leak in Stream 1.

This example illustrates the fact that least-squares estimation can give biased estimates of the parameters. This is why the estimated model parameter during normal operating conditions (0.88) was used to compute the residual, rather than the true model parameter (1). With the properly defined residual, this bias in the parameter estimate did not affect the ability of the parameter estimation method to correctly detect the fault. An FDI system based on parameter estimation should always include model validation, where the parameters are estimated using normal operating conditions. Beyond just ensuring that the parameter estimation algorithm is correctly implemented, this allows the determination of consistent biases in the parameter estimates, so that the residuals can be redefined to avoid false alarms. If the biases are too large, then an unbiased parameter estimation algorithm should be used [25, 199].

Let us further illustrate the parameter estimation method with a multi-input-multi-output example. Consider a process consisting of a centrifugal

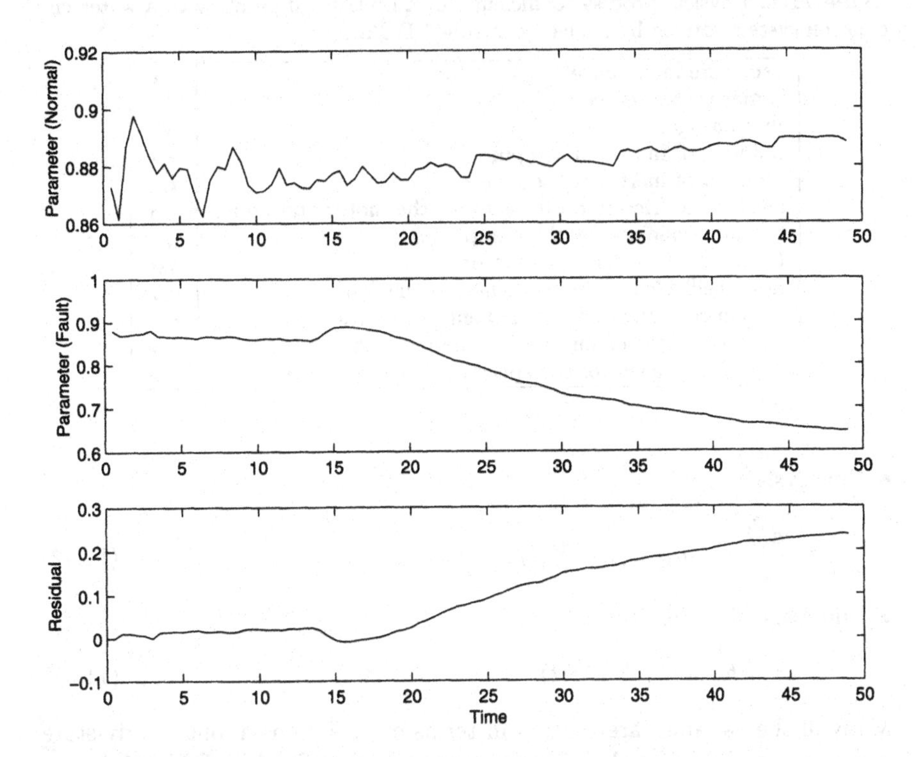

Fig. 11.5. The estimated parameter $\hat{B}$ during normal operating conditions (top plot), the estimated parameter when a leak in Stream 1 occurs at $t = 14.5$ (middle plot), and the associated residual for the gravity tank (bottom plot)

pump with a water circulation system, driven by a speed-controlled direct-current (DC) motor [96, 135]. The physical process coefficients are listed in Table 11.1. Because these coefficients are not measurable, changes in their values are determined by parameter estimation.

The first step of parameter estimation is to model the input-output relationship of the system to satisfy phenomenological relationships and the underlying physical laws such as the material, momentum, and energy balance equations. The first-principles model for this system is [135]:

- Armature circuit

$$L_1 \frac{dI_1(t)}{dt} = -R_1 I_1(t) - \Psi\omega(t) + u_1(t) \tag{11.40}$$

- Mechanics of motor and pump

$$(I_M + I_P)\frac{d\omega(t)}{dt} = \Psi I_1(t) - (g_\omega + c_{F1})\omega(t) - g_M \dot{M}(t) \tag{11.41}$$

Table 11.1. Physical process coefficients for a centrifugal pump with a water circulation system, driven by a speed-controlled DC motor

armature inductance	L_1
armature resistance	R_1
flux linkage	Ψ
moment of inertia of the pump	I_P
moment of inertia of the motor	I_M
sum of the friction coefficients for the motor and pump	c_{F1}
torque coefficient for the pump	g_ω
torque coefficient for the motor	g_M
first coefficient of the momentum equation	a_{ac}
second coefficient of the momentum equation	a_R
first coefficient of the specific energy of the pump	h_ω
second coefficient of the specific energy of the pump	h_M

- Pipe system

$$a_{ac}\frac{d\dot{M}}{dt} = -a_R\dot{M}(t) + Y(t) \tag{11.42}$$

- Pump specific energy

$$Y(t) = h_\omega \omega(t) + h_M \dot{M}(t). \tag{11.43}$$

Many of the variables are written in terms of deviations about steady-state operating conditions, and the signal variables are defined in Table 11.2.

Table 11.2. Input, state, and output variables for the centrifugal pump with a water circulation system, driven by a speed-controlled DC motor. All the variables are measurable.

armature voltage of the motor	u_1	input
armature current of the motor	I_1	state
angular velocity of the motor	ω	state
mass flow rate of the pump	$\dot{M}$	state
specific energy of the pump	Y	output

The state-space equations for the system are

$$\frac{d\mathbf{x}}{dt} = A\mathbf{x}(t) + B\mathbf{u}(t) \tag{11.44}$$

$$\mathbf{y}(t) = C\mathbf{x}(t), \tag{11.45}$$

with the manipulated variable

$$\mathbf{u}(t) = u_1(t), \tag{11.46}$$

the state vector

$$\mathbf{x}(t) = \begin{bmatrix} I_1(t) \\ \omega(t) \\ \dot{M}(t) \end{bmatrix}, \tag{11.47}$$

and the plant output vector

$$\mathbf{y}(t) = \begin{bmatrix} I_1(t) \\ \omega(t) \\ \dot{M}(t) \\ Y(t) \end{bmatrix}. \tag{11.48}$$

The state-space matrices are

$$A = \begin{bmatrix} a_{11} & a_{12} & 0 \\ a_{21} & a_{22} & a_{23} \\ 0 & a_{32} & a_{33} \end{bmatrix}, \tag{11.49}$$

$$B = \begin{bmatrix} b_1 \\ 0 \\ 0 \end{bmatrix}, \tag{11.50}$$

$$C = \begin{bmatrix} 1 & 0 & 0 \\ 0 & 1 & 0 \\ 0 & 0 & 1 \\ 0 & h_\omega & h_M \end{bmatrix}, \tag{11.51}$$

where

$$\begin{aligned} a_{11} &= -\frac{R_1}{L_1} \\ a_{12} &= -\frac{\Psi}{L_1} \\ a_{21} &= \frac{\Psi}{I_M + I_P} \\ a_{22} &= -\frac{c_{F1} + g_\omega}{I_M + I_P} \\ a_{23} &= -\frac{g_M}{I_M + I_P} \end{aligned} \tag{11.52}$$

$$a_{32} = \frac{h_\omega}{a_{ac}}$$

$$a_{33} = \frac{h_M - a_R}{a_{ac}}$$

$$b_1 = \frac{1}{L_1}.$$

The state-space matrices A and B are nonlinear in the physical model parameters (see Equation 11.52), the new model parameters a_{ij} and b_1 are defined so that A and B are linear in these parameters. This results in a parameter estimation problem that is linear in the parameters.

To estimate the a_{ij} and b_1 in (11.49) and (11.50), the process model equations for the measurable input and output signals are written in the form:

$$z_j(t) = \psi_j^T(t)\theta_j + e_j(t) \qquad j = 1, 2, 3, 4 \tag{11.53}$$

where

$$z_1(t) = \frac{dI_1(t)}{dt}; \; z_2(t) = \frac{d\omega(t)}{dt}; \; z_3(t) = \frac{d\dot{M}(t)}{dt}; \; z_4(t) = Y(t) \tag{11.54}$$

$$\psi_1^T(t) = [I_1(t), \omega(t), u_1(t)]; \quad \theta_1 = [a_{11}, a_{12}, b_1]^T \tag{11.55}$$

$$\psi_2^T(t) = [I_1(t), \omega(t), \dot{M}(t)]; \quad \theta_2 = [a_{21}, a_{22}, a_{23}]^T \tag{11.56}$$

$$\psi_3^T(t) = [\omega(t), \dot{M}(t)]; \qquad \theta_3 = [a_{32}, a_{33}]^T \tag{11.57}$$

$$\psi_4^T(t) = [\omega(t), \dot{M}(t)]; \qquad \theta_4 = [h_\omega, h_M]^T \tag{11.58}$$

The functions $\psi_j^T(t)$ are measured variables. If the measurement of $Y(t)$ is less noisy than the measurement of $\omega(t)$, then (11.57) can be replaced by

$$\psi_3^T(t) = [Y(t), \dot{M}(t)]; \quad \theta_3 = [a'_{32}, a'_{33}]^T \tag{11.59}$$

where

$$a'_{32} = \frac{1}{a_{ac}} \tag{11.60}$$

and

$$a'_{33} = -\frac{a_R}{a_{ac}}. \tag{11.61}$$

The $z_j(t)$ are determined by differentiating the measurements. The derivatives can be numerically approximated such as by backward differences, for example,

$$\frac{dI_1(t)}{dt} \approx \frac{I_1(t) - I_1(t - T_0)}{T_0} \tag{11.62}$$

where T_0 is the sampling interval. However, this method can give poor results when the measurements are noisy, which is the usual case. Filtering approaches can give better results [355].

Obtaining accurate parameter estimates for a system of this complexity requires using multiple consecutive measurements to obtain the estimates. Since the measurements of the input and output signals are made at discrete sampling instances t, (11.53) can be written as

$$z_j(t) = \psi_j^T(t)\theta_j + e_j(t), \qquad j = 1, 2, 3, 4;\ t = 0, 1, \cdots, n \tag{11.63}$$

To put this system into the standard form for parameter estimation, stack the elements of $z_j(t)$ and $\psi_j^T(t)$ into vectors

$$\mathbf{z_j} = [z_j(0), z_j(1), \cdots, z_j(n)]^T \tag{11.64}$$

and

$$\Psi_j = \begin{bmatrix} \psi_j^T(0) \\ \psi_j^T(1) \\ \vdots \\ \psi_j^T(n) \end{bmatrix}. \tag{11.65}$$

Then

$$\mathbf{z_j} = \Psi_j \theta_j + \mathbf{e_j} \tag{11.66}$$

where

$$\mathbf{e_j} = [e_j(0), e_j(1), \cdots, e_j(n)]^T \tag{11.67}$$

is the vector of the equation errors.

For each j, the estimated model parameter $\hat{\theta}_j$ is obtained by minimizing the sum-of-squared-errors $\mathbf{e_j}^T\mathbf{e_j}$. The estimated model parameters are computed by least squares:

$$\hat{\theta}_j = (\Psi_j^T \Psi_j)^{-1} \Psi_j^T \mathbf{z_j}. \tag{11.68}$$

Alternatively, the model parameters could be estimated simultaneously using (11.29).

The relationships between the estimated model parameters $\hat{\theta}_j$ and the estimated physical parameters $\hat{\mathbf{p}}$ are determined by rearranging (11.52), which gives

$$\hat{L}_1 = \frac{1}{\hat{b}_1}; \qquad \hat{R}_1 = -\frac{\hat{a}_{11}}{\hat{b}_1} \tag{11.69}$$

$$\hat{\Psi} = -\frac{\hat{a}_{12}}{\hat{b}_1}; \qquad \hat{I}_M + \hat{I}_P = -\frac{\hat{a}_{12}}{\hat{a}_{21}\hat{b}_1} \tag{11.70}$$

$$\hat{c}_{F1} + \hat{g}_\omega = \frac{\hat{a}_{22}\hat{a}_{12}}{\hat{a}_{21}\hat{b}_1}; \qquad \hat{g}_M = \frac{\hat{a}_{23}\hat{a}_{12}}{\hat{a}_{21}\hat{b}_1} \tag{11.71}$$

$$\hat{a}_{ac} = \frac{1}{\hat{a}'_{32}}; \qquad \hat{a}_R = -\frac{\hat{a}'_{33}}{\hat{a}'_{32}} \tag{11.72}$$

$$\hat{h}_\omega = \hat{h}_\omega; \qquad \hat{h}_M = \hat{h}_M. \tag{11.73}$$

While all of the coefficients that describe the linearized dynamic behavior can be determined by least squares, several of the parameters had to be lumped together so that there are only ten unique combinations of parameters. For example, the moments of inertia of the pump and the motor show up only as the sum of the two terms. While the sum of the moments of inertia can be determined by parameter estimation, their individual values could not. This lumping is usually needed in practice to result in an identifiable parameter estimation problem. Thus, a significant change in the sum of the moments of inertia of the motor and the pump $(\hat{I}_M + \hat{I}_P)$ may be due to a fault in either the motor or the pump. A significant change in most of the other physical parameters can be isolated to a particular component. For example, a significant change in the torque coefficient for the motor g_M indicates that a fault has occurred in the motor. Some faults are associated with significant changes in multiple physical parameters, in which case a historical database of parameter changes that occurred during past faults can be used to isolate the faults.

11.4 Observer-based Method

The observer-based method is appropriate if the faults are associated with changes in actuators, sensors, or unmeasurable state variables, that is, it is especially appropriate for detecting and isolating additive faults. A detailed mathematical model for the plant is required, preferably derived from first principles so that the states in the state-space equations have a physical interpretation. The unmeasured states are reconstructed from the measurable input and output variables using a Luenberger observer or Kalman filter [31, 38, 54, 153]. The observer-based method is in sharp contrast to the CVA-based method for process monitoring described in Chapter 7, in which the states are directly constructed from the process data, rather than through the use of a known process model and an observer.

For the states that are measured, a residual can be defined as the difference between the estimated state and the measured state. For states that are unmeasurable (the usual case), the residual is defined based on the difference in the estimated plant output and the measured plant output, or by some linear transformation of this difference. Based on thresholds on the residuals of the state variables or output variables, abrupt changes can be detected [341]. The main reason for preferring first-principles models is that such models add significant structure to the state-space equations, which is especially useful for modeling the effect of faults on the states and plant outputs. Also, physically-meaningful states greatly aid in isolating and diagnosing faults once thresholds on the residuals have been violated.

It is also possible to design an observer-based FDI scheme purely from an input-output point of view, which allows a frequency-based design based on the transfer functions (11.12) or (11.23) [90]. A drawback of such an approach is that relationships to any physically-meaningful states are lost. An advantage of a frequency-domain method is that model uncertainties, one of the main concerns in an FDI system, is often more conveniently modeled in the frequency domain [232, 291]. Hence a frequency-domain method can be more natural for designing FDI systems that simultaneously optimize sensitivity to faults, while minimizing sensitivity to model uncertainties.

This section focuses on state-space methods, because it is useful for both linear and nonlinear plants, and it provides a connection between the FDI system and any physically-meaningful states. Readers interested in the frequency-based design of observer-based FDI systems are referred to a rather detailed review [41]. Several more general reviews describing process monitoring methods based on the observer-based method are available [86, 135, 140]. Several papers have been published using these methods, especially in recent years [136, 137, 141, 162, 182, 220, 233, 367].

11.4.1 Full-order State Estimator

This section describes the basic idea of the observer-based method, illustrating the concepts with a simple process example.

The state vectors can be reconstructed from the measurable plant input $\mathbf{u}$ and plant output $\mathbf{y}$ using an observer. Consider a linear process with the state-space equations (11.9) and (11.10), in which the disturbance $\mathbf{d}(t)$ is lumped together with the noise term $\mathbf{n}(t)$, and the matrix D is assumed to be zero:

$$\mathbf{x}(t+1) = A\mathbf{x}(t) + B\mathbf{u}(t) + B_f\mathbf{f}(t) + B_d\mathbf{d}(t) \tag{11.74}$$

$$\mathbf{y}(t) = C\mathbf{x}(t) + D_f\mathbf{f}(t) + D_d\mathbf{d}(t). \tag{11.75}$$

The state $\hat{\mathbf{x}}(t)$ and output $\hat{\mathbf{y}}(t)$ estimated by a linear full-order observer is described by the equations:

$$\hat{\mathbf{x}}(t+1) = A\hat{\mathbf{x}}(t) + B\mathbf{u}(t) + H[\mathbf{y}(t) - \hat{\mathbf{y}}(t)] \tag{11.76}$$

$$\hat{\mathbf{y}}(t) = C\hat{\mathbf{x}}(t). \tag{11.77}$$

The observer gain H is selected to satisfy design specifications such as stability, fault sensitivity, and robustness.

With (11.74)-(11.77), the relations for the state estimation error $\Delta\mathbf{x}(t) = \mathbf{x}(t) - \hat{\mathbf{x}}(t)$ and the output estimation error $\Delta\mathbf{y}(t) = \mathbf{y}(t) - \hat{\mathbf{y}}(t)$ are

$$\Delta\mathbf{x}(t+1) = [A - HC]\Delta\mathbf{x}(t) + [B_f - HD_f]\mathbf{f}(t) + [B_d - HD_d]\mathbf{d}(t) \tag{11.78}$$

$$\Delta\mathbf{y}(t) = C\Delta\mathbf{x}(t) + D_f\mathbf{f}(t) + D_d\mathbf{d}(t). \tag{11.79}$$

The state estimation error $\Delta\mathbf{x}(t)$ and the output estimation error $\Delta\mathbf{y}(t)$ are functions of the disturbances $\mathbf{d}(t)$ and the faults $\mathbf{f}(t)$, but do not depend on the input $\mathbf{u}(t)$. If the states were measured, then $\Delta\mathbf{x}(t)$ could be used to detect and diagnose faults. Usually the states are not measured, and $\Delta\mathbf{y}(t)$ is used as the residual which forms the basis for the observer-based FDI system. This residual is usually transformed so as to increase the effect of faults and decrease the effect of disturbances on the transformed residuals. Before describing these transformations, let us first illustrate the procedure of using the full-order observer method for detecting faults on the gravity tank example introduced in Section 11.2. When Stream 1 has a leak, the measured input $\mathbf{u}(t)$ is related to the true input $\mathbf{u}^\circ(t)$ by

$$\mathbf{u}^\circ(t) = \mathbf{u}(t) + \Delta\mathbf{u}(t), \tag{11.80}$$

where $\Delta\mathbf{u}(t) = \mathbf{f}(t)$ is a negative value representing the magnitude of the leak. The state-space equations (11.32) and (11.33) become

$$\frac{d\mathbf{x}}{dt} = A\mathbf{x}(t) + B\mathbf{u}(t) + B_f\mathbf{f}(t) \tag{11.81}$$

$$\mathbf{y}(t) = C\mathbf{x}(t) \tag{11.82}$$

where $B_f = 1/A_c$. The estimated state $\hat{\mathbf{x}}(t)$ is obtained using (11.76). In this particular example, the state is the height of the liquid which is measurable. Hence in this case the state estimation error $\Delta\mathbf{x}(t)$ could be used as the residual. Since the states are unmeasurable in most practical problems, we will use the output estimation error $\Delta\mathbf{y}(t)$ from (11.79) as the residual. The measured output, the estimated output, and the associated residual during normal operating conditions are plotted in Figure 11.6, where the observer matrix $H = 0.01$. The estimated output matches fairly well with the mea-

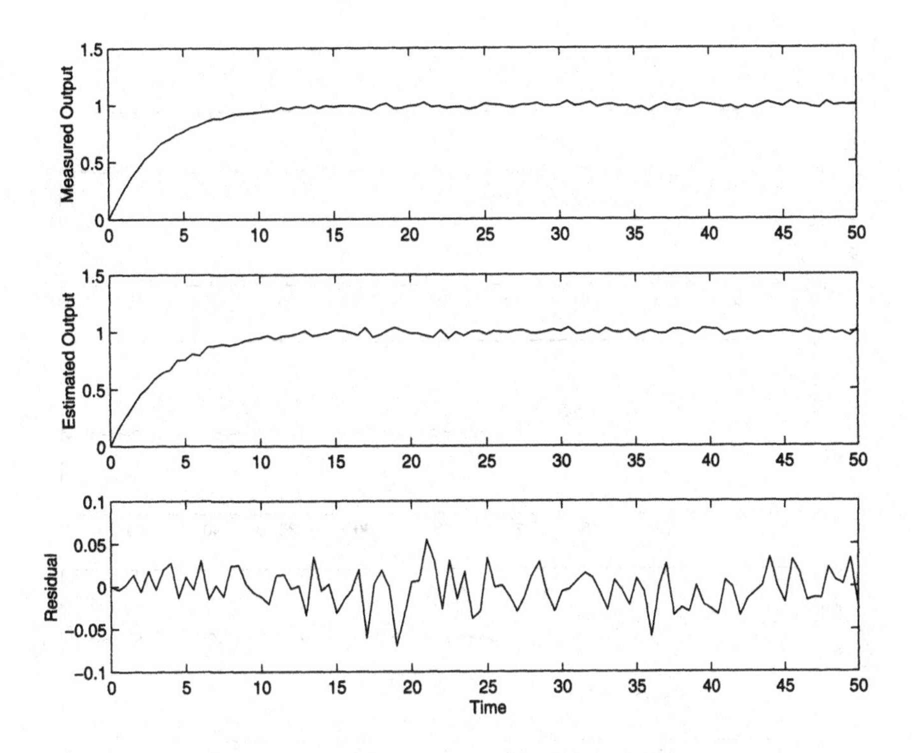

Fig. 11.6. The measured output, the estimated output, and their associated residual obtained from the full-order observer in the normal operating conditions of the gravity tank

sured output, with a fairly small residual ($-0.05 < \Delta y < 0.05$ for almost all t). The measured output, the estimated output, and the associated residual in the case where a leak occurs at $t = 14.5$ are shown in Figure 11.7. The residual deviates from zero significantly at $t = 15.5$, indicating that the fault is detected in two sampling intervals. The full-order observer was much more prompt in detecting the fault than the parameter estimation method shown in the last section.

This example illustrates the point that some faults can be modeled equally well as being additive or multiplicative. The best approach for such faults depends on performance and convenience. For this particular example, the observer-based method (which modeled the fault as being additive) had a much shorter detection delay than the parameter estimation method (which modeled the fault as being multiplicative). If all the faults are best modeled as being parametric faults except for a few faults that can be modeled as being either additive or multiplicative, then it is more convenient to model all of the faults as being multiplicative, so that the FDI system only depends on a single method.

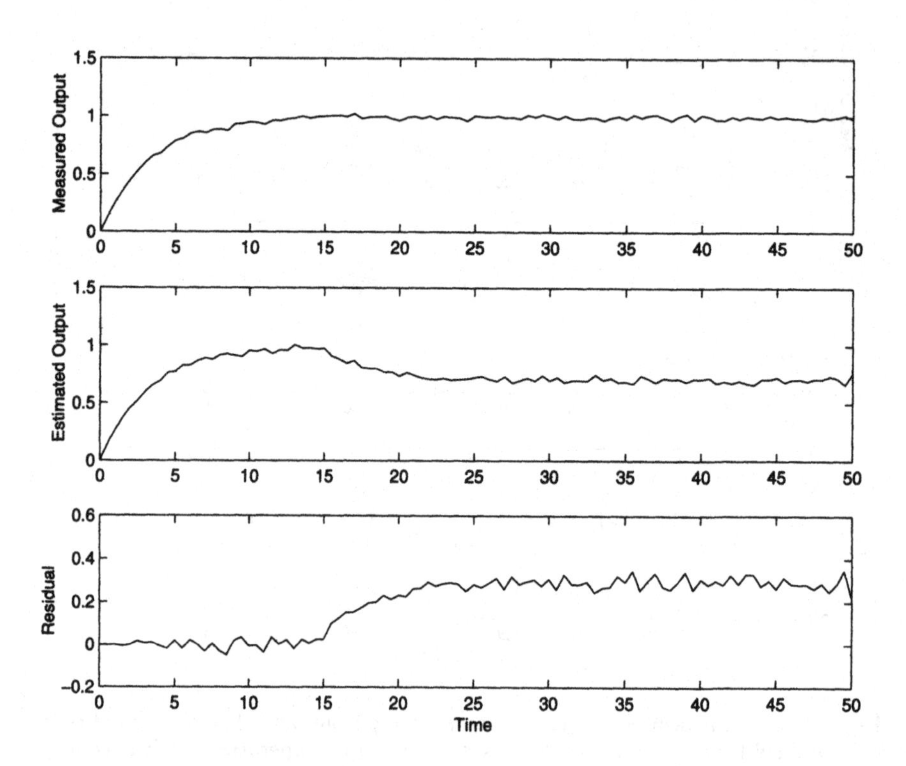

Fig. 11.7. The measured output, the estimated output, and the associated residual obtained from the full-order observer when a leak occurs in Stream 1 at $t = 14.5$, for the gravity tank

In this example disturbances were not considered and the residual was driven only by the fault and zero-mean white noise in the measured variables. In practice the output estimation error $\Delta\mathbf{y}(t)$ is driven by significant disturbances $\mathbf{d}(t)$. Also, model errors can result in an imperfect cancellation of the effect of the control inputs $\mathbf{u}(t)$ on the output estimation error.

The potential sensitivity of the output estimation error to disturbances and model errors motivates the use of a transformed output estimation error as the residual:

$$\mathbf{r}(t) = \breve{W} \Delta \mathbf{y}(t). \tag{11.83}$$

The design problem is to select the matrix $\breve{W}$ so that the residual $\mathbf{r}(t)$ is sensitive to faults while being relatively insensitive or even invariant to disturbance and model errors. Various procedures have been proposed for the design of the observer gain H and the transformation matrix $\breve{W}$ to satisfy these goals [139]. One procedure is known as eigenstructure assignment, in which the matrices are designed to zero out the effect of the disturbances on the residual $\mathbf{r}(t)$ [249, 253]. A related method is to use an unknown input observer to decouple the disturbances from the state estimation error [42, 331]. This method is described below.

11.4.2 Reduced-order Unknown Input Observer

This section derives the design equations for an **unknown input observer** (UIO), which is a fairly general method for the design of an observer-based FDI system.

As in the last section, consider a plant described by the state-space equations:

$$\mathbf{x}(t+1) = A\mathbf{x}(t) + B\mathbf{u}(t) + B_f\mathbf{f}(t) + B_d\mathbf{d}(t) \tag{11.84}$$

$$\mathbf{y}(t) = C\mathbf{x}(t) + D_f\mathbf{f}(t) + D_d\mathbf{d}(t) \tag{11.85}$$

A generalized reduced-order observer for this system is:

$$\mathbf{z}(t+1) = F\mathbf{z}(t) + G\mathbf{y}(t) + J\mathbf{u}(t) \tag{11.86}$$

$$\mathbf{r}(t) = L_1\mathbf{z}(t) + L_2\mathbf{y}(t). \tag{11.87}$$

The observer estimates a linear transformation of the state, $T\mathbf{x}(t)$, where T is a constant matrix. For a reduced-order observer, the number of rows of T is less than the number of columns. The design matrices are F, G, J, L_1, and L_2.

The estimation error is defined by

$$\mathbf{e}(t) = \mathbf{z}(t) - T\mathbf{x}(t). \tag{11.88}$$

Inserting the observer equations and the state-space equations for the plant, and grouping terms gives

$$\mathbf{e}(t+1) = F\mathbf{z}(t) + [GC - TA]\mathbf{x}(t) + [J - TB]\mathbf{u}(t) + [GD_f - TB_f]\mathbf{f}(t) + [GD_d - TB_d]\mathbf{d}(t) \tag{11.89}$$

and the residual

$$\mathbf{r}(t) = L_1\mathbf{z}(t) + L_2C\mathbf{x}(t) + L_2D_f\mathbf{f}(t) + L_2D_d\mathbf{d}(t). \tag{11.90}$$

In the unknown input observer, the observer matrices are designed so that the residual $\mathbf{r}(t)$ and estimation error $\mathbf{e}(t)$ are independent of the plant inputs $\mathbf{u}(t)$ and the disturbances $\mathbf{d}(t)$ (the "unknown inputs"). This implies that

$$J = TB \tag{11.91}$$

$$GD_d = TB_d \tag{11.92}$$

$$L_2D_d = 0. \tag{11.93}$$

This gives the simplified equations

$$\mathbf{e}(t+1) = F\mathbf{z}(t) + [GC - TA]\mathbf{x}(t) + [GD_f - TB_f]\mathbf{f}(t) \tag{11.94}$$

and the residual

$$\mathbf{r}(t) = L_1\mathbf{z}(t) + L_2C\mathbf{x}(t) + L_2D_f\mathbf{f}(t). \tag{11.95}$$

For fault detection, it is also desired for the estimation error and the residual to be independent of the plant states $\mathbf{x}(t)$. This is achieved by setting

$$GC - TA = -FT \tag{11.96}$$

and

$$L_2C = -L_1T, \tag{11.97}$$

which results in

$$\mathbf{e}(t+1) = F\mathbf{e}(t) + [GD_f - TB_f]\mathbf{f}(t) \tag{11.98}$$

and the residual

$$\mathbf{r}(t) = L_1\mathbf{e}(t) + L_2D_f\mathbf{f}(t). \tag{11.99}$$

The estimation error and the residual depend solely on the faults and are independent of the process state $\mathbf{x}(t)$, input $\mathbf{u}(t)$, and disturbances $\mathbf{d}(t)$. For stability of the estimation error, the matrix F must have its eigenvalues within the unit circle. To maximize the effect of the faults on the residual, the matrices G, T, and L_2 should be selected so that the matrix

$$\begin{bmatrix} GD_f - TB_f \\ L_2 D_f \end{bmatrix} \tag{11.100}$$

has a high rank. Maximizing the rank of $L_2 D_f$ is especially useful, since this term is a direct mapping of the faults to the residual, without being filtered by the observer dynamics (see Equation 11.99).

In the UIO method, the matrices F, G, J, L_1, and L_2 are designed so that F is stable and Equations 11.91, 11.92, 11.93, 11.96, and 11.97 are satisfied. Extra degrees of freedom are used to maximize the rank of the matrices (11.100). The extra degrees of freedom can also be used to decouple the effect of each fault on the residual. Necessary and sufficient conditions for the existence of solutions to these types of equations are available, as well as methods for computing the design matrices [41, 352].

11.5 Parity Relations

It was shown in the last section how observers can be used to generate residuals. Another popular method to generate the residuals is to use parity relations.

11.5.1 Residual Generation

The residual must be generated solely from the observations. A general equation for the residual is

$$\mathbf{r}(t) = V(q)\mathbf{u}(t) + W(q)\mathbf{y}(t) \tag{11.101}$$

where $\mathbf{r}(t)$ is the residual vector, and $V(q)$ and $W(q)$ are transfer function matrices. The residual should be zero when the unknown inputs (the faults $\mathbf{f}(t)$, disturbances $\mathbf{d}(t)$, and noise $\mathbf{n}(t)$) are zero. Substituting the system equation (11.12) into (11.101) and setting the unknown inputs to zero gives

$$V(q)\mathbf{u}(t) + W(q)P(q)\mathbf{u}(t) = 0. \tag{11.102}$$

For this to hold for all inputs $\mathbf{u}(t)$, we must have

$$V(q) = -W(q)P(q) \tag{11.103}$$

Inserting this into (11.101) gives

$$\mathbf{r}(t) = W(q)[\mathbf{y}(t) - P(q)\mathbf{u}(t)], \tag{11.104}$$

The transfer function $P(q)$ (or matrices A, B, C, and D in (11.1) and (11.2)) is assumed to be known either from first principles or from prior identification of the plant. Specifying the transfer function $W(q)$ is the main focus of the design of the FDI system.

Substituting (11.12) into (11.104) gives the residual $\mathbf{r}(t)$ in terms of the unknown inputs:

$$\mathbf{r}(t) = W(q)[P_f(q)\mathbf{f}(t) + P_d(q)\mathbf{d}(t) + P_n(q)\mathbf{n}(t)]. \tag{11.105}$$

This equation gives the dependence of the residual on the faults, disturbances, and noise. Before going into details on how $W(q)$ is designed, let us first illustrate the use of the parity relation (11.105) for detecting faults. Recall the gravity tank example in which there is a single potential fault (see Figure 11.2). Since there are no significant disturbances in this example, the design matrix $W(q)$ can be set to one.

The residual (11.104) was computed both during normal operating conditions and in the case where there is a fault (a leak in Stream 1 at $t = 14.5$). The residuals are plotted in Figure 11.8. In the normal operating conditions,

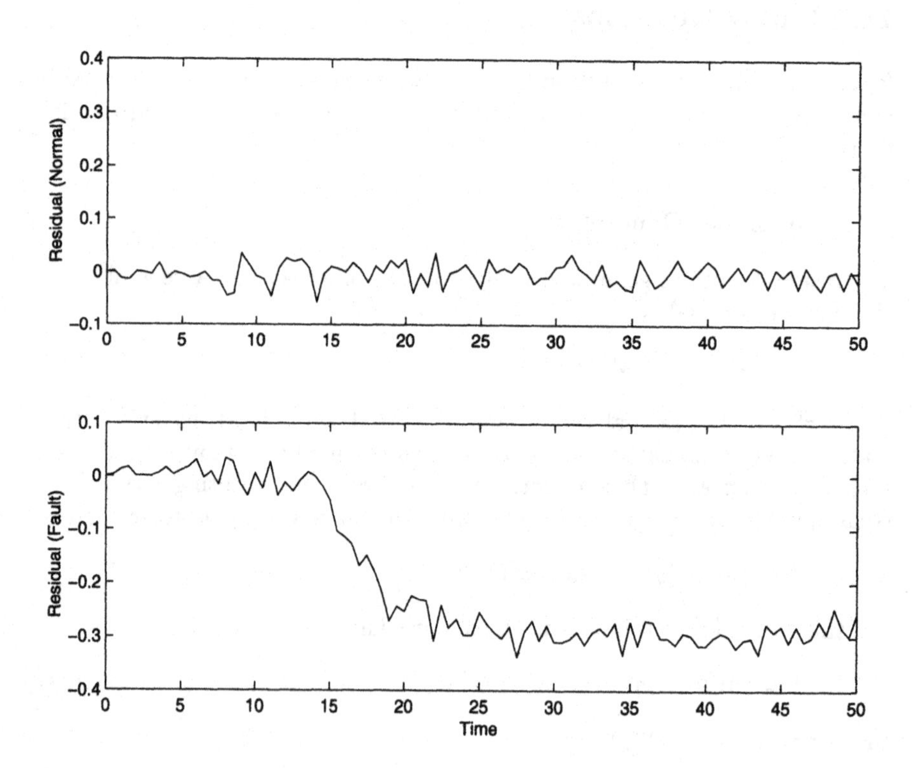

Fig. 11.8. The residual obtained from the parity relations in the normal operating conditions and the faulty condition (a leak in Stream 1 at $t = 14.5$) of the gravity tank

the residual remains close to zero, indicating that a threshold can be defined so that there is a low false alarm rate. When the fault occurs, the residual deviates from zero significantly at $t = 15$, indicating that a fault is detected. Similarly to the full-order observer, the parity relation method promptly detected the fault.

A general equation for the residual $\mathbf{r}(t)$ in terms of the unknown inputs can also be derived directly from the state-space equations (11.9) and (11.10) [47, 101]. To simplify the algebra, the presentation will neglect the disturbance and noise terms, and insert them later.

With a time delay σ, (11.9) and (11.10) become

$$\mathbf{x}(t-\sigma+1) = A\mathbf{x}(t-\sigma) + B\mathbf{u}(t-\sigma) + B_f\mathbf{f}(t-\sigma) \tag{11.106}$$

$$\mathbf{y}(t-\sigma) = C\mathbf{x}(t-\sigma) + D\mathbf{u}(t-\sigma) + D_f\mathbf{f}(t-\sigma). \tag{11.107}$$

Inserting (11.106) into (11.107) gives the expression:

$$\begin{aligned}\mathbf{y}(t-\sigma+1) = {} & CA\mathbf{x}(t-\sigma) + CB\mathbf{u}(t-\sigma) + CB_f\mathbf{f}(t-\sigma)\\ & + D\mathbf{u}(t-\sigma+1) + D_f\mathbf{f}(t-\sigma+1).\end{aligned} \tag{11.108}$$

Recursively, the following extended state equation is obtained:

$$\begin{aligned}\underbrace{\begin{bmatrix}\mathbf{y}(t-\sigma)\\ \mathbf{y}(t-\sigma+1)\\ \vdots\\ \mathbf{y}(t)\end{bmatrix}}_{\breve{\mathbf{y}}(t)} = {} & J\mathbf{x}(t-\sigma) + K\underbrace{\begin{bmatrix}\mathbf{u}(t-\sigma)\\ \mathbf{u}(t-\sigma+1)\\ \vdots\\ \mathbf{u}(t)\end{bmatrix}}_{\breve{\mathbf{u}}(t)}\\ & + L_f\underbrace{\begin{bmatrix}\mathbf{f}(t-\sigma)\\ \mathbf{f}(t-\sigma+1)\\ \vdots\\ \mathbf{f}(t)\end{bmatrix}}_{\breve{\mathbf{f}}(t)}\end{aligned} \tag{11.109}$$

where

$$J = \begin{bmatrix}C\\ CA\\ \vdots\\ CA^{\sigma}\end{bmatrix}, \tag{11.110}$$

$$K = \begin{bmatrix} D & 0 & \cdots & 0 & 0 \\ CB & D & \ddots & \ddots & 0 \\ CAB & CB & \ddots & \ddots & \vdots \\ \vdots & \vdots & \ddots & \ddots & 0 \\ CA^{\sigma-1}B & CA^{\sigma-2}B & \cdots & CB & D \end{bmatrix}, \tag{11.111}$$

and

$$L_f = \begin{bmatrix} D_f & 0 & \cdots & 0 & 0 \\ CB_f & D_f & \ddots & \ddots & 0 \\ CAB_f & CB_f & \ddots & \ddots & \vdots \\ \vdots & \vdots & \ddots & \ddots & 0 \\ CA^{\sigma-1}B_f & CA^{\sigma-2}B_f & \cdots & CB_f & D_f \end{bmatrix}. \tag{11.112}$$

Then the residual can be written as:

$$\mathbf{r}(t) = W[\check{\mathbf{y}}(t) - K\check{\mathbf{u}}(t)] \tag{11.113}$$

where $W \in \mathcal{R}^{p \times (\sigma+1)m_y}$ is the transformation matrix, and p is the dimension of the residual vector. This equation can be used to compute the residual from the measured inputs and outputs of the plant. Inserting (11.109) into this equation gives

$$\mathbf{r}(t) = W[J\check{\mathbf{x}}(t-\sigma) + L_f\check{\mathbf{f}}(t)]. \tag{11.114}$$

The dependence of the state vector $\check{\mathbf{x}}(t-\sigma)$ can be eliminated by choosing a transformation W such that

$$WJ = 0. \tag{11.115}$$

For an appropriately large σ, it follows from the Cayley-Hamilton theorem [153] that the solution for W always exists [47]. Then the residual is only a function of the faults

$$\mathbf{r}(t) = WL_f\check{\mathbf{f}}(t). \tag{11.116}$$

For a particular fault to be detectable, W must be selected so that the appropriate columns of WL_f are not equal to the zero vector. For a residual to be affected by at least one fault, W must be selected so that none of the rows of WL_f are equal to the zero vector.

The length of the data window, σ, is a design parameter. A sufficiently large σ guarantees that there is a large number of degrees of freedom in W

for satisfying the above detectability conditions. However, a low value of σ is preferred to simplify the design and implementation of the FDI system. The smallest value of σ such that (11.115) can be satisfied is given by the inequalities [41, 226, 227]:

$$\frac{\text{rank}(\mathcal{O})}{\text{rank}(C)} \leq \sigma_{min} \leq \text{rank}(\mathcal{O}) - \text{rank}(C) + 1 \tag{11.117}$$

where $\mathcal{O}$ is the observability matrix [153] (which is closely related to J)

$$\mathcal{O} = \begin{bmatrix} C \\ CA \\ \vdots \\ CA^a \end{bmatrix} \tag{11.118}$$

and a is the number of states. If the system is observable and the rows of the matrix C are linearly independent, then the inequality can be written as

$$\frac{a}{m_y} \leq \sigma_{min} \leq a - m_y + 1 \tag{11.119}$$

To consider additive noise and disturbances in the system, the vector $\breve{\mathbf{n}}(t)$ and $\breve{\mathbf{d}}(t)$ are defined similarly to $\breve{\mathbf{f}}(t)$, and its accompanying matrices L_n and L_d can be computed. This more general form of the residual (11.116) is

$$\mathbf{r}(t) = W[L_f\breve{\mathbf{f}}(t) + L_n\breve{\mathbf{n}}(t) + L_d\breve{\mathbf{d}}(t)]. \tag{11.120}$$

While this state-space form for the transformation matrix can be used, more insights can be obtained by using the transfer function $W(q)$ in (11.104). Hence the rest of this chapter will use the transfer function form.

11.5.2 Detection Properties of the Residual

Ideally, the transformation matrix $W(q)$ is designed so that non-zero residuals occur only when faults occur. However, the residuals can also be affected by measurement noise, model uncertainty, and disturbances. The simplest approach to reduce the effect of noise is low-pass filtering of the measured signals. More sophisticated Kalman filtering can be used for more complicated noise signals [12]. Quantifying the contribution of the measurement noise on the residuals based on (11.105) is rather straightforward provided that the noise is modeled stochastically. The noise will continue to have some effect on the residuals, so a threshold must be used to determine whether a fault has occurred.

Characterizing the model uncertainties and quantifying their effect on the residuals are more difficult. The larger the model uncertainty, the more difficult it is to detect and diagnose faults using residuals. Much attention has been focused on improving the robustness of analytical redundancy methods to model uncertainty. Two of the more popular methods in the literature include robust residual generators [88, 108, 322], and structured residuals with an unknown input observer [91, 248, 286]. The simplest approach is to model the uncertainties as additional disturbances. Then, when the transformation matrix $W(q)$ is designed so that the residual is insensitive to this larger set of disturbances, the residual is also insensitive to the uncertainties. This approach is possible when the total number of disturbances and uncertainties is small [101].

An ideal residual would be sensitive to each fault in the system. The **triggering limit** is a useful measure of the sensitivity of the residual with respect to faults [101]. Recall the general equation for the residual (11.105) as a function of the faults, disturbances, and noise:

$$\mathbf{r}(t) = W(q)[P_f(q)\mathbf{f}(t) + P_d(q)\mathbf{d}(t) + P_n(q)\mathbf{n}(t)]. \quad (11.121)$$

The relationship between the jth fault $f_j(t)$ and the ith residual induced by the fault is

$$r_i(t|f_j) = \mathbf{w}_i^T(q)\mathbf{p}_{fj}(q)f_j(t) \quad (11.122)$$

where $\mathbf{w}_i^T(q)$ is the ith row of $W(q)$ and $\mathbf{p}_{fj}(q)$ is the jth column of $P_f(q)$. The time response of the ith residual depends on the time response of the jth fault, which is not usually precisely known. If the time response $f_j(t)$ is not known, then it is simplest to assume that it is a unit step function $H(t)$ [245, 296]. Then the absolute value of the steady-state value for the ith residual is

$$\lim_{t\to\infty} |r_i(t|H(t))| = \left|\mathbf{w}_i^T(q)\mathbf{p}_{fj}(q)\right|_{q=1} \quad (11.123)$$

from the final value theorem for discrete-time systems (a similar equation holds for continuous-time systems). The triggering limit is defined as

$$TL_{ij} = \frac{k_i}{\left|\mathbf{w}_i^T(q)\mathbf{p}_{fj}(q)\right|_{q=1}} \quad (11.124)$$

where k_i is the threshold for $r_i(t)$. A small triggering limit indicates a high fault sensitivity.

If the nominal magnitude, f_{jo}, of the jth fault is known, then it is useful to define a normalized triggering limit

$$TL_{Nij} = \frac{k_i}{f_{jo}\left|\mathbf{w}_i^T(q)\mathbf{p}_{fj}(q)\right|_{q=1}}. \quad (11.125)$$

A normalized triggering limit TL_{Nij} greater than one indicates that the fault does not bring the residual to its threshold at steady-state, clearly an undesirable situation. A normalized triggering limit TL_{Nij} less than one is desired.

Alternative definitions of the triggering limits can be useful in certain applications. If the time response of the jth fault $f_j(t)$ is known, then the time response can be used to define the triggering limit instead of the unit step function. Also, the maximum of the ith residual can be used instead of the steady-state value.

11.5.3 Specification of the Residuals

Recall that the main design consideration for an FDI system based on parity relations is the design of the transformation matrix $W(q)$ in (11.121). The approach to the design of $W(q)$ is similar to the design of feedforward controllers as taught in an undergraduate process control course [245]. This approach is to specify the desired transfer functions between the inputs and outputs, and then compute $W(q)$ that gives the desired transfer functions. The inputs are the disturbances and faults, and the outputs are the residuals.

Denote the response of the ith element of the residual to the fault $f_j(t)$ as $r_i(t|f_j)$ and its response to the disturbance $d_j(t)$ as $r_i(t|d_j)$. For additive faults and disturbances, the response specifications are given in the form of transfer functions that incorporate all the desired behavior:

$$r_i(t|f_j) = z_{fij}(q)f_j(t) \tag{11.126}$$

and

$$r_i(t|d_j) = z_{dij}(q)d_j(t) \tag{11.127}$$

where $z_{fij}(q)$ and $z_{dij}(q)$ are scalar transfer functions. The response specification for a scalar residual $r_i(t)$ can be written in terms of the vector of additive faults $\mathbf{f}(t)$ and the vector of additive disturbances $\mathbf{d}(t)$:

$$r_i(t) = \mathbf{z}_{\mathbf{fi}}^T(q)\mathbf{f}(t) + \mathbf{z}_{\mathbf{di}}^T(q)\mathbf{d}(t) \tag{11.128}$$

where $\mathbf{z}_{\mathbf{fi}}^T(q) = [z_{fi1} \; z_{fi2} \; \cdots \; z_{fim_f}]$ and $\mathbf{z}_{\mathbf{di}}^T(q) = [z_{di1} \; z_{di2} \; \cdots \; z_{dim_d}]$ are vectors of the individual transfer functions and m_f and m_d are the numbers of faults and disturbances, respectively. The response of the full residual vector $\mathbf{r}(t)$ can be written in terms of the vector of additive faults $\mathbf{f}(t)$ and the vector of additive disturbances $\mathbf{d}(t)$:

$$\mathbf{r}(t) = Z_f(q)\mathbf{f}(t) + Z_d(q)\mathbf{d}(t) \tag{11.129}$$

where $Z_f = [\mathbf{z}_{\mathbf{f1}} \; \mathbf{z}_{\mathbf{f2}} \; \cdots \; \mathbf{z}_{\mathbf{fp}}]^T$ and $Z_d = [\mathbf{z}_{\mathbf{d1}} \; \mathbf{z}_{\mathbf{d2}} \; \cdots \; \mathbf{z}_{\mathbf{dp}}]^T$ are transfer function matrices, and p is the number of residuals.

For disturbance decoupling, the response to the disturbances is specified as zero (that is, $r_i(t|d_j) = 0$ or $z_{dij}(q) = 0$ in Equation 11.127). For the faults, either zero or specific non-zero responses are specified for each $z_{fij}(q)$.

11.5.4 Implementation of the Residuals

Ignoring noise (which is assumed to be addressed by filtering as discussed in Section 11.5.2), the single residual $r_i(t)$ from (11.105) is

$$r_i(t) = \mathbf{w}_\mathrm{i}^T(q)[P_f(q)\mathbf{f}(t) + P_d(q)\mathbf{d}(t)]. \tag{11.130}$$

Comparing (11.130) with the specification (11.128) reveals that

$$\mathbf{w}_\mathrm{i}^T(q)P_{fd}(q) = \mathbf{z}_\mathrm{i}^T(q) \tag{11.131}$$

where

$$P_{fd}(q) = [P_f(q) \ P_d(q)] \tag{11.132}$$

and

$$\mathbf{z}_\mathrm{i}^T(q) = \left[\mathbf{z}_\mathrm{fi}^T(q) \ \mathbf{z}_\mathrm{di}^T(q)\right]. \tag{11.133}$$

The transfer function $P_{fd}(q)$ is governed by the plant, and is assumed known. The $\mathbf{z}_\mathrm{i}^T(q)$ are specifications on the residuals, which are set by the engineer. Equation 11.131 relates the rows of the transformation matrix $W(q)$ with the specifications. If a transformation matrix $W(q)$ can be computed that satisfies (11.131), then the desired specifications on the residuals will be achieved.

One objective of the design is to obtain an appropriate transformation $\mathbf{w}_\mathrm{i}(q)$ such that its elements are rational functions or polynomials in the shift operator. The transformation $\mathbf{w}_\mathrm{i}(q)$ also needs to be causal and stable. Actually, both $W(q)$ and $W(q)P(q)$ must be stable and implementable in (11.104). This implies that $W(q)$ must cancel any unstable poles of the plant $P(q)$. It is also desired for $W(q)$ to be of low complexity.

If $P_{fd}(q)$ is a square matrix and it has a stable inverse, then setting the ith row of W as

$$\mathbf{w}_\mathrm{i}^T(q) = \mathbf{z}_\mathrm{i}^T(q)P_{fd}^{-1}(q) \tag{11.134}$$

satisfies (11.131). If the inverse of $P_{fd}(q)$ exists but is not stable, then the specifications in $\mathbf{z}_\mathrm{i}^T(q)$ can be modified so that $\mathbf{w}_\mathrm{i}^T(q)$ consists of stable transfer functions. If there are multiple solutions to (11.131), then some elements of $\mathbf{w}_\mathrm{i}^T(q)$ can be fixed so that the resulting system has a unique solution. The transformation matrix $W(q)$ is constructed by stacking up its rows $\mathbf{w}_\mathrm{i}^T(q)$.

The above procedure looks at each residual $r_i(t)$ individually. Alternatively, the equations can be written in terms of the vector residual $\mathbf{r}(t)$. Stacking the equations (11.131) gives the design condition on the transformation matrix $W(q)$:

$$W(q)[P_f(q) \ P_d(q)] = [Z_f(q) \ Z_d(q)]. \tag{11.135}$$

If $P_{fd}(q)$ is a square matrix and it has a stable inverse, then

$$W(q) = [Z_f(q) \;\; Z_d(q)]P_{fd}^{-1}(q) \tag{11.136}$$

satisfies the specifications on the residuals.

An alternative method to design $W(q)$ is to use an observer. This relationship between the observer-based method and parity equations is made clear in the next section. But first, let us consider an example.

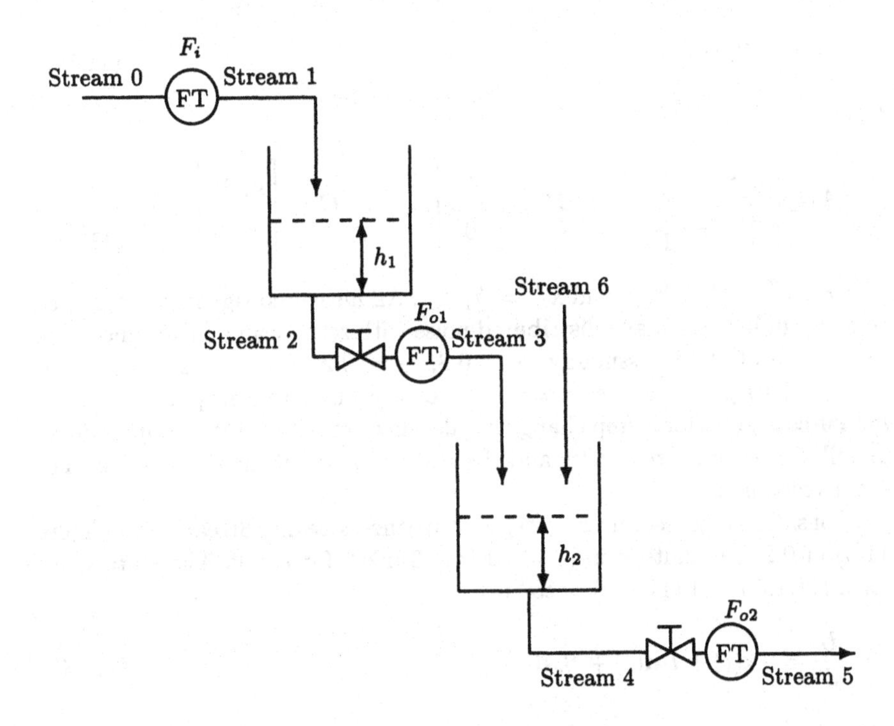

Fig. 11.9. Two non-interacting flow tanks in series. The system has one measured input F_i, two measured outputs F_{o1} and F_{o2}, and two measured states h_1 and h_2. The FT is standard nomenclature for a flow transmitter [134].

A system consisting of two non-interacting flow tanks in series is used to illustrate the use of parity relation for disturbance decoupling (see Figure 11.9) [245]. The system in the disturbance-free and fault-free case is governed by the material balance equations:

$$A_{c1}\frac{dh_1}{dt} = F_i - c_1 h_1 \tag{11.137}$$

and

$$A_{c2}\frac{dh_2}{dt} = c_1 h_1 - c_2 h_2 \tag{11.138}$$

where A_{c1} and A_{c2} are the cross-sectional areas of Tanks 1 and 2, h_1 and h_2 are the liquid levels for Tanks 1 and 2, c_1 and c_2 are constants which depend on the valves, and F_i is the measured inlet flow rate. The outlet flow rates F_{o1} and F_{o2} are measured, and are nominally equal to $c_1 h_1$ and $c_2 h_2$, respectively. Equations 11.137 and 11.138 can be written in state-space form

$$\frac{d\mathbf{x}}{dt} = A\mathbf{x}(t) + B\mathbf{u}(t) \tag{11.139}$$

$$\mathbf{y}(t) = C\mathbf{x}(t) \tag{11.140}$$

where

$$A = \begin{bmatrix} \frac{-c_1}{A_{c1}} & 0 \\ \frac{c_1}{A_{c2}} & \frac{-c_2}{A_{c2}} \end{bmatrix}, \qquad B = \begin{bmatrix} \frac{1}{A_{c1}} \\ 0 \end{bmatrix}, \qquad C = \begin{bmatrix} c_1 & 0 \\ 0 & c_2 \end{bmatrix}, \tag{11.141}$$

$\mathbf{u} = F_i$, $\mathbf{y}^T = [F_{o1}\ F_{o2}]$, and $\mathbf{x}^T = [h_1\ h_2]$. All measured signals are assumed to have additive normally distributed noise with zero mean and variance with magnitude of 10^{-4}. Assuming $c_1 = 0.3$, $c_2 = 0.2$, $A_{c1} = 1$, and $A_{c2} = 1$, Figure 11.10 plots the measured input $\mathbf{u}$, the measured output $\mathbf{y}$, and the measured state $\mathbf{x}$ for a step change in inlet flow rate. The state x_1 and output y_1 follow first-order responses and the state x_2 and output y_2 follow second-order responses.

Consider the case where a step disturbance stream (Stream 6 in Figure 11.9) of 0.2 flow units is introduced into Tank 2 for $t \geq 0$. The state-space form (11.139) and (11.140) becomes

$$\frac{d\mathbf{x}}{dt} = A\mathbf{x}(t) + B\mathbf{u}(t) + B_d\mathbf{d}(t) \tag{11.142}$$

$$\mathbf{y}(t) = C\mathbf{x}(t) \tag{11.143}$$

where $\mathbf{d} = d$ represents the disturbance stream, and $B_d^T = [0\ 1/A_{c2}]$. Figure 11.11 plots the measured input $\mathbf{u}$, the measured output $\mathbf{y}$, and the measured state $\mathbf{x}$ for a step change in inlet flow rate. Comparison between Figures 11.10 and 11.11 indicates that the disturbance stream increases x_2 and y_2 by roughly 20% and that it does not affect x_1 and y_1.

Now consider the case where a leak (a 30% drop in Stream 1) occurs in Stream 1 for $t \geq 14.5$ (see Figure 11.9 for definition of Stream 1), in addition to the disturbance stream introduced into Tank 2 for $t \geq 0$. The state-space form (11.142) and (11.143) becomes

$$\frac{d\mathbf{x}}{dt} = A\mathbf{x}(t) + B\mathbf{u}(t) + B_d\mathbf{d}(t) + B_f\mathbf{f}(t) \tag{11.144}$$

$$\mathbf{y}(t) = C\mathbf{x}(t) \tag{11.145}$$

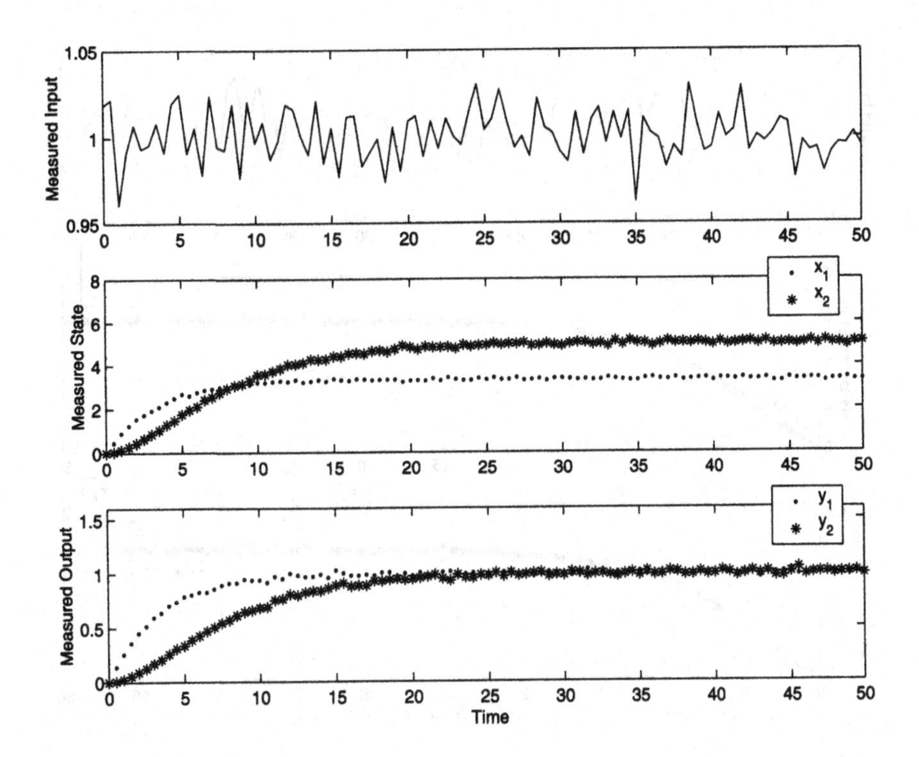

Fig. 11.10. The measured input (inlet flow rate), the measured state (liquid height), and the measured output (outlet flow rate) of two non-interacting flow tanks in series during normal operating conditions

where $\mathbf{f} = f$ represents the leak in Stream 1, and $B_f^T = [1/A_{c1}\ 0]$. Figure 11.12 plots the measured input $\mathbf{u}$, the measured output $\mathbf{y}$, and the measured state $\mathbf{x}$ for a step change in inlet flow rate. Comparison between Figures 11.11 and 11.12 shows that the leak in Stream 1 caused a decrease in x_1, x_2, y_1, and y_2.

The residual (11.104) was computed during normal operating conditions, in the case where there is a disturbance, and in the case where there are a disturbance and a fault (a leak in Stream 1 at $t = 14.5$). For illustration purposes, first use the transformation $\mathbf{w}(s) = [1\ 1]^T$ where the residuals are plotted in Figure 11.13. Similarly to Figure 11.8, the residual remains close to zero in the normal operating conditions, indicating that a threshold can be defined so that there is a low false alarm rate. With the chosen transformation $\mathbf{w}(s)$, the disturbance and the fault both cause the residual to deviate significantly from zero.

To determine the appropriate transformation such that the disturbance is decoupled from the residual, (11.134) is used in the discrete-time case. In

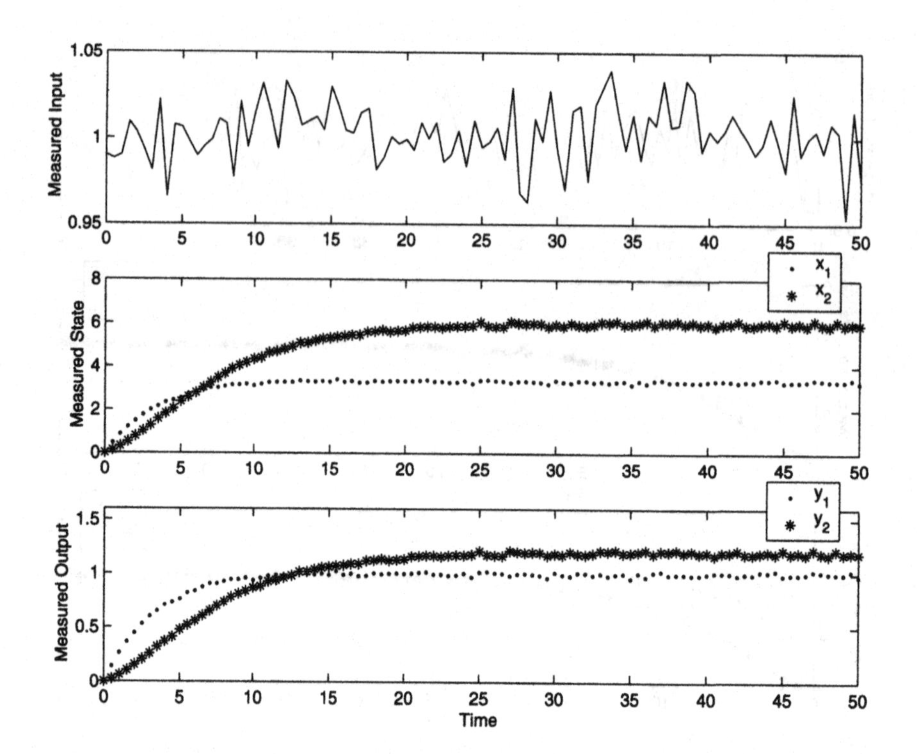

Fig. 11.11. The measured input (inlet flow rate), the measured state (liquid height), and the measured output (outlet flow rate) of two non-interacting flow tanks in series when a step disturbance is introduced into Tank 2 at $t = 0$

the continuous-time case, the equation becomes

$$\mathbf{w}^T(s) = \mathbf{z}^T(s)P_{fd}^{-1}(s). \tag{11.146}$$

In this example, we have

$$P_{fd}(s) = \begin{bmatrix} \frac{c_1}{c_1+A_{c1}s} & 0 \\ \frac{c_1c_2}{(c_1+A_{c1}s)(c_2+A_{c2}s)} & \frac{c_2}{c_2+A_{c2}s} \end{bmatrix}. \tag{11.147}$$

With the specification $\mathbf{z}^T(s) = [\frac{c_1}{c_1+A_{c1}s}\ 0]$, the transformation $\mathbf{w}^T(s) = [1\ 0]$. With the proper choice of transformation, the residual is decoupled from the disturbance, but remains driven by the fault (see Figure 11.14).

11.5.5 Connection Between the Observer and Parity Relations

Recall the equations for a full-order observer:

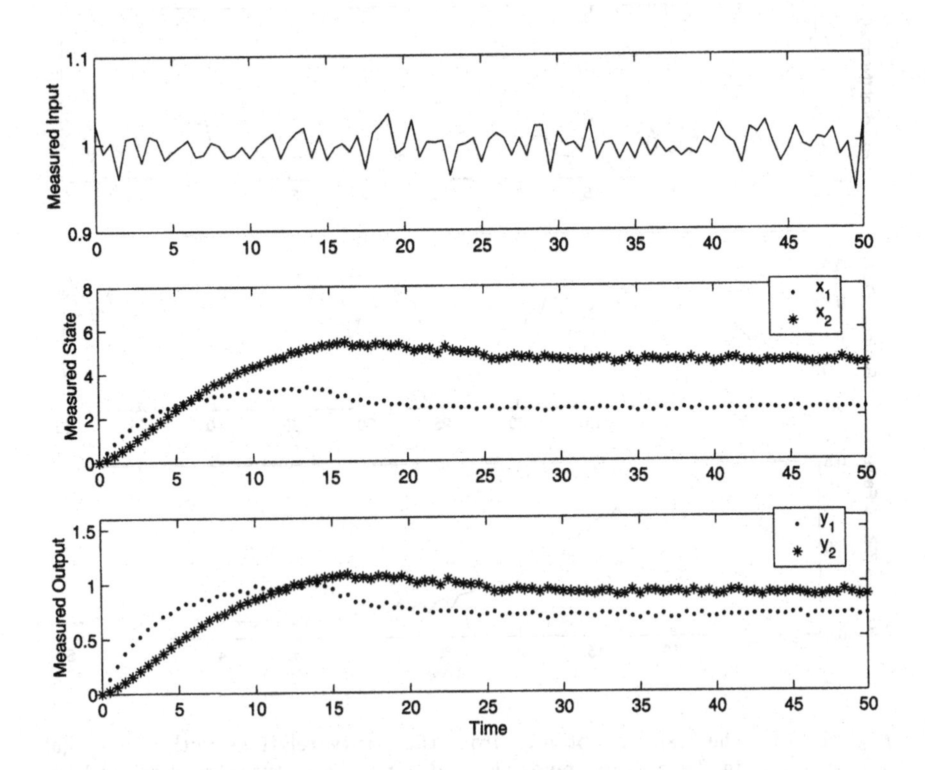

Fig. 11.12. The measured input (inlet flow rate), the measured state (liquid height), and the measured output (outlet flow rate) of two non-interacting flow tanks in series when a leak in Stream 1 occurs at $t = 14.5$ and a step disturbance is introduced into Tank 2 at $t = 0$

$$\hat{\mathbf{x}}(t+1) = A\hat{\mathbf{x}}(t) + B\mathbf{u}(t) + H[\mathbf{y}(t) - \hat{\mathbf{y}}(t)] \tag{11.148}$$

$$\hat{\mathbf{y}}(t) = C\hat{\mathbf{x}}(t) + D\mathbf{u}(t). \tag{11.149}$$

Inserting (11.149) into (11.148), introducing the shift operator, and solving for the state estimates gives

$$\hat{\mathbf{x}}(t) = (qI - A + HC)^{-1}[H\mathbf{y}(t) - HD\mathbf{u}(t) + B\mathbf{u}(t)]. \tag{11.150}$$

Inserting this into (11.149) gives the output estimates

$$\hat{\mathbf{y}}(t) = C(qI - A + HC)^{-1}[H\mathbf{y}(t) - HD\mathbf{u}(t) + B\mathbf{u}(t)] + D\mathbf{u}(t). \tag{11.151}$$

Recall the definition of the output estimation error:

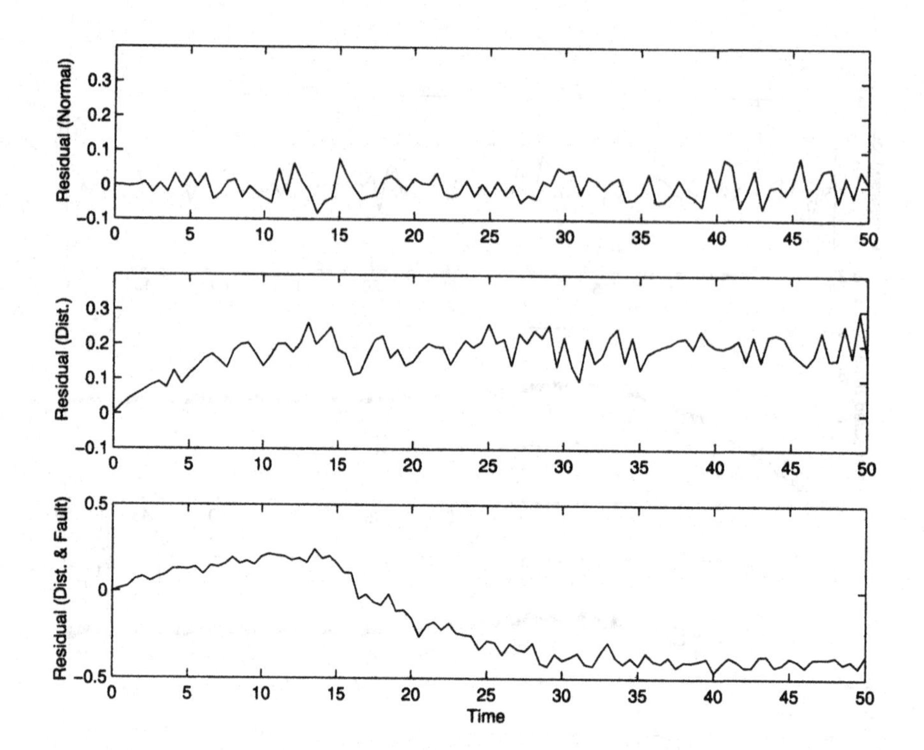

Fig. 11.13. The residual obtained from the parity relations with the weight $\mathbf{w}^T(s) = [1\ 1]$ in the normal operating conditions, the disturbance case, and the faulty condition of the two non-interacting tanks in series

$$\Delta \mathbf{y}(t) = \mathbf{y}(t) - \hat{\mathbf{y}}(t). \tag{11.152}$$

Inserting (11.151) gives the transfer function relationship for the output estimation error

$$\begin{aligned}\Delta \mathbf{y}(t) = {} & [I - C(qI - A + HC)^{-1}H]\mathbf{y}(t) \\ & + [(C(qI - A + HC)^{-1}(HD - B) - D]\mathbf{u}(t).\end{aligned} \tag{11.153}$$

Some matrix algebra simplifies this to

$$\Delta \mathbf{y}(t) = [I - C(qI - A + HC)^{-1}H][\mathbf{y}(t) - (C(qI - A)^{-1}B + D)\mathbf{u}(t)]. \tag{11.154}$$

Since $P(q) = C(qI - A)^{-1}B + D$, we have

$$\Delta \mathbf{y}(t) = [I - C(qI - A + HC)^{-1}H][\mathbf{y}(t) - P(q)\mathbf{u}(t)]. \tag{11.155}$$

In the observer-based method, the output estimation error is multiplied by a transformation matrix

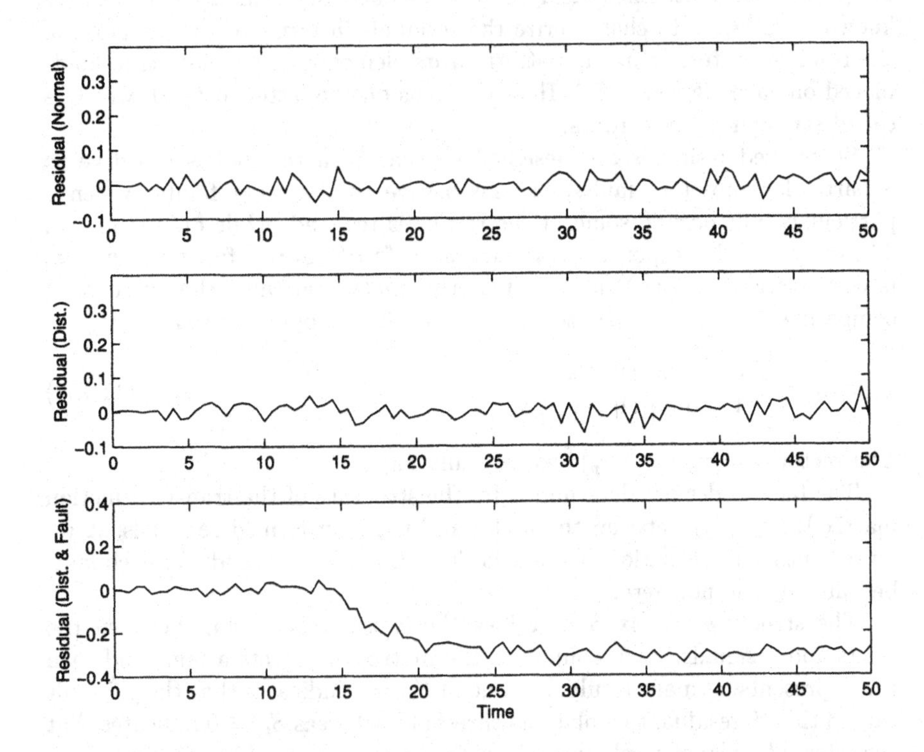

Fig. 11.14. The residual obtained from the parity relations with weight $\mathbf{w}^T(s) = [1\ 0]$ (determined using Equation 11.134) in the normal operating conditions, the disturbance case, and the faulty condition of the two non-interacting tanks in series

$$\mathbf{r}(t) = \check{W}\Delta\mathbf{y}(t). \tag{11.156}$$

Equation 11.156 can be written in terms of the generic form for the residual (11.104), by setting

$$W(q) = \check{W}[I - C(qI - A + HC)^{-1}H]. \tag{11.157}$$

Hence any observer implementation can be replaced by a direct implementation using parity relations in accordance with (11.157). The residuals obtained from the direct implementation are identical. More detailed discussions of the connections between the observer-based method and parity relations are available [101].

11.5.6 Isolation Properties of the Residual

A vector of residuals is required for isolating faults. To distinguish among faults, it is desirable for the residuals corresponding to a given fault to be unique for that fault. Since the dynamics of a particular fault are not precisely known, it is useful to characterize the residuals in terms of the direction of the residual vector, or in the pattern of its elements that violate thresholds placed on each element [101]. Here we focus on the latter method, which is called **structured residuals**.

Structured residuals are designed so that each residual is sensitive to a particular subset of faults, but insensitive to the other faults. When a particular fault occurs, some of the residuals respond, while others do not. The pattern of the response set, known as the **fault code** or **fault signature**, is characteristic of the fault. To state this mathematically, the outcome of comparing the residual $r_i(t)$ to its threshold k_i is a binary variable $\gamma_i(t)$:

$$\gamma_i(t) = \begin{cases} 0 & \text{if } |r_i(t)| < k_i, \\ 1 & \text{if } |r_i(t)| \geq k_i. \end{cases} \tag{11.158}$$

The vector $\gamma = [\gamma_1 \, \gamma_2 \, \cdots \, \gamma_p]^T$ is the fault code.

The fault codes are determined by the structure of the transfer function matrix $W(q)P_f(q)$ between the faults and the transformed residuals. A requirement for the isolation of single faults is that the fault code for each fault be different and non-zero.

The structure matrix S describes the causal relationship between the faults and residuals. Each column of the matrix represents a fault and each row represents a binary result. A value of $S_{ij} = 1$ indicates that the jth fault caused the ith residual to violate its threshold, whereas $S_{ij} = 0$ indicates that the ith residual is relatively insensitive to the occurrence of the jth fault. The structure matrix is defined by the residual specifications (see Section 11.5.3). The columns of S must be distinct to be able to distinguish among all the faults.

For fault isolation, the fault code $\gamma(t)$ is computed from the observations and compared with the columns of the structure matrix S. If the observed fault code satisfies

$$\gamma(t) = \mathbf{s_j}, \tag{11.159}$$

where $\mathbf{s_j}$ is the jth column of the structure matrix S, then the jth fault is indicated as having occurred. For simple implementation, the number of residuals p should be kept low while the number of "0" elements in each column should be made high.

To illustrate the procedure of designing structured residuals for isolating faults, the non-interacting tanks system is used (see Figure 11.9). To simplify the algebra, this example will neglect the disturbance and consider only two faulty cases, where there is a leak in Stream 1 for $t \geq 14.5$ (denoted as Fault 1) and that the flow transmitter for F_{o2} gives a biased reading for $t \geq 7$ (denoted as Fault 2). When the faults occur, the state-space equations are

$$\frac{d\mathbf{x}}{dt} = A\mathbf{x}(t) + B\mathbf{u}(t) + B_f\mathbf{f}(t) \tag{11.160}$$

$$\mathbf{y}(t) = C\mathbf{x}(t) + D_f\mathbf{f}(t) \tag{11.161}$$

where

$$A = \begin{bmatrix} \frac{-c_1}{A_{c1}} & 0 \\ \frac{c_1}{A_{c2}} & \frac{-c_2}{A_{c2}} \end{bmatrix}, \quad B = \begin{bmatrix} \frac{1}{A_{c1}} \\ 0 \end{bmatrix}, \quad B_f = \begin{bmatrix} \frac{1}{A_{c1}} & 0 \\ 0 & 0 \end{bmatrix}, \tag{11.162}$$

$$C = \begin{bmatrix} c_1 & 0 \\ 0 & c_2 \end{bmatrix}, \quad D_f = \begin{bmatrix} 0 & 0 \\ 0 & 1 \end{bmatrix},$$

$\mathbf{u} = F_i$, $\mathbf{y}^T = [F_{o1}\ F_{o2}]$, $\mathbf{x}^T = [h_1\ h_2]$, $\mathbf{f}^T = [f_1\ f_2]$, f_1 represents the leak in Stream 1, and f_2 represents the bias in the F_{o2} measurement.

All measured signals are assumed to have additive normally distributed noise with zero mean and variance with magnitude of 10^{-4}. Assuming $c_1 = 0.3$, $c_2 = 0.2$, $A_{c1} = 1$, and $A_{c2} = 1$, the measured input $\mathbf{u}$, the measured output $\mathbf{y}$, and the measured state $\mathbf{x}$ are plotted in Figures 11.15 and 11.16 during occurrences of Faults 1 and 2. Using (11.24), we have

$$P_f(s) = \begin{bmatrix} \frac{c_1}{c_1 + A_{c1}s} & 0 \\ \frac{c_1 c_2}{(c_1 + A_{c1}s)(c_2 + A_{c2}s)} & 1 \end{bmatrix}. \tag{11.163}$$

The dimension of the residual vector $\mathbf{r}(t)$ is set to 2, so that there is enough dimensionality to distinguish between the faults based on the structure of the residuals. For convenience, the structure matrix S is set to the identity matrix. In other words, the residuals $r_1(t)$ and $r_2(t)$ are driven by Faults 1 and 2, respectively. This suggests that the transfer function matrix $Z_f(s)$ for the residual specification should have the form:

$$Z_f(s) = \begin{bmatrix} z_{f11}(s) & 0 \\ 0 & z_{f22}(s) \end{bmatrix}. \tag{11.164}$$

Equation 11.136 can be used to determine the transformation $W(q)$ in the discrete-time case. In the continuous-time and disturbance-free case, the equation becomes

$$W(s) = Z_f(s)P_f^{-1}(s). \tag{11.165}$$

Therefore, the transformation matrix $W(s)$ can be determined as

$$W(s) = \begin{bmatrix} z_{f11}(s)(1 + \frac{A_{c1}}{c_1}s) & 0 \\ z_{f22}(s)\frac{-c_2}{c_2 + A_{c2}s} & z_{f22}(s) \end{bmatrix}. \tag{11.166}$$

The design parameter $z_{f11}(s)$ was set as $\frac{c_1}{c_1 + A_{c1}s}$ and $z_{f22}(s)$ was set as one and the transformation matrix becomes

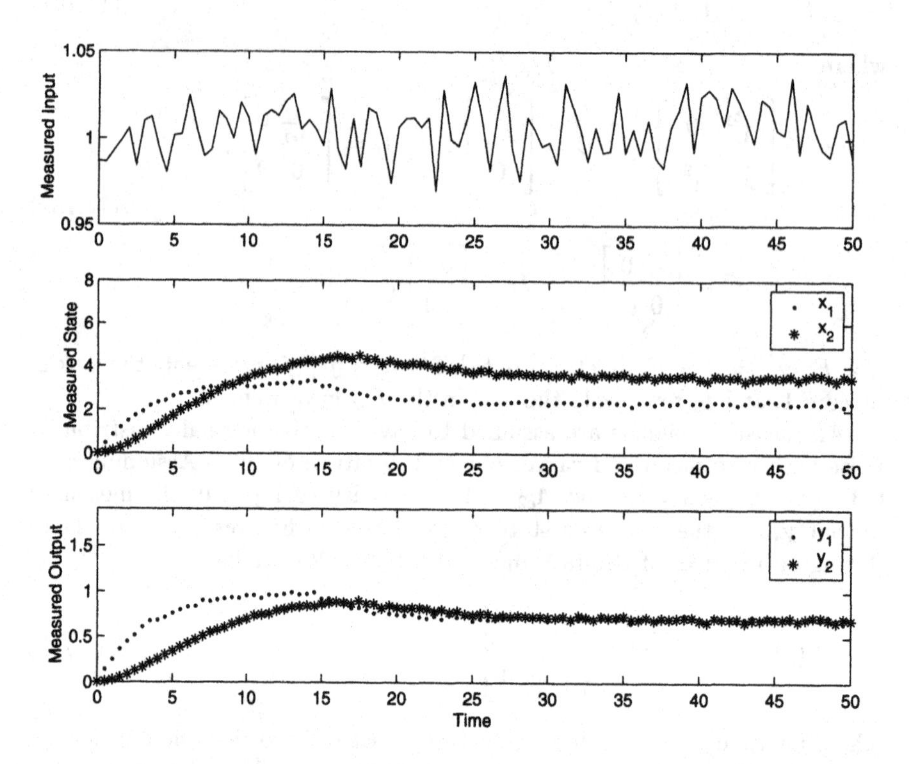

Fig. 11.15. The measured input (inlet flow rate), the measured state (liquid height), and the measured output (outlet flow rate) of two non-interacting flow tanks in series when Fault 1 (a leak in Stream 1) occurs at $t = 14.5$

$$W(s) = \begin{bmatrix} 1 & 0 \\ \frac{-c_2}{c_2 + A_{c2}s} & 1 \end{bmatrix}. \tag{11.167}$$

The residuals $r_1(t)$ and $r_2(t)$ are plotted in Figures 11.17 and 11.18, respectively. The residual $r_1(t)$ remains close to zero in the normal operating conditions and in the case when Fault 2 occurs. As suggested by the design specification, $r_1(t)$ is driven only by Fault 1. The residual $r_2(t)$ remains close to zero in the normal operating conditions and in the case when Fault 1 occurs. As suggested by the design specification, $r_2(t)$ is driven only by Fault 2. These indicate that the transformation matrix $W(s)$ was robust and sensitive.

11.5.7 Residual Evaluation

After the residuals are computed, the resulting residual is used as feature inputs to fault detection and diagnosis through logical, causal, or pattern

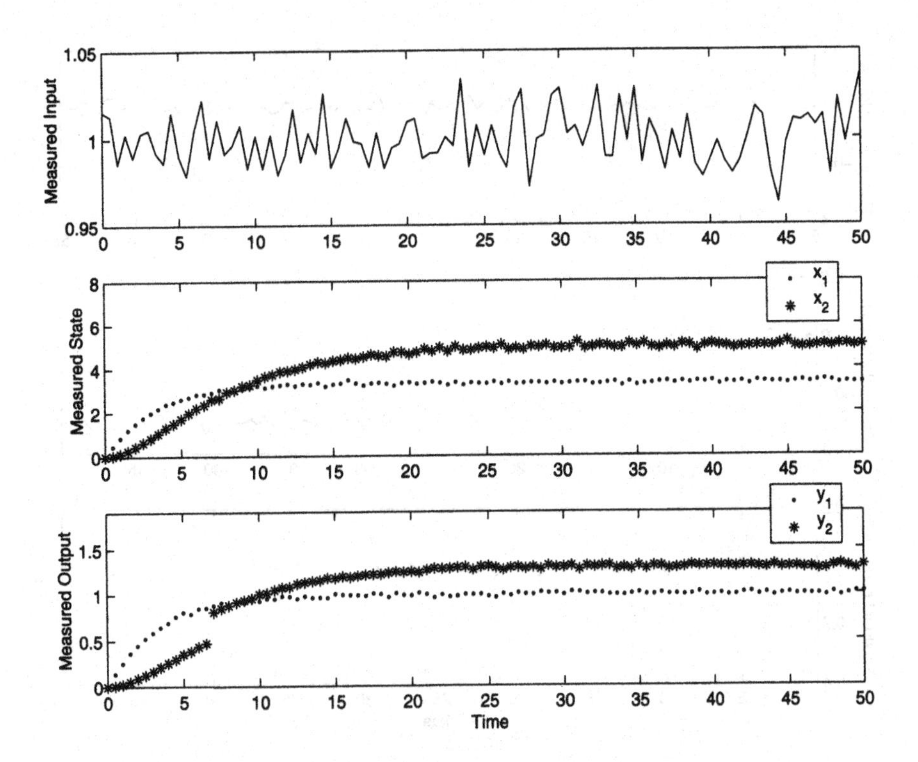

Fig. 11.16. The measured input (inlet flow rate), the measured state (liquid height), and the measured output (outlet flow rate) of two non-interacting flow tanks in series when Fault 2 (F_{o2} gives a biased reading) occurs at $t = 7$

recognition techniques. When the disturbances and model uncertainty are decoupled from the residuals (see Section 11.5.2), then only the noise and the faults contribute to the residuals:

$$\mathbf{r}(t) = \mathbf{r_f}(t) + \mathbf{r_n}(t) \tag{11.168}$$

where $\mathbf{r_f}(t) = W(q)P_f(q)\mathbf{f}(t)$ is the fault-induced part of the residual, while $\mathbf{r_n}(t) = W(q)P_n(q)\mathbf{n}(t)$ is the noise-induced part of the residuals. Although unknown, the faults are assumed to be deterministic. With the assumption that the noise has zero mean, the residual has a time-varying mean contributed entirely by the faults

$$\mu_{\mathbf{r}}(t) = \mathbf{r_f}(t). \tag{11.169}$$

The noise is assumed to be stochastic. If the faults are not stochastic, then the covariance of the residual is entirely due to the noise:

$$\mathrm{Cov}(\mathbf{r}(t), \mathbf{r}(t-\tau)) = \mathrm{Cov}(\mathbf{r_n}(t), \mathbf{r_n}(t-\tau)). \tag{11.170}$$

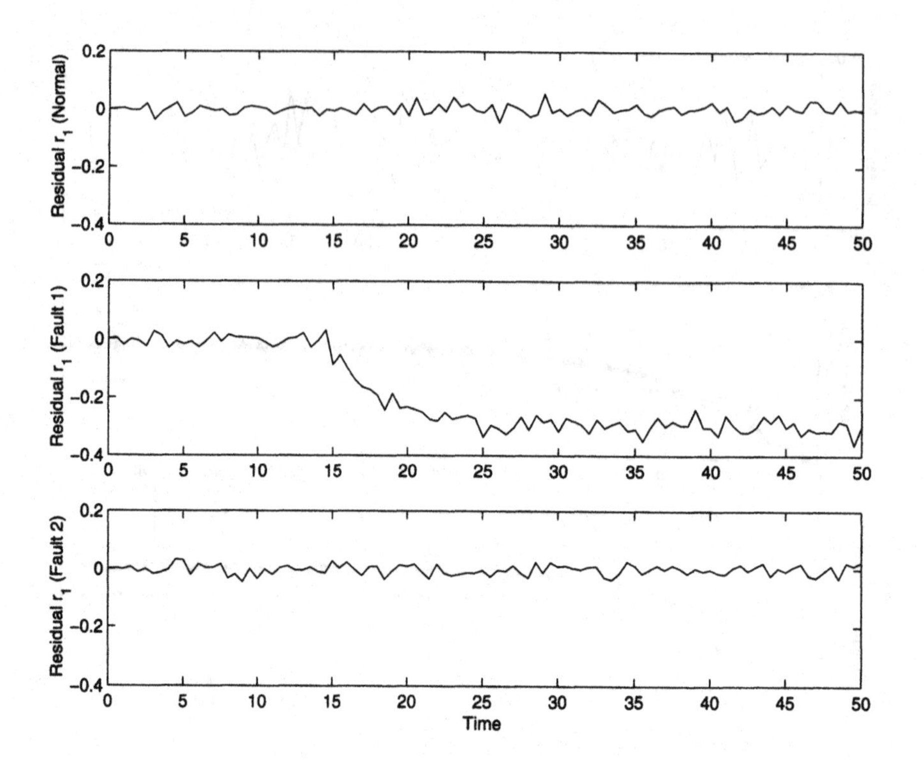

Fig. 11.17. The residual $r_1(t)$ obtained from the parity relations with the weight $\mathbf{w}_1^T(s) = [1\ 0]$ in the normal operating conditions and in the cases when Faults 1 and 2 occur

For fault detection, the null hypothesis is that the residual has zero mean. The T^2 statistics in Chapter 2 can be used to define thresholds on the residuals used for determining whether a fault has occurred. Each residual can be tested separately, as done in univariate control charts, or the residual vector can be tested using a single threshold defined by multivariate statistics. The process monitoring procedure can be made more sensitive to slow drifts by taking window averages, by applying exponential moving averages, or by using cumulative sums on the residuals. The methods of dealing with temporal correlation discussed in Part III, such as time histories, can also be applied. These methods also apply to fault isolation.

One way to diagnose faults is to apply pattern classification techniques, as discussed in Chapter 3, on the residuals. Discriminant analysis can be used to select the fault class which maximizes the *a posteriori* probability. This allows the direct incorporation of prior fault probabilities to improve fault diagnosis. A closely related approach is the **generalized likelihood ratio** technique. In this approach, conditional estimates of the residual means are computed

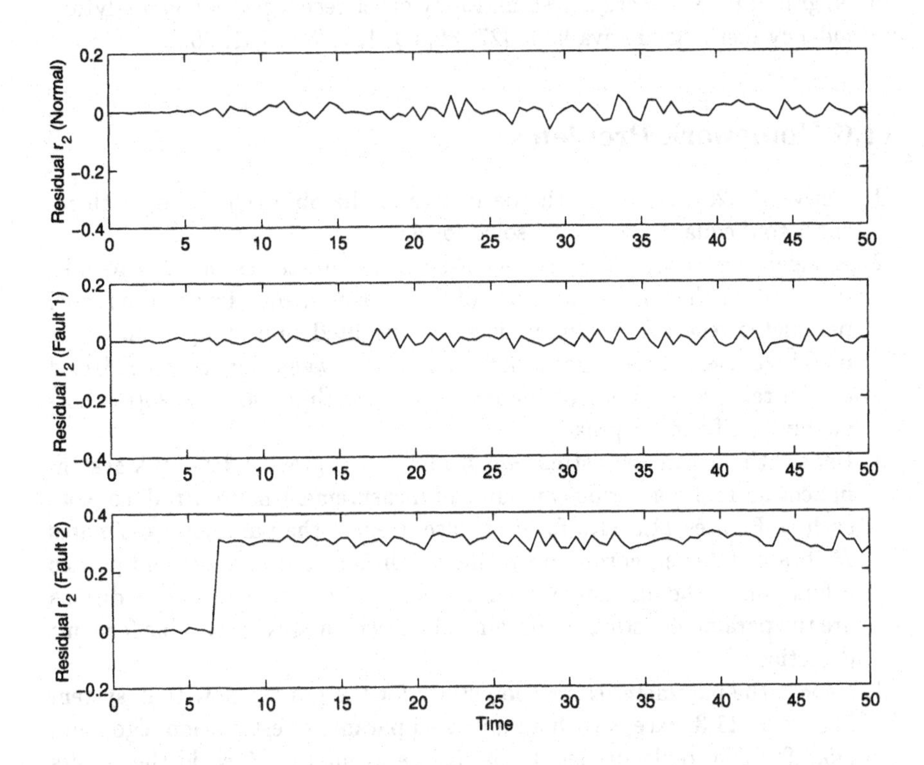

Fig. 11.18. The residual $r_2(t)$ obtained from the parity relations with the weight $\mathbf{w}_2^T(s) = [\frac{-c_2}{c_2+A_{c2}s} \; 1]$ in the normal operating conditions and in the cases when Faults 1 and 2 occur

with the various fault hypotheses, under the geometric constraints posed by the respective structure, and the likelihood functions obtained with those means are compared [101]. This allows the directionality and structuring of residuals to be included rather explicitly.

Gomez *et al.* [105] suggested using operating point computation, Scheffee's statistic, and Hotelling's statistic to detect the normality of the residuals. The results are then formulated as a fuzzy logic rule for detecting and diagnosing faults. Frank and Kiupel [92] evaluated the residual based on fuzzy logic incorporated with either adaptive thresholds or fuzzy inference with the assistance of a human operator. Garcia and Frank [94] proposed a method to integrate the observer-based method with the parameter estimation approach. The observer-based residual is used for fault detection; when the signals are sufficiently rich, the parameter identification residual is then used for fault diagnosis. Ding and Guo [68] suggested that integrating the generation of the residual with its evaluation may improve the ability to detect and diagnose faults. They proposed a frequency domain method to design an integrated fault detection system. Many other recent papers on analytical redundancy methods are available [27, 36, 69, 197, 242, 251, 252].

11.6 Homework Problems

1. Derive (11.29). Hint: Set the derivative of the objective ($\mathbf{e}^T\mathbf{e}$) with respect to θ equal to zero, and solve for θ.
2. A weighting matrix Ω is usually used in parameter estimation to take into account the relative accuracy of the measurements. Derive the model parameter vector $\hat{\theta}$ that minimizes the weighted sum of equation errors $\mathbf{e}^T\Omega^T\Omega\mathbf{e}$. Describe an approach to defining the weighting matrix Ω based on the relative accuracy of the measurements (hint: see [22, 25]). Thoroughly justify the approach.
3. Repeat the parameter estimation study for the gravity flow tank system in Section 11.3, for various amounts of measurement noise. Produce plots such as Figures 11.3-11.5 for each case. Report the parameter estimates $\hat{B}$ obtained during normal operating conditions in each case, and plot as a function of the amount of measurement noise. Under what conditions are the parameter estimates obtained by least squares acceptable for fault detection?
4. Repeat the parameter estimation study for the gravity flow tank system in Section 11.3, except with an unbiased parameter estimation algorithm [25, 199, 355] replacing the least-squares estimates. How do the results change when the measurement noise is increased by a factor of 5?
5. Repeat the parameter estimation study for the gravity flow tank system in Section 11.3, except with the parameters estimated using only the past 10 consecutive time intervals. Comment on the accuracy of the parameter estimates obtained after the process dynamics have settled out. Discuss the importance of having persistent excitation when applying the parameter estimation method. Change the operating conditions for the process so that there is persistent excitation, and reapply the parameter estimation method. Comment on the accuracy of the parameter estimates, and discuss how this affects the ability to detect and diagnose faults. Using the same change in operating conditions, reapply the parameter estimation method using 30 consecutive time intervals. Comment on the relationships between the number of consecutive time intervals used in the parameter estimation method, the time it takes to detect the fault, and the accuracy of the parameter estimates.
6. Repeat the parameter estimation study for the gravity flow tank system in Section 11.3, except with the leak occurring immediately after the valve at the exit of the tank.
7. Repeat the parameter estimation study for the gravity flow tank system in Section 11.3, except where leaks at four locations are considered: (i) Stream 1, (ii) the tank, (iii) Stream 2, and (iv) between the valve and

the second flow transmitter. Can the parameter estimation algorithm distinguish between leaks at the four locations? Thoroughly justify your answer.

8. Consider the pump example in Section 11.3, except where only a steady-state model for the process is used. Formulate the parameter estimation problem. Specify the model parameters θ_j. How many model parameters are there? Derive the relationship between the θ_j and the p_j. How many physical parameters can be estimated? Which of the physical parameters can be estimated uniquely, and which physical parameters are lumped?
9. Consider the pump example in Section 11.3, but with the valve closed, so that $\dot{M}(t) = 0$. The measured variables are $u_1(t)$, $I_1(t)$, and $\omega(t)$. Formulate the parameter estimation problem. Specify the model parameters θ_j. How many model parameters are there? Derive the relationship between the θ_j and the p_j. How many physical parameters can be estimated? Which of the physical parameters can be estimated uniquely, and which physical parameters are lumped?
10. For Problem 8, how would the answers change if only a steady-state model for the process was used?
11. Consider the pump example in Section 11.3, except with $Y(t)$ not measurable. Formulate the parameter estimation problem. Specify the model parameters θ_j. How many model parameters are there? Derive the relationship between the θ_j and the p_j. How many physical parameters can be estimated? Which of the physical parameters can be estimated uniquely, and which physical parameters are lumped?
12. For Problem 10, how would the answers change if only a steady-state model for the process was used?
13. Consider the pump example in Section 11.3, except with $\dot{M}(t)$ not measurable. Formulate the parameter estimation problem. Specify the model parameters θ_j. How many model parameters are there? Derive the relationship between the θ_j and the p_j. How many physical parameters can be estimated? Which of the physical parameters can be estimated uniquely, and which physical parameters are lumped?
14. For Problem 12, how would the answers change if only a steady-state model for the process was used? Note: Problems 7-13 are patterned after an experimental study [96, 137].
15. Consider a chemical reaction where the reactant A forms products B and C on a catalyst surface. Consider the estimation of the kinetic rate constant k in the rate law, $r_A = kC_A$, where C_A is the molar concentration of species A. Assume that the experiments are carried out in a well-mixed batch reactor with initial concentration C_{A0}, and that the volume and temperature remain constant throughout the reaction. Assume that the concentration of A can be measured once a minute.

a) Solve for C_A as a function of the initial concentration of A, the kinetic rate constant k, and time t. [Hint: the molar balance equation for species A is $\frac{dC_A}{dt} = -kC_A$.]
b) Write out the least-squares objective function for the estimation of the kinetic rate constant k. Clearly define each variable and its dimensions. Simplify as much as possible. Explain in words how to compute the best fit model parameter k. Is it possible to derive an analytical expression for the best fit k?
c) Write out the least-squares objective function for the estimation of the kinetic rate constant k as above, except with the assumption that the logarithm of the concentration of A can be measured directly (this happens, for example, when a pH probe is used to measure hydrogen ion concentration). Derive an analytical expression for the best fit kinetic rate constant k as a function of the time at the sampling instances and the measurement of the logarithm of the concentration of A at each sampling instant.
d) During a batch run, changes in the kinetic rate constant can occur due to deactivation of the catalyst used in the reaction. Explain how you would determine the threshold on the change in the kinetic rate constant which would signal when catalyst deactivation has occurred.

16. Repeat the full-order observer study for the gravity flow tank system in Section 11.4.1, for various amounts of measurement noise, where the initial estimated state is 0. Plot the state and output estimates obtained during normal operating conditions and during fault conditions in each case. How does the tuning of H depend on the noise level? Repeat the problem for the case where the initial estimated state is 0.3. Discuss how to tune H depending on the noise level and the accuracy of the initial estimated state. Repeat the problem for the case where the values for A and B in the observer equations are 20% larger than the A and B in the state-space equations for the process (this represents model uncertainties). Discuss how to tune H depending on the level of model uncertainty.
17. Repeat the full-order observer study for the gravity flow tank system in Section 11.4.1, except with the leak occurring immediately after the valve at the exit of the tank.
18. Repeat the full-order observer study for the gravity flow tank system in Section 11.4.1, except where leaks at four locations are considered: (i) Stream 1, (ii) the tank, (iii) Stream 2, and (iv) between the valve and the second flow transmitter. Can the fault detection algorithm distinguish between leaks at the four locations? Thoroughly justify your answer.
19. Propose a method to blend the observer-based method with canonical variate analysis (CVA) as described in Chapter 7. Thoroughly justify your method, while listing both its advantages and disadvantages over the

CVA-based measures in Chapter 7 and the pure observer-based method discussed in this chapter.

20. Rederive the equations in Section 11.4.2 for continuous-time systems. Compare with a published derivation [89]. Which derivation is more general? Note: there is a typographical error in Equation 23 of [89].
21. It is stated in Section 11.5.1 that (11.115) can always be satisfied for sufficiently large σ. Prove this statement.
22. Derive (11.117). Hint: see [226, 227].
23. Derive (11.119) from (11.117).
24. Derive the expressions for $\breve{\mathbf{n}}(t)$, $\breve{\mathbf{d}}(t)$, L_n, and L_d in (11.120). Comment on the design of WL_d so that the disturbances do not affect the transformed residuals.
25. Derive (11.154) from (11.153).
26. Recall the full-order observer study for the gravity flow tank system in Section 11.4.1. Compute the transformation matrix $W(q)$ for the equivalent fault detection system based on parity relations. Then compute the associated specification on the residual $Z_f(q)$. Are the dynamics in these transfer functions what you would expect? Does this provide some insight into the suitability of the observer design? Thoroughly justify your answers. Hint: derive $P_f(q)$, and use (11.135) and (11.157).
27. Read one of the following papers: [53, 88, 139, 151, 154, 155, 178, 238, 249, 311, 312, 331, 332]. Write a summary report. Compare the method described in the paper with the methods described in this chapter. Which methods are more general? Which types of faults are best handled by each method? Thoroughly justify your answers.
28. A technique that has been applied in the process industries is **data reconciliation**. Read one of the following papers on data reconciliation: [6, 43, 57, 121, 214, 275, 276, 306, 337]. Write a summary report. Compare the method described in the paper with the methods described in this chapter. Which approaches are more general? Which types of faults are best handled by each method? Thoroughly justify your answers.
29. In the generalized observer scheme, an observer dedicated to a certain sensor is driven by all outputs except that of the respective sensor. This allows the detection and isolation of a single fault in any sensor [85]. Write a summary report based on [85]. Would such an method be expected to give better results for faults in single sensors than the observer-based methods described in this chapter? Does this answer depend on the characteristics of the plant? Thoroughly justify your answers.
30. In contrast to a single observer, a bank of observers can also be used, in which each observer is excited by all outputs [54]. For fault isolation, multiple hypotheses testing can be applied, in which each of the estimators is designed for a different fault hypothesis. The hypotheses are tested in terms of likelihood functions (e.g., Bayesian decision theory) [341]. Read the papers [54, 341] which describe this method, and write a

report describing the method in some detail. Compare and contrast with the single observer method. What are the advantages and disadvantages of each method? Thoroughly justify your answers.

31. For a stochastic process, the innovations (prediction errors) of a Kalman filter can be used to detect faults. In the fault-free case, the innovations are white noise with zero mean and known covariance matrix. A fault is detected when the character of zero mean white noise with known covariance has changed. Read the papers [221, 341, 342] which describe this method, and write a report describing the method in some detail. Compare and contrast with the observer-based method described in this chapter. What are the advantages and disadvantages of each method? Thoroughly justify your answers.
32. For nonlinear processes, nonlinear observers can be used to estimate the state [1, 18, 86, 88, 89]. Write a report based on [89], which discusses the differences between the unknown input observer design for linear and nonlinear systems.

12. Knowledge-based Methods

12.1 Introduction

As discussed in Chapter 11, the analytical approach requires a detailed quantitative mathematical model in order to be effective. For large-scale systems, such information may not be available or may be too costly and time-consuming to obtain. An alternative method for process monitoring is to use knowledge-based methods such as causal analysis, expert systems, and pattern recognition. These techniques are based on qualitative models, which can be obtained through causal modeling of the system, expert knowledge, a detailed description of the system, or fault-symptom examples. Causal analysis techniques are based on the causal modeling of fault-symptom relationships. Qualitative and semi-quantitative relationships in these causal models can be obtained without using first principles. Causal analysis techniques including signed directed graphs and symptom trees are primarily used for diagnosing faults. These techniques are described in Section 12.2.

Expert systems are used to imitate the reasoning of human experts when diagnosing faults. The experience from a domain expert can be formulated in terms of rules, which can be combined with the knowledge from first principles or a structural description of the system for diagnosing faults. Expert systems are able to capture human diagnostic associations that are not readily translated into mathematical or causal models. A description of expert systems is provided in Section 12.3.

Pattern recognition techniques use associations between data patterns and fault classes without explicit modeling of internal process states or structure. Examples include artificial neural networks and self-organizing maps. These techniques are related to the data-driven techniques (PCA, PLS, FDA, and CVA) described in Chapters 4 to 7 in terms of modeling the relationships between data patterns and fault classes. The data-driven techniques are dimensionality reduction techniques based on rigorous multivariate statistics, whereas neural networks and self-organizing maps are black box methods that learn the patterns based entirely from training sessions. Section 12.4 provides a description of these pattern recognition techniques.

Each of the data-driven, analytical, and knowledge-based approaches have strengths and limitations. Incorporating several techniques for process monitoring can be beneficial in many applications. Many of these approaches can be combined with fuzzy logic. Section 12.5 discusses various combinations of process monitoring techniques.

12.2 Causal Analysis

Approaches based on causal analysis use the concept of **causal modeling** of **fault-symptom** relationships. Causal analysis is primarily used for diagnosing faults. Several recent papers that use causal analysis are available [127, 196, 228, 229, 303, 304, 320].

12.2.1 Signed Directed Graph

The **signed directed graph** (SDG) is a qualitative model-based approach for fault diagnosis that incorporates causal analysis [133, 287, 310]. It is a map showing the relationship of the process variables and it also reflects the behavior of the equipment involved as well as general system topology. A SDG for the gravity flow tank system in Figure 11.2 is shown in Figure 12.1. Nodes can depict process variables, sensors, system faults, component

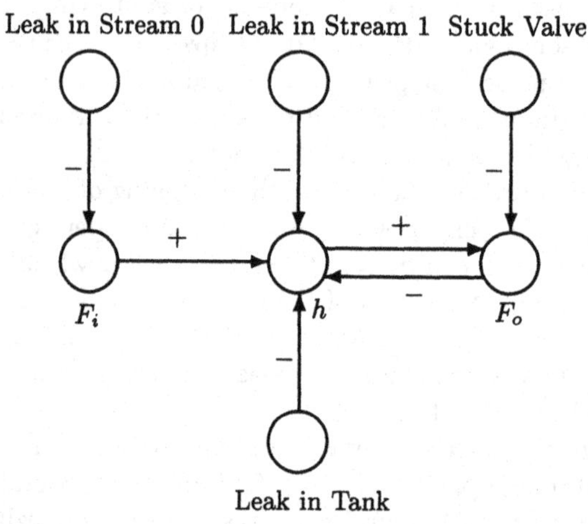

Fig. 12.1. A signed directed graph for the gravity tank system

failures, or subsystem failures. To use a SDG for diagnosing faults, high and low thresholds for each variable are first defined. A node takes the value of 0 when its measure variable is normal. A node takes a value of + when its measured variable is larger than the high threshold or the event as indicated by the node occurs. A node takes a value of – when its measured variable is smaller than the low threshold. Relationships between the cause nodes to effect nodes are embodied in the direct arcs between the nodes. These arcs may be conditional upon other events. Arc signs associated with each directed arc can take values of + and − representing whether the cause and effect change in the same direction or the opposite direction, respectively. A − sign can also be taken when the occurrence of an event in the cause node causes the negative deviation in the event in the effect node. For example, when "Leak in Stream 1" occurs, it will decrease the liquid height h. Therefore, a − sign is taken. Similarly, a + sign can also be taken when the occurrence of an event in the cause node causes the positive deviation in the event in the effect node. The goal of utilizing a SDG for diagnosing faults is to locate the possible root nodes representing the system faults based on the observed symptoms. To achieve this, the measured node deviations are propagated from effect nodes to cause nodes via consistent arcs until the root nodes are identified. An arc is **consistent** if the sign of the cause node times the sign of the arc times the sign of the effect node is positive.

Assuming that a single fault affects only a single root node and that the fault does not change other causal pathways in the SDG, the causal linkages will connect the fault origin to the observed symptoms of the fault. The gravity flow tank (see Figure 11.2) is used to illustrate the procedure of diagnosing faults using a SDG. The first step of developing a SDG is to connect the nodes in the fault-free case. The second step is the **fault modeling** step, which determines the initial effects of the fault on the SDG. The following faults are considered in this example: (i) leak in Stream 0, (ii) leak in Stream 1, (iii) leak in tank, and (iv) valve is stuck in the closed position. The corresponding SDG is shown in Figure 12.1.

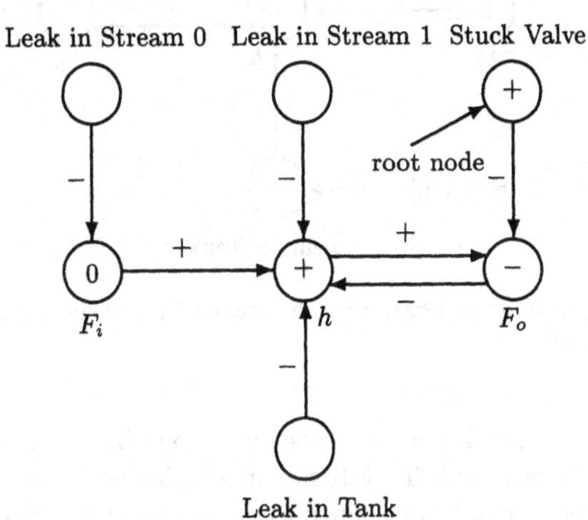

Fig. 12.2. A signed directed graph for the gravity tank system with the symptoms "h is increasing, while F_o is decreasing"

Consider the case where the observed symptoms are that the liquid level h is increasing, while the output flow rate F_o is decreasing. These symptoms indicate that the nodes F_i, h, and F_o take the values of 0, +, and −, respectively. Based on a consistent path check (see Figure 12.2), the fault is determined uniquely as "the valve is stuck in the closed position". A + sign in any of the other unmarked nodes in Figure 12.2 results in an inconsistent arc.

Now consider the case where the observed symptoms are that h and F_o are decreasing. The nodes h and F_o now take values of −, while the node F_i takes a value of 0. Based on a consistent path check (see Figure 12.3), the possible root nodes responsible for the symptoms are identified as "Leak in Tank" and "Leak in Stream 1". Simulations of these two faults are shown in Figures 11.4 and 12.4, respectively. The simulations indicate that these two faults share the same symptoms. The SDG narrows down the search for the possible faults, but it can produce more than one fault candidate. To determine the *exact* cause of the symptoms, expert knowledge is often needed. Alternatively, taking additional measurements of the process at different locations may reveal different symptoms for the faults "Leak in Tank" and "Leak in Stream 1".

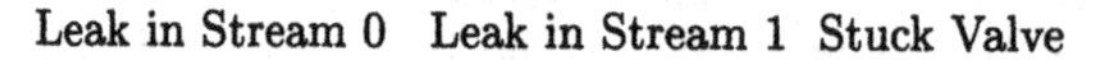

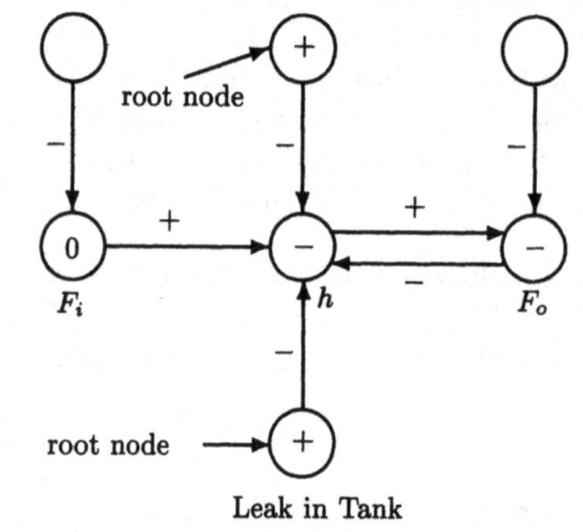

Fig. 12.3. A signed directed graph for the gravity tank system with the symptoms "h and F_o are decreasing"

The SDG shown in Figure 12.1 was developed based on knowledge from observation and analysis of the system. For complex and large-scale systems, a SDG for the process can be developed from the model equations of individual units in the process [235]. Alternatively, the SDG can also be developed based on the knowledge of the process from a domain expert or historical data. The SDG is able to provide a list of possible fault candidates. Expert knowledge is often needed to deduce the most likely fault candidates from the list.

There are some drawbacks of using this basic version of the SDG. These include the lack of resolution, potentially long computing times, and the single

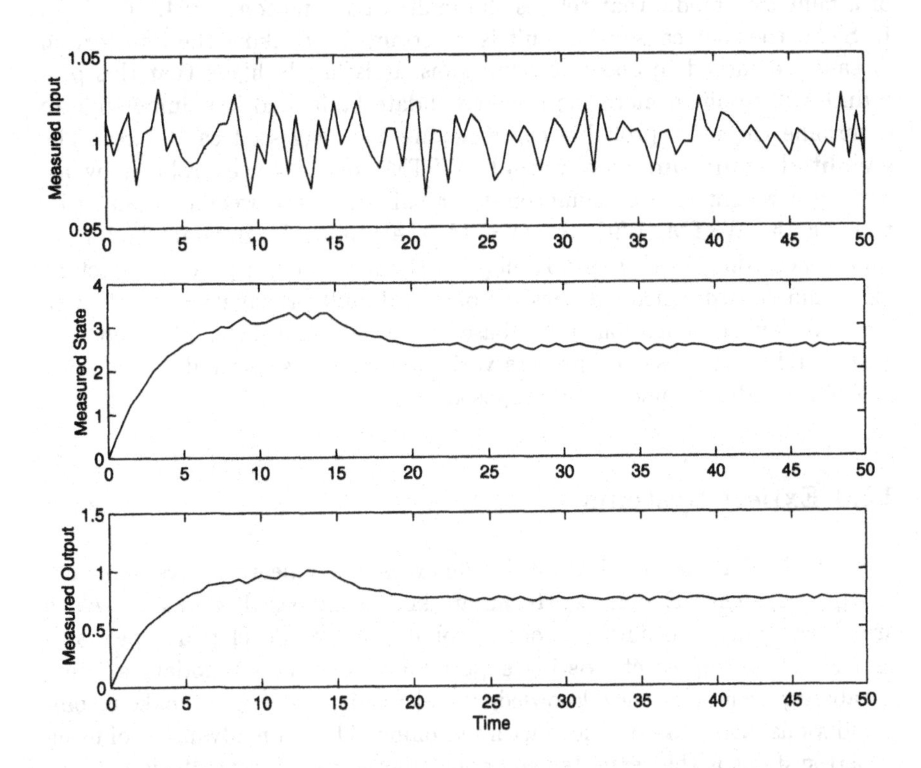

Fig. 12.4. The measured input (inlet flow rate), measured state (liquid height), and measured output (outlet flow rate) of a gravity tank in faulty case (Tank is leaking)

fault assumption. The resolution of a SDG has been improved by using extra information on the reliability of equipment, the infeasibility of certain root nodes, and equipment maintenance schedules [319]. This knowledge base is used to eliminate physically improbable nodes.

The computing time required for using the SDG can be reduced by compiling the SDG into rules [173]. The SDG has been extended to handle variables with compensatory response and inverse response [81, 247]. A digraph-based diagnosis reasoning approach known as the **possible cause-effect graph** can reduce the search space [339, 340]. The SDG has also been extended to multiple fault diagnosis by assuming that the probability of occurrence of a multiple fault decreases with an increasing number of faults [319].

12.2.2 Symptom Tree Model

A closely related representation to the SDG that can be used in causal analysis is the **symptom tree model** (STM). The STM is a real-time version of a fault tree model that relates the faults and symptoms [354, 362, 363]. In STM, the root cause of a fault is determined by taking the intersection of causes attached to observed symptoms. It is highly likely that this procedure will result in more than one candidate fault, and it is impossible to determine the most probable cause among the suggested candidates. The **weighted symptom tree model** (WSTM) resolves the problem by attaching a weight to each symptom-fault pair, with the weight obtained by training the WSTM. With the WSTM, the proposed candidate faults are ranked according to their probability. In the next step, a pattern matching algorithm is used which matches the observed fault propagation trends with standard fault propagation trends based on training set [246]. The fault that best matches the observed process variable changes is selected as the most probable candidate among the proposed ones.

12.3 Expert Systems

Many fault diagnosis applications in the areas of engineering have made use of expert systems. **Expert systems** are knowledge-based techniques which are closer in style to human problem solving. A well-developed expert system is able to represent existing expert knowledge, accommodate existing databases, accumulate new knowledge, make logical inferences, make recommendations, and make decisions with reasoning. The main advantage of using expert systems is that experts need not be present for a consultation.

The basic components of an expert system include a knowledge base, an inference engine, and a human/expert system interface. The knowledge base can be obtained via **shallow knowledge** (based on heuristics and expert testimony) and/or **deep knowledge** (based on structural, behavioral, or mathematical models) [257, 366]. Various types of knowledge representation schemes can be used including production rules, frames, and semantic networks. The correctness and completeness of the information stored in the knowledge base specifies the performance achievable by the expert system. To benefit from new experience and knowledge, the knowledge base also needs to be updated periodically. The inference engine directs the use of the knowledge base. Inference mechanisms include forward-backward chaining, hypothesis/test methods, heuristic search, meta-rules, and artificial neural networks [23]. The human/expert system interface must translate user input into computer language and presents conclusions and explanations to the user in an easy-to-understand form.

Early work on expert systems was focused primary on medical diagnostic systems [32, 52]. Efforts have been made to expand the applications to equipment maintenance and diagnostics, science, engineering, agriculture, business, and finance [34, 175, 198, 211, 284, 314]. Here we provide an introduction to expert systems. Many references provide a more detailed description

[37, 107, 131, 161, 335]. Several recent papers describing applications of expert systems are available [14, 33, 35, 265, 359].

12.3.1 Shallow-Knowledge Expert System

An experienced engineer or domain expert is capable of diagnosing faults in a much shorter time than an inexperienced operator because the experienced personnel have accumulated knowledge and experience. To assist the personnel to diagnose faults, expert experience can be formulated as a set of IF-THEN rules, which can be used to build an expert system. This is referred to as a **shallow-knowledge expert system** (also known as experiential knowledge and empirical reasoning expert systems) [174, 198]. The method does not depend on a functional understanding of the mechanism or physics of the system.

Advantages of shallow-knowledge expert systems are that they are flexible and their conclusions can be easily verified and explained. Shallow-knowledge expert systems map the observations to conclusions directly; therefore, shallow knowledge can also be applied to areas where fundamental principles or complete descriptions of the systems are lacking, but heuristic solutions are available. For example in medical diagnosis where detailed and reliable models of the subjects are lacking, rules have been formulated to relate sets of symptoms to possible diseases [32, 52].

The results from a shallow-knowledge expert system depend strongly on the adequacy of the knowledge incorporated into the expert system. However, heuristics do not guarantee any solution to the fault diagnosis problems, especially for situations in which the domain experts have not encountered before (*i.e.*, knowledge outside of the domain of expertise). At a minimum, a well-developed shallow-knowledge expert system should be able to offer solutions which are good enough most of the time [60].

The main difficulty of applying shallow-knowledge expert systems is in the **knowledge acquisition** step, which is the step of collecting adequate knowledge from domain experts and translating it into computer programs. First, domain experts may not be available for unique operating scenarios and for new or retrofitted plants. Second, when domain experts are available, they may not understand or be able to explain clearly how they solve a problem [115, 174, 293, 309, 327]. Each expert system is application specific. Developing an effective expert system from scratch can be time-consuming and costly for a large-scale system.

12.3.2 Deep-Knowledge Expert Systems

In contrast to shallow-knowledge expert systems, deep-knowledge expert systems are based on a model such as engineering fundamentals, a structural description of a system, or a complete behavioral description of its components in faulty and normal cases. Deep-knowledge expert systems are also known as model-based, functional reasoning, or diagnosis-from-first-principles expert systems. For novel or unique situations, deep-knowledge expert systems often provide useful information for diagnosing faults. Deep knowledge is often needed when a particularly difficult problem is confronted or an explanation to the diagnostic process is required [273].

Deep knowledge involves using reasoning on causal and functional information. Knowledge of the principles which govern the process can be used in a deep-knowledge expert system. Governing equations based on physical laws provide a set of constraints on the values of process variables. Significant violations of these constraints are an indication of process faults. Each constraint is associated with the set of faults which cause violation of the constraint [174].

Another method to develop a deep-knowledge expert system is to use causal reasoning via a SDG [173, 174, 301, 356]. One rule can be produced for each possible fault origin in the SDG; combining these rules produces all viable fault candidates [174].

Similarly to the analytical techniques which rely heavily on first principles, a deep-knowledge expert system is also hard to develop for a complex large-scale system whose mathematical model may not be available.

12.3.3 Combination of Shallow-Knowledge and Deep-Knowledge Expert Systems

An experienced engineer uses a combination of techniques for diagnosis, including a familiarity with the system documentation, a functional understanding of the system components, an understanding of the system interrelationships, knowledge of the failure history of the device, along with numerous heuristics [254]. This suggests that shallow knowledge and deep knowledge should be combined in an expert system. Deep knowledge reasoning is often needed to supplement the shallow knowledge.

Although it is costly to obtain a first-principles model for a large-scale system, models of individual components are usually available [216]. Such information can be combined with shallow knowledge in order to effectively diagnose faults. One method to combine shallow and deep knowledge is to convert the deep knowledge into production rules [174].

12.3.4 Machine Learning Techniques

As mentioned in Section 12.3.1, the main difficulty of using shallow knowledge is in knowledge acquisition. Experts are usually better at collecting and archiving cases than in expressing the experience and cases explicitly into production rules [293, 327].

One way to solve this problem is to use machine learning techniques, in which knowledge is automatically extracted from data [21, 200, 201]. Symbolic information can be integrated into an artificial neural network learning algorithm [156, 298]. Such a learning system allows for knowledge extraction and background knowledge encoded in the form of rules. Fuzzy rules can also be used to extract knowledge from the data [156, 327].

12.3.5 Knowledge Representation

The simplest form of knowledge representation in an expert system is to use a series of IF-THEN rules to represent the expert knowledge in the system. The majority of industrial expert systems use a rule-based system, which is composed of a rule base, a working memory, and a rule interpreter [107, 309]. The **rule base** is often partitioned into groups of rules, called **rule clusters**. Each rule cluster encodes the knowledge required to perform a certain task. A **working memory** is a database holding input data, inferred hypotheses, and internal information about the program. A **rule interpreter** is the mechanism to select rules and evaluate rules. Advantages of rule-based systems are that they enforce a homogeneous representation of knowledge, allow incremental knowledge growth through the addition of rules, and allow unplanned but useful interactions [107, 293].

In addition to the rule-based systems to represent expert knowledge, a semantic network can also be used. A semantic network is a method of knowledge representation in which concepts are represented as nodes in the network and relations are represented as directed arcs (see Figure 12.5). There must be a way of associating meaning with the network. One way to do this is to associate a set of programs that operate on descriptions in the representation [293, 294].

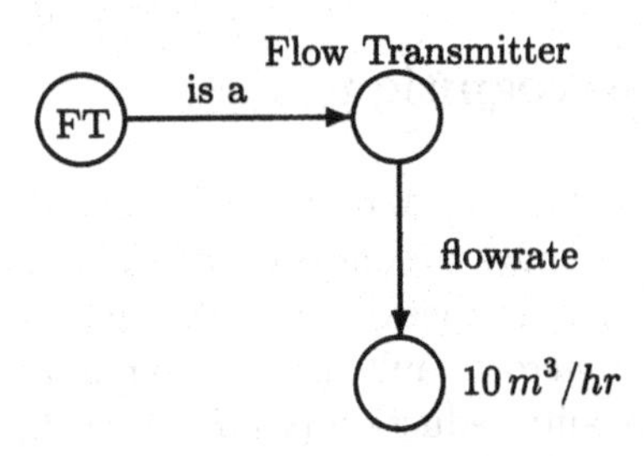

Fig. 12.5. A semantic network representing the knowledge "FT is a flow transmitter with a flow rate of 10 m^3/hr"

Alternatively, frames can be used for knowledge representation. A frame is a collection of semantic net nodes that together provides a structured representation of an object, act, or event [293]. Frames may be linked in hierarchies to show the relationships between domain objects, while rules can only indirectly describe the objects which comprise the domain [174, 293].

12.3.6 Inference Engine

An **inference engine** uses an inference mechanism to gather the information needed (from the knowledge base or the user) to draw inferences or conclusions for the processes involved, and presents these inferences or conclusions with explanations or bases. The most common approach used in an inference mechanism is backward/forward chaining. In backward chaining, the system works backward from tentative conclusions or goals to find supporting evidence. Backward chaining starts with selecting a particular hypothesis, the rules are examined to see if the hypothesis is a consequence. If so, the premise (also called a condition, pattern, or antecedent) forms the next set of hypotheses. The procedure is continued until some hypotheses are false or all hypotheses are true based on the data.

In forward chaining the system reasons forward from a set of known facts to infer the conclusions [293, 309]. System design is a forward-chaining application where the expert system starts with the known requirements, investigates the possible arrangements, and makes a recommendation. A combination of forward and backward chaining are common in many applications [309].

The hypothesis/test method is patterned closely to human diagnostic reasoning. This method first generates a hypothesis based on observations. The effects of the hypothetical fault on the process are determined and compared with the actual measurements. If the hypothesis cannot be verified, another hypothesis is checked. The procedure is repeated until all hypotheses are exhausted.

12.4 Pattern Recognition

Many data-driven, analytical, and knowledge-based methods incorporate pattern recognition techniques to some extent. For example, Fisher discriminant analysis is a data-driven process monitoring method based on pattern classification theory. Numerous fault diagnosis approaches described in Part III combined dimensionality reduction (via PCA, PLS, FDA, or CVA) with discriminant analysis, which is a general approach from the pattern recognition literature. Other uses of pattern recognition in process monitoring are discussed in Section 12.2.

Some pattern recognition methods for process monitoring use the relationship between the data patterns and fault classes without modeling the internal process states or structure explicitly. These approaches include **artificial neural networks** (ANN), and **self-organizing maps**. Since pattern recognition approaches are based on inductive reasoning through generalization from a set of stored or learned examples of process behaviors, these techniques are useful when data are abundant, but when expert knowledge is lacking. Recent reviews of pattern recognition approaches are available [192, 224, 282]. The goal here is to describe artificial neural networks and self-organizing maps, as these are two of the most popular pattern recognition approaches, and they are representative of other approaches.

12.4.1 Artificial Neural Networks

The **artificial neural network** (ANN) was motivated from the study of the human brain, which is made up of millions of interconnected neurons. These interconnections allow humans to implement pattern recognition computations. The ANN was developed in an attempt to mimic the computational structures of the human brain.

An ANN is a nonlinear mapping between input and output which consists of interconnected "neurons" arranged in layers. The layers are connected such that the signals at the input of the neural net are propagated through the network. The choice of the neuron nonlinearity, network topology, and the weights of connections between neurons specifies the overall nonlinear behavior of the neural network. Many books are available that provide an introduction to neural networks [24, 28, 61, 156, 309, 358]. Numerous papers are available which apply ANNs to fault detection and diagnosis; many of these techniques were derived from the pattern recognition perspective [17, 26, 50, 49, 106, 118, 119, 124, 210, 250, 297, 313, 333, 334, 361, 365].

Of all the configurations of ANNs, the three-layer feedforward ANN is the most popular (see Figure 12.6). The network consists of three components: an input layer, a hidden layer, and an output layer. Each layer contains neurons (also called nodes). The input layer neurons correspond to input variables and the output layer neurons correspond to output variables. Each neuron in the hidden layer is connected to all input layer neurons and output layer neurons. No connection is allowed within its own layer and the information flow is in one direction only.

One common way to use a neural network for fault diagnosis is to assign the input neurons to process variables and the output neurons to fault indicators. The number of output neurons is equal to the number of different fault classes in the training data. The jth output neuron is assigned to '1' if the input neurons are associated with fault j, and '0' otherwise.

Each neuron j in the hidden and output layers receives a signal from the neurons of the previous layer $\mathbf{v}^T = [v_1\ v_2\ \cdots\ v_r]$, scaled by the weight $\mathbf{w_j}^T = [w_{1j}\ w_{2j}\ \cdots\ w_{rj}]$. The strength of connection between two linked neurons is represented in the weights, which are determined via the training process. The jth neuron computes the following value:

$$s_j = \mathbf{w_j}^T \mathbf{v} + b_j \tag{12.1}$$

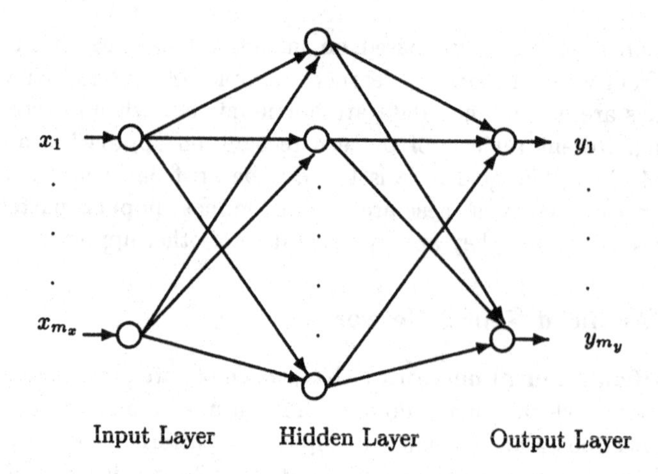

Fig. 12.6. Basic structure of a three-layer feedforward artificial neural network

where b_j is the optional bias term of the jth neuron. Adding a bias term provides an offset to the origin of the activation function and hence selectively inhibits the activity of certain neurons [282, 309, 353]. The bias term b_j can be regarded as an extra weight term w_{0j} with the input fixed at one. Therefore, the weight becomes $\mathbf{w}_j^T = [w_{0j}\ w_{1j}\ w_{2j}\ \cdots\ w_{rj}]$. The input layer neuron uses a linear activation function and each input layer neuron j receives only one input signal x_j.

The quantity s_j is passed through an activation function resulting in an output o_j. The most popular choice of the activation function is to use a **sigmoid** function, which satisfies the following properties:

1. The function is bounded, usually in the range [0,1] or [-1,1].
2. The function is monotonically nondecreasing.
3. The function is smooth and continuous (*i.e.*, differentiable everywhere in its domain).

A common choice of sigmoid function is the **logistic** function:

$$o_j = \frac{1}{1+e^{-s_j}}. \tag{12.2}$$

The logistic function has been a popular choice of activation function because many ANN training algorithms use the derivative of the activation function, and the logistic function has a simple derivative, $\frac{\partial o_j}{\partial s_j} = o_j(1-o_j)$. Another choice of sigmoid function is the **bipolar logistic** function:

$$o_j = \frac{1-e^{-s_j}}{1+e^{-s_j}} \tag{12.3}$$

which has a range of [-1,1]. Another common sigmoid function is the **hyperbolic tangent**:

$$o_j = \frac{e^{s_j} - e^{-s_j}}{e^{s_j} + e^{-s_j}}. \tag{12.4}$$

Also, radial basis functions (Gaussian, bell-shaped functions) can be used in place of or in addition to sigmoid functions [44, 130, 231, 262].

The training session of the network uses the error in the output values to update the weights $\mathbf{w_j}$ of the neural network, until the accuracy is within the tolerance level. An error quantity based on the difference between the correct decision made by the domain expert and the one made by the neural network is generated, and used to adjust the neural network's internal parameters to produce a more accurate output decision. This type of learning is known as **supervised learning**. Mathematically, the objective of the training session is to minimize the total **mean square error** (MSE) for all the output neurons in the network and all the training data:

$$E = \frac{1}{Mm_y} \sum_{m=1}^{M} \sum_{j=1}^{m_y} (\hat{y}_j^{(m)} - y_j^{(m)})^2 \tag{12.5}$$

where M is the number of training data patterns, m_y is the number of neurons in the output layer, $\hat{y}_j^{(m)}$ is the prediction for the jth output neuron for the given m^{th} training sample, and $y_j^{(m)}$ is the target value of the jth output neuron for the given m^{th} training sample.

The backpropagation training algorithm is a commonly used steepest descent method which searches for optimal solutions for the input layer-hidden layer weights $\mathbf{w_i^h}$ and hidden layer-output layer weights $\mathbf{w_j^o}$ for (12.5). The general procedure for training a three-layer feedforward ANN is [156, 353]:

1. Initialize the weights (this is iteration $t = 0$).
2. Compute the output $\hat{y}_j(t)$ for an input $\mathbf{x}$ from the training data. Adjust the weights between the ith hidden layer neuron and the jth output neuron using the **delta rule** [156]

$$w_{ij}^o(t+1) = w_{ij}^o(t) + \Delta w_{ij}^o(t+1) \tag{12.6}$$

where

$$\Delta w_{ij}^o(t+1) = \eta \delta_j(t) o_i^h(t) + \alpha \Delta w_{ij}^o(t), \tag{12.7}$$

η is the **learning rate**, α is the coefficient of **momentum term**, $o_i^h(t)$ is the output value of the ith hidden layer neuron at iteration t, and $\delta_j(t) = y_j - \hat{y}_j(t)$ is the output error signal between the desired output value y_j and the value $\hat{y}_j(t)$ produced by the jth neuron at iteration t. Alternatively, the **generalized delta rule** can be used:

$$\Delta w_{ij}^o(t+1) = \eta \delta_j(t) o_i^h(t) \frac{do_j^o}{ds_j^o}(s_j^o(t)) + \alpha \Delta w_{ij}^o(t), \tag{12.8}$$

where $o_j^o(s_j^o)$ is the activation function, and

$$s_j^o = \sum_i w_{ij}^o o_i^h(s_i^h) + b_j^o \tag{12.9}$$

is the combined input value from all of the hidden layer neurons to the jth output neuron. When the activation function o_j^o is the logistic function (12.2), the derivative becomes

$$\frac{do_j^o(s_j^o)}{ds_j^o} = o_j^o(1 - o_j^o) = \hat{y}_j(1 - \hat{y}_j). \tag{12.10}$$

3. Calculate the error e_i for the ith hidden layer neuron:

$$e_i = \sum_{j=1}^{m_y} \delta_j w_{ij}^o \frac{do_j^o}{ds_j^o}. \tag{12.11}$$

4. Adjust the weights between the kth input layer neuron and the ith hidden neuron:

$$w_{ki}^h(t+1) = w_{ki}^h(t) + \Delta w_{ki}^h(t+1). \tag{12.12}$$

When the delta rule (12.7) is used in Step 2, $\Delta w_{ki}^h(t+1)$ is calculated as

$$\Delta w_{ki}^h(t+1) = \eta e_i(t) x_k(t) + \alpha \Delta w_{ki}^h(t), \tag{12.13}$$

where x_k is the kth input variable. When the generalized delta rule (12.8) is used in Step 2, $\Delta w_{ki}^h(t+1)$ is calculated as

$$\Delta w_{ki}^h(t+1) = \eta e_i(t) x_k(t) \frac{do_i^h}{ds_i^h}(s_i^h(t)) + \alpha \Delta w_{ki}^h(t), \tag{12.14}$$

where

$$s_i^h = \sum_{k=1}^{n_x} w_{ki}^h x_k + b_i^h \tag{12.15}$$

is the combined input value from all of the input layer neurons to the ith hidden neuron.

Steps 2 to 4 are repeated for an additional **training cycle** (also called an **iteration** or **epoch**) with the same training samples until the error E in (12.5) is sufficiently small, or the error no longer diminishes significantly.

The backpropagation algorithm is a gradient descent algorithm, indicating that the algorithm can stop at a local minimum instead of the global minimum. In order to overcome this problem, two methods are suggested [156]. One method is to randomize the initial weights with small numbers in an interval $[-1/n, 1/n]$, where n is the number of the neuronal inputs. Another method is to introduce noise in the training patterns, synaptic weights, and output values.

The training of the feedforward neural networks requires the determination of the network topology (the number of hidden neurons), the learning rate η, the momentum factor α, the error tolerance (the number of iterations), and the initial values of weights. It has been shown that the proficiency of neural networks depends strongly on the selection of the training samples [50].

The learning rate η sets the step size during gradient descent. If $0 < \eta < 1$ is chosen to be too high (e.g., 0.9), the weights oscillate with a large amplitude, whereas a small η results in slow convergence. The optimal learning rate has been shown to be inversely proportional to the number of hidden neurons [156]. A typical value for the learning rate is taken to be 0.35 for many applications [327]. The learning rate η is usually taken to be the same for all neurons. Alternatively, each connection weight can have its individual learning rate (known as the **delta-bar-delta rule** [146]). The learning rate should be decreased when the weight changes alternate in sign and it should be increased when the weight change is slow.

The degree to which the weight change $\Delta w_{ij}^{o}(t+1)$ depends on the previous weight change $\Delta w_{ij}^{o}(t)$ is indicated by the coefficient of momentum term α. The term can accelerate learning when η is small and suppress oscillations of the weights when η is big. A typical value of α is taken to be 0.7 $(0 < \alpha < 1)$.

The number of hidden neurons depends on the nonlinearity of the problem and the error tolerance. The number of hidden neurons must be large enough to form a decision region that is as complex as required by a given problem. However, the number of hidden neurons must not be so large that the weights cannot be reliably estimated from available training data patterns. A practical method is to start with a small number of neurons and gradually increase the number. It has been suggested that the minimum number should be greater than $(M-1)/(m_x+2)$ where m_x is the number of inputs of the network, and M is the number of training samples [156].

In [156] a (4,4,3) feedforward neural network (*i.e.*, 4 input neurons, 4 hidden neurons, and 3 output neurons) was used to classify Fisher's data set (see Figure 4.2 and Table 4.1) into the three classes. The network was trained based on 120 samples (80% of Fisher's data). The rest of the data was used for testing. A mean square error (MSE) of 0.0001 was obtained for the training process and all of the testing data were classified correctly.

To compare the classification performance of neural networks with the PCA and FDA methods, 40% of Fisher's data (60 samples) were used for training, while the rest of the data was used for testing. The MATLAB Neural Network Toolbox [65] was used to train the network to obtain a MSE of 0.0001 using the backpropagation algorithm. The input layer-hidden layer weights w^h_{ki} and the hidden layer-output layer weights w^o_{ij} are listed in Table 12.1. The hidden neuron biases b^h_i and the output neuron biases b^o_j are listed in Table 12.2. For example, w_{21} is 1.783 according to Table 12.1. This means that the weight between the second input neuron and the first hidden neuron is 1.783.

Table 12.1. The weights of the neural network for Fisher's data [45, 82]

w^h_{ki}	1	2	3	4	w^o_{ij}	1	2	3
1	3.714	-0.2953	1.253	0.0536	1	0.0001	1.698	-1.726
2	1.783	2.178	0.656	-0.0421	2	2.206	-2.811	-0.0002
3	-18.89	-3.908	-3.261	0.0187	3	0	1.112	0.0002
4	-9.644	-1.767	-1.513	0.2086	4	3.031	0.120	1.834

Table 12.2. The bias weights of the neural network for Fisher's data [45, 82]

	b^h_i	b^o_j
1	15.04	-0.8244
2	-3.252	-0.1191
3	0.0059	-0.1069
4	4.368	–

The misclassification rates for Fisher's data are shown in Table 12.3. The overall misclassification rate for the testing set is 0.033, which is the same as the best classification performance using the PCA or FDA methods (see Table 5.3). This suggests that using a neural network is a reasonable approach for this classification problem.

Table 12.3. Misclassification rate of Fisher's data from [45, 82] using the neural network method

	Class 1	Class 2	Class3	Overall
Training	0	0	0	0
Testing	0	0.10	0	0.033

The training time for a neural network using one of the variations of backpropagation can be substantial (hours or days). For a simple 2-input 2-output system with 50 training samples, 100,000 iterations are not uncommon [50]. In the Fisher's data example, the computation time required to train the neural network is noticeably longer than the time required by the data-driven methods (PCA and FDA). For a large-scale system, the memory and computation time required for training a neural network can exceed the hardware limit. Training a neural network for a large-scale system can be a bottleneck in developing a fault diagnosis algorithm.

To investigate the dependence of the size of the training set on the proficiency of classification, 120 observations (instead of 60 observations) were used for training and the rest of Fisher's data were used for testing. A MSE of 0.002 was obtained and the network correctly classified all the observations in the testing set, which is consistent with the performance obtained by the PCA and FDA methods (see Table 5.5).

Recall that the training of neural networks is based entirely on the available data. Neural networks can only recall an output when presented with an input *consistent* with the training data. This suggests that the neural networks need to be retrained when there is a slight change of the normal operating conditions (e.g., a grade change in a paper machine).

Neural networks can represent complex nonlinear relationships and are good at classifying phenomena into preselected categories used in the training process. However, their reasoning ability is limited. This has motivated research on using expert systems or fuzzy logic to improve the performance of neural networks (this is discussed in Section 12.5).

12.4.2 Self-Organizing Map

Neural network models can also be used for **unsupervised learning** using a **self-organizing map** (SOM) (also known as a **Kohonen self-organizing map**), in which the neural network learns some internal features of the input vectors $\mathbf{x}$ [156, 164, 165, 166]. A SOM maps the nonlinear statistical dependencies between high-dimensional data into simple geometric relationships, which preserve the most important topological and metric relationships of the original data. This allows the data to be clustered without knowing the class memberships of the input data.

As shown in Figure 12.7, a SOM consists of two layers; an input layer and an output layer. The output layer is also known as the **feature map**, which represents the output vectors of the output space. The feature map can be n-dimensional, but the most popular choice of the feature map is two-dimensional. The topology in the feature map can be organized in a rectangular grid, a hexagonal grid, or a random grid. The number of the neurons in the feature map depends on the complexity of the problem. The number of neurons must be chosen large enough to capture the complexity of the problem, but the number must not be so large that too much training time is required.

The weight $\mathbf{w}_j$ connects all the m_x input neurons to the jth output neuron. The input values may be continuous or discrete, but the output values are binary. A particular implementation of a SOM training algorithm is outlined below [7, 156]:

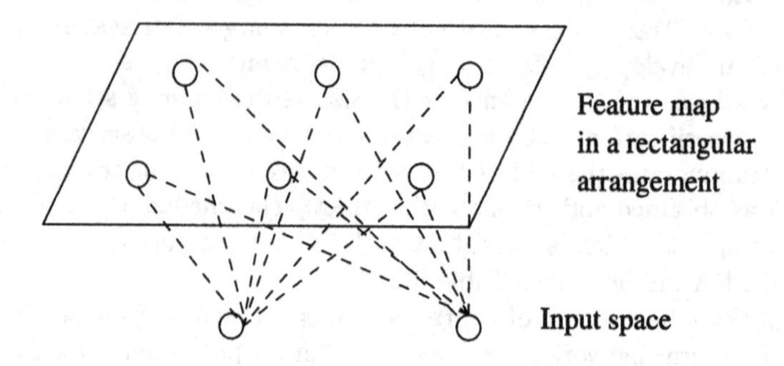

Fig. 12.7. A self-organizing map with two inputs and a two-dimensional output map

1. Assign small random numbers to the initial weight vector $\mathbf{w_j}$ for each neuron j from the output map (this is iteration $t = 0$).
2. Retrieve an input vector $\mathbf{x}$ from the training data, and calculate the Euclidean distance between $\mathbf{x}$ and each weight vector $\mathbf{w_j}$:

$$\|\mathbf{x} - \mathbf{w_j}\|. \tag{12.16}$$

3. The neuron closest to $\mathbf{x}$ is declared as the **best matching unit** (BMU). Denote this as neuron k.
4. Each weight vector is updated so that the BMU and its topological neighbors are moved closer to the input vector in the input space. The update rule for neuron j is:

$$\mathbf{w_j}(t+1) = \begin{cases} \mathbf{w_j}(t) + \alpha(t)[\mathbf{x}(t) - \mathbf{w_j}(t)] & j \in N_k(d) \\ \mathbf{w_j}(t) & j \notin N_k(d) \end{cases} \tag{12.17}$$

   where $N_k(d)$ is the neighborhood function around the winning neuron k and $0 < \alpha(t) < 1$ is the learning coefficient. Both the neighborhood function and learning coefficient are decreasing functions of iteration number t. In general, the neighborhood function $N_k(d)$ can be defined to contain the indices for all of the neurons that lie within a radius d of the winning neuron k.

Steps 2 to 4 are repeated for all the training samples until convergence. The final accuracy of the SOM depends on the number of the iterations. A "rule of thumb" is that the number of iterations should be at least 500 times the number of network units; over 100,000 iterations are not uncommon in applications [166].

To illustrate the principle of the SOM, Fisher's data set (see Table 4.1 and Figure 4.2) is used. The MATLAB Neural Network Toolbox [65] was used to train the SOM, in which 60 observations are used and 15 by 15 neurons in a rectangular arrangement are defined in the feature map. The feature map of the training set after 2,000 iterations is shown in Figure 12.8. Each marked neuron ('x', 'o', and '*') represents the BMU of an observation in the training set. The activated neurons form three clusters. The SOM organizes the neurons in the feature map such that observations from the three classes can be separated.

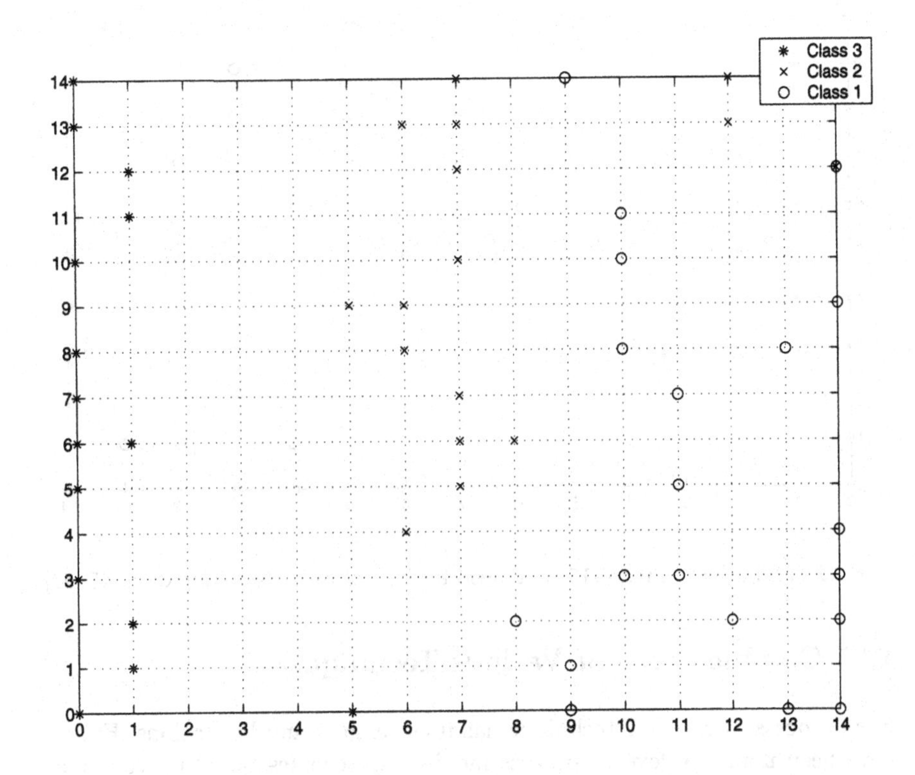

Fig. 12.8. The feature map (15 by 15 grid format) of Fisher's training data [45, 82]

The feature map of a testing set is shown in Figure 12.9. The positions of the 'x', 'o', and '*' occupy the same regions as in Figure 12.8. This suggests that the SOM has a fairly good recall ability when applied to new data. An increase in the number of neurons and the number of iterations would improve the clustering of the three classes.

The SOM has been successfully applied in fault diagnosis [289, 290]. For fault detection, a SOM is trained to form a mapping of the input space during normal operating conditions; a fault can be detected by monitoring the distance between the observation vector and the BMU [7].

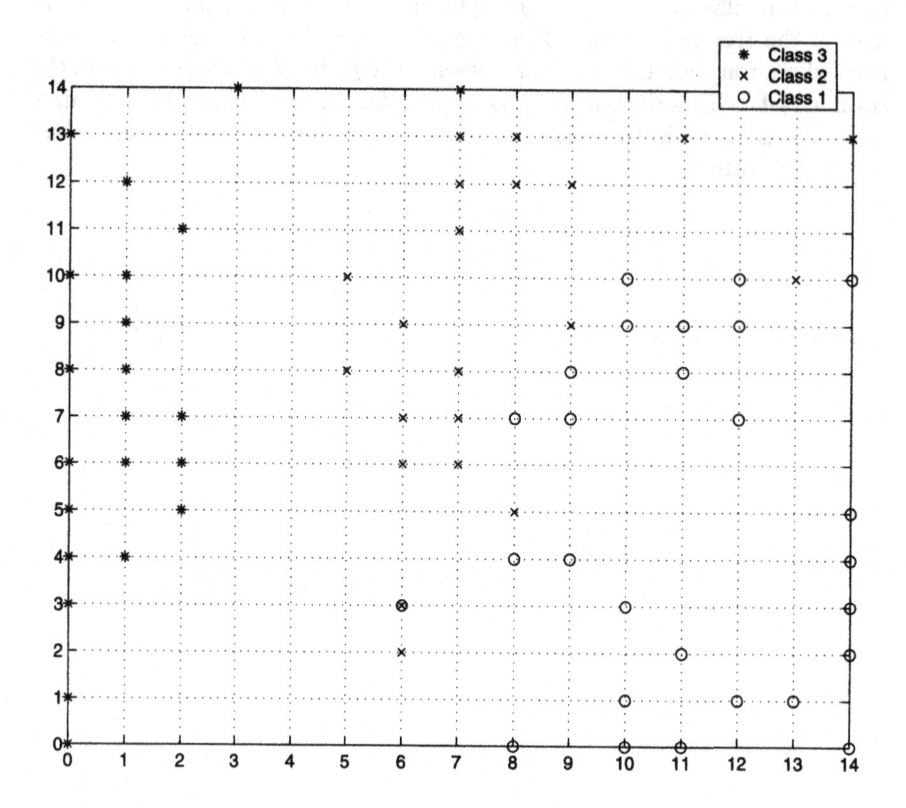

Fig. 12.9. The feature map (15 by 15 grid format) of Fisher's testing data [45, 82]

12.5 Combinations of Various Techniques

Each process monitoring technique has its strengths and limitations. Efforts have been made to develop process monitoring schemes based on combinations of techniques from knowledge-based, analytical, and data-driven approaches [51, 78, 95, 236, 323, 324]. Results show that combining multiple approaches can result in better process monitoring performance for many applications.

12.5.1 Neural Networks and Expert Systems

Most of the knowledge-based methods can be used in conjunction with each other. For example, neural networks and expert systems have been combined and used in industrial applications [329, 330]. As shown in Section 12.3, the strength of expert systems is their ability to mimic human reasoning on solving fault diagnosis problems and the weakness is the knowledge acquisition bottleneck. As shown in Section 12.4, the strength of neural networks is their ability to recognize patterns based on training examples and the weakness is their lack of ability to explain the results.

The most direct application to using neural networks for improving expert systems is to have a neural network serve as the knowledge base for an expert system. This allows the expert system to acquire knowledge from data. The training may be on line or performed during an initialization period. Knowledge bases may also contain models of systems which produce real-time results or certain learning systems via neural networks to provide new knowledge.

Expert systems can be used to improve neural networks as well. One application is to use an expert system as an interpreter of neural networks to execute fault diagnosis and evaluate the results [309, 366]. An expert system can also be used to retrain the neural network to adapt to challenging situations. A combined neural network and expert system tool was developed for transformer fault diagnosis [329, 330]. Results were that a tool which combines an artificial neural network and an expert system provided better performance than using either of the individual components.

12.5.2 Fuzzy Logic

Fuzzy logic was first developed in the mid-1960s for representing uncertain and imprecise knowledge [357]. Fuzzy logic provides an approximate but effective means of describing complex ill-defined systems by using graded statements rather than ones that are strictly true or false. Fuzzy logic has been widely applied to many areas of engineering in recent years [2, 11, 48, 149, 148, 323]. There are many books on fuzzy logic (e.g., [156, 309, 364]).

Descriptions commonly used in engineering systems such as "big or small" or "high or low" are inherently fuzzy. The fuzzy description is a conceptualization of numerical values that can be qualitative and meaningful to operators. A process variable can be translated to fuzzy concepts via a **membership function** $\mu_A(x)$, which maps every element x of the set X to the interval [0,1]. Mathematically, it can be defined as:

$$\mu_A(x) : X \rightarrow [0, 1] \tag{12.18}$$

where A is a **fuzzy subset** of X. Each value of the membership function is called a **membership degree**. A membership degree of 0 indicates no membership, while a membership degree of 1 indicates full membership in the set A. A set defined in classic logic (commonly referred to as a **crisp set**) is a special case of fuzzy set, in which only two membership degrees 0 and 1 are allowed. A fuzzy set A defined on X may be written as a collection of ordered pairs

$$A = \bigcup_{x \in X} (x, \mu(x)) \tag{12.19}$$

where each pair $(x, \mu(x))$ is called a **singleton**. If the set X is discrete, a membership function can be defined by a finite set:

$$A = \bigcup_k (x_k, \mu(x_k)). \tag{12.20}$$

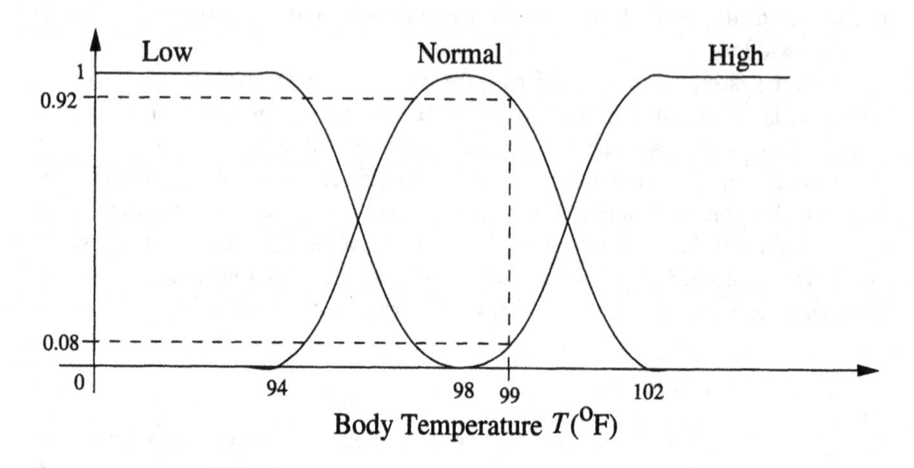

Fig. 12.10. Membership functions representing three fuzzy sets for the linguistic variable "body temperature"

Fuzzy logic allows the representation of variables and relationships in linguistic terms. A **linguistic variable** is a variable which takes fuzzy values and has a linguistic meaning. Linguistic variables can be based on *quantitative* variables in the process, for example, the linguistic variable *body temperature*, which can take the fuzzy values of "Low", "Normal", and "High". Each fuzzy value may be modeled as shown in Figure 12.10. For example, a body temperature of 99°F takes a fuzzy value of "Normal" and a membership degree of 0.92 via $\mu_{Normal}(T)$. It also takes a fuzzy value of "High" and a membership degree of 0.08 via $\mu_{High}(T)$. Linguistic variables can also be *qualitative*, for example, the linguistic variable *certainty* which can take fuzzy values such as "Highly Certain" or "Not Very Certain". The process of representing a linguistic variable into a set of fuzzy values is called **fuzzy quantization.**

The membership functions shown in Figure 12.10 are defined based on statistical data. The membership functions for "Low", "Normal", and "High" are represented by a Z-function (which is 1 minus a sigmoid function), bell-shaped function, and sigmoid function, respectively. Other types of membership functions including the trapezoidal, triangular, and single-valued functions can also be used [156].

Fuzzy logic systems address the imprecision of the input and output variables directly by defining them with fuzzy numbers and fuzzy sets that can be expressed in linguistic terms. Complex process behavior can be described in general terms without precisely defining the complex phenomena involved. However, it is difficult and time consuming to determine the correct set of rules and membership functions for a reasonably complex system. Fine tuning a fuzzy solution takes a large amount of time. To resolve some of the issues, neural networks can be used to learn the best membership function through training.

12.5.3 Fuzzy Expert Systems

It has been observed that the number of IF-THEN rules required to define an expert system tends to grow exponentially as the complexity of the system increases. As the number of IF-THEN rules becomes larger than 200, it is virtually impossible to write a meaningful rule that does not conflict with the existing rules [309]. This has motivated recent research in incorporating fuzzy logic into expert systems in an attempt to reduce the number of rules required. Several recent papers based on fuzzy expert systems are available [35, 152, 323].

A **fuzzy expert system** (also known as a **fuzzy system**) is defined in the same way as an ordinary expert system as described in Section 12.3, except that fuzzy logic is used. Fuzzy expert systems use fuzzy data, fuzzy rules, and a fuzzy inference mechanism which may include fuzzification and defuzzification. Input and output data can be fuzzy (as described in Section 12.5.2) or exact (crisp).

When the input data and output values are crisp, then the "fuzzification, fuzzy rule, and defuzzification" inference method is applied. **Fuzzification** is the process of finding the membership function $\mu_A(x)$ so that input data x belong to the fuzzy set A. **Rule evaluation** deals with single values of the membership function $\mu_A(x)$ and produces the output membership function. **Defuzzification** is the process of calculating single-output numerical values for a fuzzy output variable on the basis of the inferred membership function for this variable.

The fuzzy rules and the membership functions form the system knowledge base. **Fuzzy rules** deal with fuzzy values. The most popular rule is the IF-THEN rule. Fuzzy IF-THEN rules are conditional statements that describe the dependence of one or more linguistic variable on another. The number of different implication relations is over 40 [194, 195]. The simplest form is the Zadeh-Mamdani's fuzzy rule:

$$\text{IF} \, (\text{“}x \text{ is } A\text{”}), \; \text{THEN} \, (\text{“}y \text{ is } B\text{”}) \tag{12.21}$$

where x and y are fuzzy variables, A and B are fuzzy sets and ("x is A") and ("y is B") are fuzzy propositions. The fuzzy rules can be generated based on clustering of data into groups [156, 326, 327]. To illustrate this idea, Fisher's data (see Table 4.1 and Figure 4.2) is used to generate the fuzzy rules [156, 326, 327]:

1. As shown in Section 4.2, Fisher's data set contains 3 groups, with each group containing four measurements and 50 observations. The sepal length, sepal weight, petal length, and petal width, are fuzzified into 4, 3, 6, and 3 fuzzy regions, respectively. Each region is represented by a membership function (see Figure 12.11). Triangular functions are used for intermediate intervals with the center of a triangular membership function placed at the center of the interval and the other two vertexes placed at the middle points of the neighboring intervals. Trapezoidal membership functions are used for the end intervals.
2. The four measurement variables are fuzzified. For example, the first observation of Class 3 is ($SL = 5.1, SW = 3.5, PL = 1.4$, and $PW = 0.2$), the variables can be fuzzy-quantized using the membership functions (see Equation 12.11) and the results are shown in Table 12.4.

Table 12.4. Fuzzy-quantizing of an observation of Fisher's data [45, 82]

	Measurement	Fuzzy Value	Membership Degree
Sepal Length (SL)	5.1	M1	0.6
Sepal Width (SW)	3.5	M	1
Petal Length (PL)	1.4	S1	1
Petal Width (PW)	0.2	S	1
Iris Setosa (Class 3)	–	–	0.6

3. Each observation is represented by one fuzzy rule attached with a degree of confidence, which is calculated by multiplying the membership degrees of the condition elements by one another. For example, the first observation of Class 3 results in the following fuzzy rule:

$$\begin{aligned}&\text{IF ("}SL \text{ is } M1\text{") AND ("}SW \text{ is } M\text{") AND ("}PL \text{ is } S1\text{")}\\ &\quad\text{AND ("}PW \text{ is } S\text{"),} \quad \text{THEN ("Class 3")}\end{aligned} \tag{12.22}$$

with a degree of confidence of 0.6 ($0.6 \times 1 \times 1 \times 1 = 0.6$).

One weakness of the fuzzy approach shown above is the relatively large number of fuzzy rules generated. To reduce the number of rules required to describe a complex system, a genetic algorithm optimization can be used [156, 309]. Alternatively, a statistical-based processor can analyze the situation and give the contribution of each rule to the solution [309].

Fuzzy inference takes inputs, applies fuzzy rules, and produces outputs. Fuzzy inference is an inference method that uses fuzzy implication relations (e.g., the IF-THEN rule), fuzzy composition operators (e.g, MIN, MAX), and

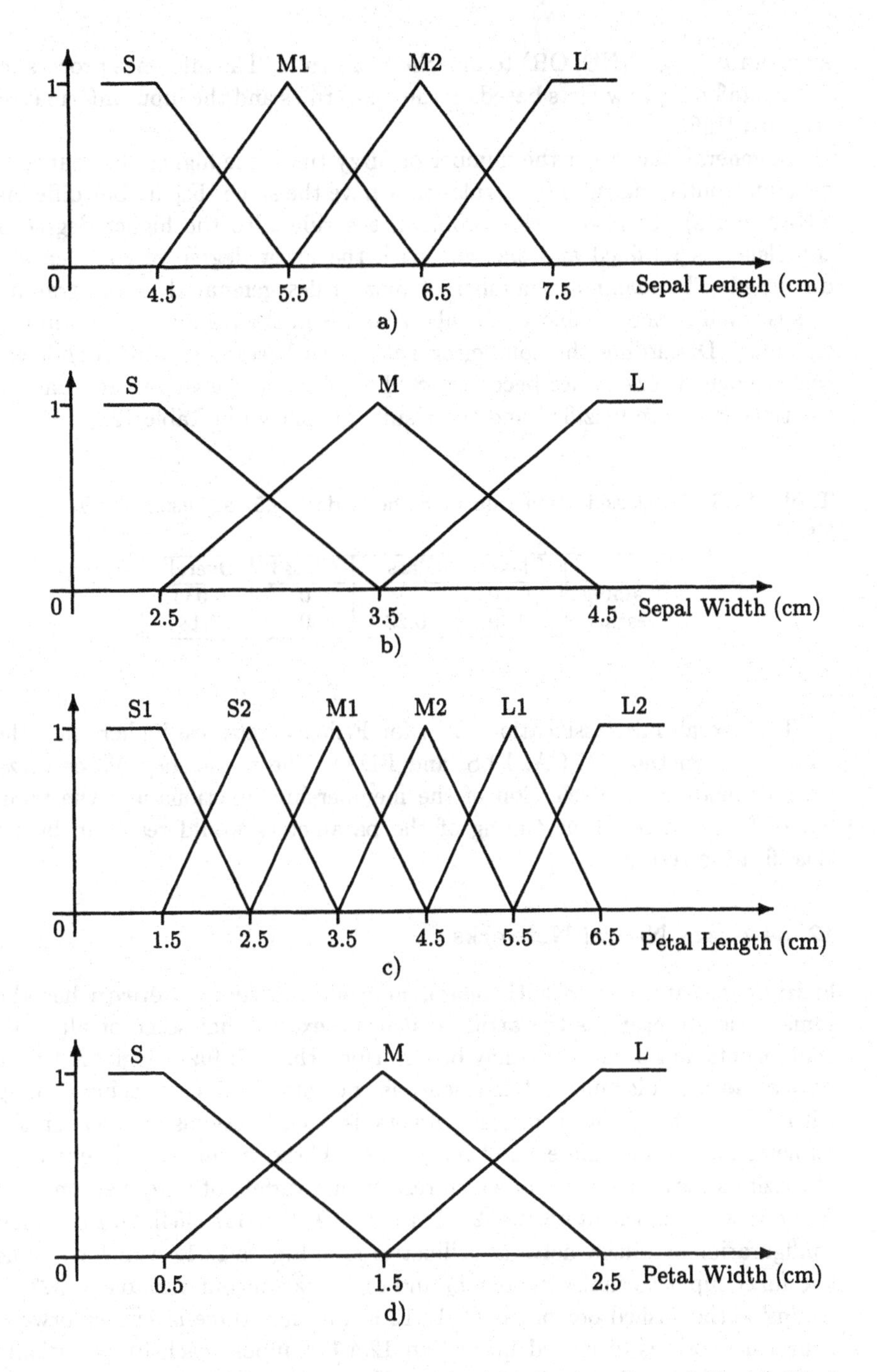

Fig. 12.11. Divisions of the input spaces into fuzzy regions for variables a) Sepal Length, b) Sepal Width, c) Petal Length, and d) Petal Width

an operator (e.g., AND, OR) to link the fuzzy rules. The inference process results in inferring new facts based on the fuzzy rules and the input information supplied [156].

In general, the larger the number of fuzzy rules, the higher the chance to generate conflicting rules (*i.e.*, rules that have the same IF part but different THEN parts). To resolve this problem, the rule with the higher degree of confidence is retained and the rule with the lower degree of confidence is discarded. The maximum number of fuzzy rules generated in the training sets is equal to the number of the observations in the training set (60 in this example). Discarding the conflicting rules with lower degree of confidence, the number of fuzzy rules becomes 58. The observations of Fisher's data in the testing set are fuzzified and the results are shown in Table 12.5.

Table 12.5. Misclassification rate of Fisher's data [45, 82] using the fuzzy set method

	Class 1	Class 2	Class3	Overall
Training	0.10	0	0	0.033
Testing	0.30	0.23	0	0.18

The overall misclassification rates for Fisher's data are higher than the data-driven methods (PCA, PLS, and FDA). The proficiency of the fuzzy rules depends on the selection of the membership functions and the number of fuzzy values. Fine tuning of the parameters would result in better classification results.

12.5.4 Fuzzy Neural Networks

Fuzzy logic can be used with neural networks. A **fuzzy neuron** has the same basic structure as the artificial neuron, except that some or all of its components and parameters may be described through fuzzy logic. A **fuzzy neural network** is built on fuzzy neurons or on standard neurons but dealing with fuzzy data. A fuzzy neural network is a connectionist model for the implementation and inference of fuzzy rules. There are many different ways to fuzzify an artificial neuron, which results in a variety of fuzzy neurons and fuzzy networks in the literature [2, 11, 15, 48, 59, 156, 327, 364]. One common configuration of a fuzzy network is illustrated in Figure 12.12, which contains two fuzzy input variables x_1 and x_2 and one fuzzy output variable y [327].

Inside the dashed box of Figure 12.12 is a normal three-layer feedforward neural network as discussed in Section 12.4.1. Suppose each fuzzy variable takes three fuzzy values: "High", "Normal", and "Low", then the membership degrees of the fuzzy values corresponding to the variables x_1 and x_2 are the input layer neurons, and the membership degrees of the fuzzy values corresponding to the variable y are the output layer neurons. The configuration of

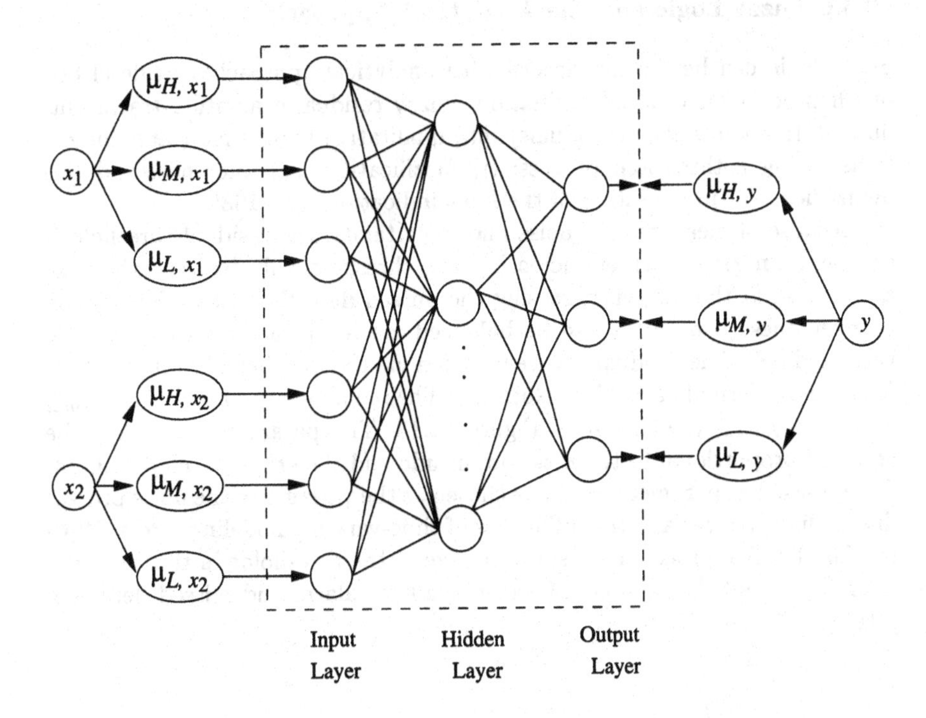

Fig. 12.12. A fuzzy three-layer feedforward neural network [327]

this fuzzy neural network increases the size of the network dramatically and increases the computational load. An alternative approach is to split each input layer neuron into two; one for describing the fuzzy value and the other for representing the membership value [327].

12.5.5 Fuzzy Signed Directed Graph

As shown in Section 12.2.1, the traditional signed directed graph (SDG) can take one of three values (−, +, 0) for each node or branch. This can give ambiguous solutions in complicated fault diagnosis problems. Fuzzy logic can be combined with the signed directed graph [128, 302, 304, 327]. A fuzzy set can be defined for a finite set of nodes and the relationship between two nodes can be represented by a fuzzy relationship [128, 327].

Each node in the fuzzy SDG takes a fuzzy variable with its fuzzy value determined by a membership function. Unlike the arcs in a traditional SDG that only have + or − sign, the arcs in a fuzzy SDG also have a weight representing the strength of the connection. The weight can be calculated based on the value range and the sensitivity of the connecting nodes. A more sophisticated method to represent the effect between two nodes is to use a single layer perceptron [327].

12.5.6 Fuzzy Logic and the Analytical Approach

Fuzzy logic can be used in accord with analytical approaches as described in Chapter 11 for residual evaluation. Fuzzy residual evaluation transforms quantitative knowledge (residuals) into qualitative knowledge (fault indications) using a three-step process: (i) fuzzification, (ii) inference, and (iii) defuzzification (presentation of the fault indication) [90, 168].

Because of measurement noise and uncertainty, the residual threshold is greater than zero. Further increasing the threshold will decrease the false alarm rate, at the cost of increasing the missed detection rate. The tradeoff between these two effects can be balanced via fuzzification on the residual threshold [90]. The residual can be fuzzified via the membership functions for fuzzy sets "Normal" and "Not Normal". The membership functions μ_{Normal} and $\mu_{Not\,Normal}$ are shown in Figure 12.13. The parameter a_0 has to be assigned proportional to the noise amplitude and the effects of modeling uncertainties. The parameter δ can be chosen as the variance of the noise process due to disturbances and the influences of time-varying modeling errors. With the fuzzification procedure, a small change of the thresholds in the fuzzy domain $[a_0, a_0 + \delta]$ has a small effect on the false alarm and missed detection rate.

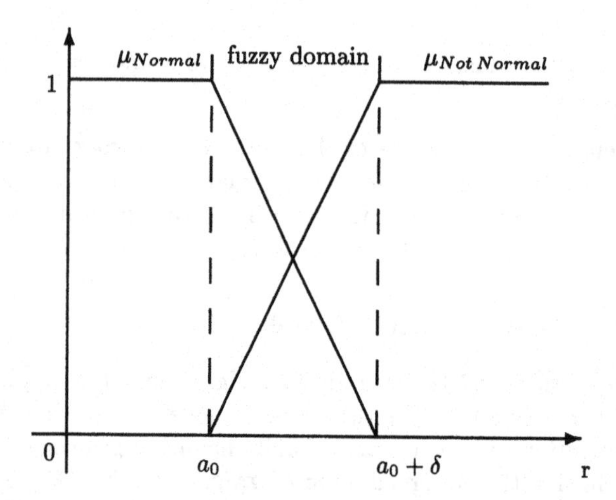

Fig. 12.13. Composition of the fuzzy set "Normal" and "Not Normal"

Similarly to the analytical approaches, the faults of interest are first defined. In the fuzzification step, each residual r_i is fuzzified into the fuzzy sets "Normal" and "Not Normal". Mathematically, it is described by

$$r_i \rightarrow r_{i0} \circ r_{i1} \tag{12.23}$$

where $\circ$ is the fuzzy composition operator, r_{i0} describes the fuzzy set "Normal" of the ith residual, and r_{i1} describes the fuzzy set "Not Normal" of the ith residual.

The inference phase is to determine the indication signals for the faults from the given rule base. The inference mechanism uses a series of IF-THEN rules to map the residual (defined by their fuzzy sets) onto the faults, for example

$$\begin{aligned}&\text{IF } (\mathit{effect} = r_{i0}) \text{ AND } (\mathit{effect} = r_{j1}) \cdots \\ &\quad \text{THEN } (\mathit{cause} = f_k)\end{aligned} \tag{12.24}$$

where f_k represents the kth fault of the system.

Two faults are distinguishable if they have at least one different definition in the premise of the rule. If all premises of two fault descriptions f_k and f_l have the same fuzzy values, a distinction is not possible. To resolve such an inconsistency, one or more fuzzy sets have to be subdivided into at least two fuzzy sets [168]. For example, the fuzzy set "Fault" can be subdivided into "Strongly deviating" and "Slightly deviating" such that the residuals of these two fuzzy sets are different for faults f_k and f_l. From the definition of the fuzzy sets and the faults defined, the number of rules is determined.

12.5.7 Neural Networks and the Analytical Approach

The neural network can replace the analytical model (e.g., observer, parity relations) describing the process under normal operating conditions. The residual is taken as the difference between the actual output and the estimated output from the neural network. It is useful to apply this approach when no exact or complete analytical or knowledge-based model can be produced, but a large amount of measurement data is available [90].

For residual evaluation, a residual database and a corresponding fault signature database can be used to train the neural networks. The residual database can be generated from another neural and/or other analytical methods such as parity relations or an observer. One difficulty of applying this approach is the lack of analytical information on the performance, stability, and robustness of the neural network; on-line approximators and learning algorithms have been proposed to resolve this problem [261].

12.5.8 Data-driven, Analytical, and Knowledge-based Approaches

The previous sections describe some efforts to combine ideas from more than one approach to process monitoring. Many of the knowledge-based approaches (e.g., the SDG, expert systems) are well suited for diagnosing faults because of their ability to incorporate reasoning. On the other hand, data-driven approaches are based on rigorous statistical development that is able to capture the most important information onto a lower-dimensional space. As such, data-driven techniques are well suited for detecting faults for large-scale industrial applications. When a detailed first-principles and other mathematical model is available, the analytical approach can incorporate physical understanding into the process monitoring scheme. For these reasons, a combined data-driven, analytical, and knowledge-based process monitoring scheme will play an important role in industrial systems for detecting, isolating, and diagnosing faults in upcoming years.

12.6 Homework Problems

1. Compare and contrast the SDG and the symptom tree model. Which method is expected to perform better for fault diagnosis? Justify your answers.
2. Read an article on the use of the SDG for diagnosing multiple faults (e.g., [196, 319]) and write a report describing in detail how the technique is implemented and applied. What are the strengths and weaknesses of the technique?
3. Which of the following expert systems (deep knowledge, shallow knowledge, or a combination of shallow knowledge and deep knowledge) is more popular in industrial applications? Why? Justify your answers, and support them with at least ten journal articles from a literature search.
4. Read the article [34] on the use of meta-knowledge architecture to accommodate both shallow and deep reasoning mechanisms in an expert system. Write a report describing in detail how the method is implemented and applied. How does the meta-knowledge architecture place in the context with methods described in Section 12.3.6?
5. Read one of the following articles [103, 280, 281, 288] on the use of discrete-event models for fault diagnosis. Write a report describing in detail how the method is implemented and applied. How do the discrete-event models compare with the methods described in Section 12.3?
6. Investigate the effects of (i) the number of hidden layer neurons, (ii) the learning rate, (iii) the coefficient of the momentum term, (iv) different types of sigmoid functions, and (e) the bias terms on the proficiency of classification using Fisher's data set. Comment on your findings.
7. Compare and contrast [156, 224, 285, 309]: (i) feedforward neural networks, (ii) adaptive neural networks, (iii) radial basis function neural networks, (iv) time-delay neural networks, (v) recurrent neural networks, and (vi) autoassociative neural networks. Find an industrial application of each type of neural network in the literature, and write a few sentences summarizing the results of each application.
8. Compare and contrast the following algorithms [65, 156, 309]: (i) variable learning rate, (ii) Rprop, (iii) scaled conjugate gradient, (iv) Fletcher-Powell conjugate gradient, (v) Polak-Ribiere conjugate gradient, (vi)

Powell-Beale conjugate gradient, (vii) one-step-secant, (viii) BFGS quasi-newton, and (ix) Levenberg-Marquardt. Run Fisher's data set on a three-layer feedforward neural network using all of the training algorithms and compare the computation speeds. Discuss your results.

9. Derive the generalized delta rule (12.8) and (12.14) based on the method of gradient descent. Hint: Write the error for a single observation $\mathbf{x}$ as

$$E = \frac{1}{2}\sum_{j=1}^{m_y}(\hat{y}_j - y_j)^2 \tag{12.25}$$

Gradient descent sets the changes in weights by

$$\Delta w_{ij}^o = -\eta \frac{dE}{dw_{ij}^o} \tag{12.26}$$

$$\Delta w_{ki}^h = -\eta \frac{dE}{dw_{ki}^h} \tag{12.27}$$

The expressions for dE/dw_{ij}^o and dE/dw_{ki}^h can be derived using the chain rule.

10. Describe in detail the idea of learning vector quantization when used with the SOM [166, 167]. Apply the technique using Fisher's data set and compare with results shown in this book. Comment on your findings.
11. Re-run Fisher's data set using a self-organizing map with the following changes: (i) use a 25 × 25 rectangular map, (ii) use 150 observations in the training set and 30 observations in the test set, (iii) use a 15 × 15 hexagonal map instead of rectangular map, (iv) set the number of iterations to 100. Comment on your findings.
12. Write a summary report that reviews a book on the use of fuzzy logic in engineering applications (e.g., [156, 309, 364]). What are the strengths and weaknesses of fuzzy logic? Find three industrial applications which use fuzzy logic in the literature, and discuss the application results.
13. Re-run Fisher's data set using the fuzzy rules with the following changes: (i) use different membership functions for each fuzzy variable, (ii) use 150 observations for training and 30 observations for testing, (iii) use different fuzzy values for each fuzzy variable, and (iv) use the fuzzy rules as suggested on page 219 of [156]. Comment on your findings.
14. Compare and contrast different types of fuzzy neural networks [2, 11, 48, 59, 156, 327, 364]. Pick two fuzzy neural networks and apply them to Fisher's data set. Compare your results with the results shown in this book. Comment on your findings.
15. Compare and contrast different types of fuzzy SDGs [128, 302, 304, 327]. What are the advantages and disadvantages of the fuzzy SDG compared to the traditional SDG?

16. Read an article on the use of SDG with PCA (e.g., [320]) and write a report describing in detail how the technique is implemented and applied. What are the advantages and disadvantages of using this technique compared to using PCA and SDG alone? Justify your answers.
17. Read an article on the use of parity equations with PCA (e.g., [100]) and write a report describing in detail how the technique is implemented and applied. What are the advantages and disadvantages of using this technique compared to using PCA and parity equation alone? Justify your answers.

References

1. H. K. Adjallah, G. Schreier, and J. Ragot. Nonlinear observers–application to state estimation of a chemical process. In *Proc. of the 4th IEEE Conf. on Control Applications*, pages 113–118, Piscataway, New Jersey, 1995. IEEE Press.
2. R. K. Aggarwal, Q. Y. Xuan, A. T. Johns, F. Li, and A. Bennett. Novel approach to fault diagnosis in multicircuit transmission lines using Fuzzy ARTmap neural networks. *IEEE Trans. on Neural Networks*, 10:1214–1221, 1999.
3. H. Akaike. Stochastic theory of minimal realization. *IEEE Trans. on Automatic Control*, 19:667–674, 1974.
4. H. Akaike. Markovian representation of stochastic processes by canonical variables. *SIAM J. of Control*, 13:162–173, 1975.
5. H. Akaike. Canonical correlation analysis of time series and the use of an information criterion. In R. K. Mehra and D. G. Jainiotis, editors, *System Identification: Advances and Case Studies*, pages 27–96. Academic Press, New York, 1976.
6. J. S. Albuquerque and L. T. Biegler. Data reconciliation and gross-error detection for dynamic systems. *AIChE J.*, 42:2841–2856, 1996.
7. E. Alhoniemi, J. Hollmen, O. Simula, and J. Vesanto. Process monitoring and modeling using the self-organizing map. *Integrated Computer-aided Engineering*, 6:3–14, 1999.
8. D. J. Allen. Digraphs and fault trees. *Ind. Eng. Chem. Fund.*, 23:175–180, 1984.
9. B. K. Alsberg, R. Goodacre, J. J. Rowland, and D. B. Kell. Classification of pyrolysis mass spectra by fuzzy multivariate rule induction-comparison with regression, K-nearest neighbour, neural and decision-tree methods. *Analytica Chimica Acta*, 348:389–407, 1997.
10. F. B. Alt. Multivariate quality control. In S. Kotz and N. L. Johnson, editors, *Encyclopedia of Statistical Sciences*. John Wiley & Sons, New York, 1985.
11. S. Altug, M. Chow, and H. J. Trussell. Fuzzy inference systems implemented on neural architectures for motor fault detection and diagnosis. *IEEE Trans. on Industrial Electronics*, 46:1069–1079, 1999.
12. B. D. O. Anderson and J. B. Moore. *Optimal Filtering*. Prentice-Hall, Englewood Cliffs, New Jersey, 1979.
13. T. W. Anderson. *Introduction to Multivariate Statistical Analysis*. John Wiley & Sons, New York, 1958.
14. C. Angeli. Online expert system for fault diagnosis in hydraulic systems. *Expert Systems*, 16:115–120, 1999.
15. A. Aoyama, F. J. Doyle III, and V. Venkatasubramanian. Fuzzy neural network systems techniques and their applications to nonlinear chemical process

control systems. In C. T. Leondes, editor, *Fuzzy Theory Systems: Techniques and Applications*, pages 485–526. Academic Press, New York, 1999.

16. H. B. Aradhye, B. R. Bakshi, and R. Strauss. Process monitoring by PCA, dynamic PCA, and multiscale PCA – Theoretical analysis and disturbance detection in the Tennessee Eastman process. In *AIChE Annual Meeting*, 1999. Paper 224g.
17. T. Asakura, T. Kobayashi, and S. Hayashi. A study of fault diagnosis system using neural networks. In *Proc. of the 29th ISCIE Int. Symp. on Stochastic Systems Theory and Its Applications*, pages 19–24, Tokyo, Japan, 1998.
18. S. A. Ashton, D. N. Shields, and S. Daley. Fault detection in pipelines using nonlinear observers. In *UKACC Int. Conf. on CONTROL*, volume 1, pages 135–140, Piscataway, New Jersey, 1998. IEEE Press.
19. K. J. Åström and B. Wittenmark. *Computer Controlled Systems: Theory and Design.* Prentice-Hall, Inc., Englewood Cliffs, New Jersey, 1984.
20. C. Bakiotis, J. Raymond, and A. Rault. Parameter and discriminant analysis for jet engine mechanical state diagnosis. In *Proc. of the IEEE Conf. on Decision and Control*, pages 151–155, Piscataway, New Jersey, 1979. IEEE Press.
21. B. R. Bakshi and G. Stephanopoulos. Representation of process trends, IV Induction of real-time patterns from operating data for diagnosis and supervisory control. *Computers & Chemical Engineering*, 18:303–332, 1994.
22. D. M. Bates and D. G. Watts. *Nonlinear Regression Analysis and Its Applications.* John Wiley & Sons, New York, 1988.
23. F. A. Batzias and G. Kopsidas. An expert system design for fault diagnosis in electrochemical operations - A quantitative approach. In M. G. Singh, K. S. Hindi, G. Schmidt, and S. Tzafestas, editors, *Fault Detection and Reliability; Knowledge Based and Other Approaches*, pages 93–102. Pergamon Press, Oxford, 1987.
24. D. R. Baughman. *Neural Networks in Bioprocessing and Chemical Engineering.* Academic Press, New York, 1995.
25. J. V. Beck and K. J. Arnold. *Parameter Estimation in Engineering and Science.* Wiley, New York, 1977.
26. H. Benkhedda and R. J. Patton. B-spline network integrated qualitative and quantitative fault detection. In *Proc. of the 13th IFAC World Congress*, volume N, pages 163–168, Piscataway, New Jersey, 1996. IEEE Press.
27. G. Betta and A. Pietrosanto. Instrument fault detection and isolation: State of the art and new research trends. In *Proc. of the IEEE Instrumentation and Measurement Technology*, volume 1, pages 483–489, Piscataway, New Jersey, 1998.
28. R. Bharath and J. Drosen. *Neural Network Computing.* McGraw-Hill, New York, 1994.
29. C. Bishop. *Neural Networks for Pattern Recognition.* Clarendon Press, New York, 1995.
30. G. E. P. Box, W. G. Hunter, and J. S. Hunter. *Statistics for Experimenters - An Introduction to Design, Data Analysis, and Model Building.* Wiley, New York, 1978.
31. W. L. Brogan. *Modern Control Theory.* Prentice-Hall, New Jersey, 1991.
32. B. G. Buchanan and E. H. Shortliffe. *Rule-based Expert Systems: The MYCIN Experiments of the Stanford Heuristic Programming Project.* Addison-Wesley, Reading, Massachusetts, 1984.
33. P. Burrell and D. Inman. Expert system for the analysis of faults in an electricity supply network: Problems and achievements. *Computers in Industry*, 37:113–123, 1998.

34. J. Caviedes, J. B. A. Brodersen, P. Osborne, A. Ross, J. D. Schaffer, and G. Bengtson. A meta-knowledge architecture for planning and explanation. In S. G. Tzafestas, editor, *Knowledge-based System Diagnosis, Supervision, and Control*, pages 11–26. Plenum Press, New York, 1989.
35. C. S. Chang, J. M. Chen, A. C. Liew, D. Srinivasan, and F. S. Wen. Fuzzy expert system for fault diagnosis in power systems. *Int. J. of Engineering Intelligent Systems for Electrical Engineering & Communications*, 5:75–81, 1997.
36. I. Chang, C. Yu, and C. Liou. Model-based approach for fault diagnosis. 1. principles of deep model algorithm. *Ind. Eng. Chem. Res.*, 33:1542–1555, 1994.
37. E. Charniak and D. McDermott. *Introduction to Artificial Intelligence*. Addison-Wesley, New York, 1985.
38. C.-T. Chen. *Linear System Theory and Design*. Harcourt Brace College Publishers, Orlando, Florida, 1984.
39. G. Chen and T. J. McAvoy. Predictive on-line monitoring of continuous processes. *J. of Process Control*, 8:409–420, 1997.
40. G. Chen, T. J. McAvoy, and M. J. Piovoso. Multivariate statistical controller for on-line quality improvement. *J. of Process Control*, 8(2):139–149, 1998.
41. J. Chen and R. J. Patton. *Robust Model-based Fault Diagnosis for Dynamic Systems*. Kluwer Academic Publishers, Boston, Massachusetts, 1999.
42. J. Chen, R. J. Patton, and H. Y. Zhang. Design of unknown input observers and robust fault-detection filters. *Int. J. of Control*, 63:85–105, 1996.
43. J. Chen and J. A. Romagnoli. Strategy for simultaneous dynamic data reconciliation and outlier detection. *Computers & Chemical Engineering*, 22:559–562, 1998.
44. S. Chen, C. F. N. Cowan, and P. M. Grant. Orthogonal least squares learning algorithm for radial basis function networks. *IEEE Trans. on Neural Networks*, 2(2):302–309, 1991.
45. Y. Q. Cheng, Y. M. Zhuang, and J. Y. Yang. Optimal Fisher discriminant analysis using the rank decomposition. *Pattern Recognition*, 25:101–111, 1992.
46. L. H. Chiang, E. L. Russell, and R. D. Braatz. Fault diagnosis in chemical processes using Fisher discriminant analysis, discriminant partial least squares, and principal component analysis. *Chemometrics and Intelligent Laboratory Systems*, 50:243–252, 2000.
47. E. Y. Chow and A. S. Willsky. Analytical redundancy and the design of robust failure detection systems. *IEEE Trans. on Automatic Control*, 29:603–614, 1984.
48. M. Chow. *Methodologies of Using Neural Network and Fuzzy Logic Technologies for Motor Incipient Fault Detection*. World Scientific, Singapore, 1997.
49. M. Chow, R. N. Sharpe, and J. C. Hung. On the application and design consideration of artificial neural network fault detectors. *IEEE Trans. on Industrial Electronics*, 40:181–198, 1993.
50. M. Y. Chow, B. Li, and G. Goddu. Intelligent motor fault detection. In L. C. Jain, R. P. Johnson, Y. Takefuji, and L. A. Zadeh, editors, *Knowledge-based Intelligent Techniques in Industry*, pages 191–223. CRC Press, New York, 1999.
51. A. Cinar, E. Tatara, J. DeCicco, R. Raj, N. Aggarwal, M. Chesebro, J. Evans, M. Shah-Khan, and A. Zloza. Automated patient monitoring and diagnosis assistance by integrating statistical and artificial intelligence tools. *Proc. of the Annual Conf. on Engineering in Medicine & Biology*, 2:700–710, 1999.
52. W. J. Clancey and E. H. Shortliffe, editors. *Readings in Medical Artificial Intelligence: The First Decease*. Addison-Wesley, Reading, Massachusetts, 1984.

53. R. N. Clark. A simplified instrument detection scheme. *IEEE Trans. on Aerospace and Electronic Systems*, 14:558–563, 1978.
54. R. N. Clark, D. C. Fosth, and V. M. Walton. Detecting instrument malfunctions in control systems. *IEEE Trans. on Aerospace and Electronic Systems*, 11:465–473, 1975.
55. T. M. Cover and P. E. Hart. Nearest neighbor pattern classification. *IEEE Trans. on Information Theory*, 30:21–27, 1967.
56. R. B. Crosier. Multivariate generalizations of cumulative sum quality-control schemes. *Technometrics*, 30:291–303, 1988.
57. C. M. Crowe. Data reconciliation - Progress and challenges. *J. of Process Control*, 6:89–98, 1996.
58. D. T. Dalle-Molle and D. M. Himmelblau. Fault detection in an evaporator via parameter estimation in real time. In M. G. Singh, K. S. Hindi, G. Schmidt, and S. Tzafestas, editors, *Fault Detection and Reliability; Knowledge Based and Other Approaches*, pages 131–138. Pergamon Press, Oxford, 1987.
59. T. J. Dasey and E. Micheli-Tzanakou. Fuzzy neural networks. In E. Micheli-Tzanakou, editor, *Supervised and Unsupervised Pattern Recognition*, pages 135–162. CRC Press, New York, 1999.
60. J. David and J. Krivien. Three artificial intelligence issues in fault diagnosis: declarative programming, expert systems, and model-based reasoning. In M. G. Singh, K. S. Hindi, G. Schmidt, and S. Tzafestas, editors, *Fault Detection and Reliability; Knowledge Based and Other Approaches*, pages 19–28. Pergamon Press, Oxford, 1987.
61. J. E. Dayhoff. *Neural Network Architectures: An Introduction*. Van Nostrand Reinhold, New York, 1990.
62. S. de Jong. An alternative approach to partial least squares regression. *Chemometrics and Intelligent Laboratory Systems*, 18:251–263, 1993.
63. J. C. Deckert, M. N. Desai, J. J. Deyst, and A. S. Willsky. F-8 DFBW sensor failure identification using analytic redundancy. *IEEE Trans. on Automatic Control*, 22:795–803, 1977.
64. M. Defernez and E. K. Kemsley. The use and misuse of chemometrics for treating classification problems. *Trends in Analytical Chemistry*, 16:216–221, 1997.
65. H. Demuth and M. Beale. *Neural Network Toolbox: For Use with MATLAB*. The MathWorks, Inc., Natick, Massachusetts, 1998.
66. L. Desborough and T. Harris. Performance assessment measures for univariate feedback control. *Can. J. of Chem. Eng.*, 70:262–268, 1992.
67. W. R. Dillon and M. Goldstein. *Multivariate Analysis, Methods and Applications*. John Wiley & Sons, New York, 1984.
68. X. Ding and L. Guo. Observer-based fault detection optimized in the frequency domain. In *Proc. of the 13th IFAC World Congress*, volume N, pages 157–162, Piscataway, New Jersey, 1996. IEEE Press.
69. X. Ding and L. Guo. Observer-based optimal fault detector. In *Proc. of the 13th IFAC World Congress*, volume N, pages 187–192, Piscataway, New Jersey, 1996. IEEE Press.
70. N. Doganaksoy, F. W. Faltin, and W. T. Tucker. Identification of out of control quality characteristics in a multivariate manufacturing environment. *Commun. Stat.- Theory Methods*, 20:2775, 1991.
71. D. Dong and T. J. McAvoy. Nonlinear principal component analysis: based on principal curves and neural networks. *Computers & Chemical Engineering*, 20:65–78, 1996.
72. J. J. Downs and E. F. Vogel. A plant-wide industrial-process control problem. *Computers & Chemical Engineering*, 17:245–255, 1993.

73. R. J. Doyle, L. Charest, N. Rouquette, J. Wyatt, and C. Robertson. Causal modeling and event-driven simulation for monitoring of continuous systems. *Computers in Aerospace*, 9:395–405, 1993.
74. R. O. Duda and P. E. Hart. *Pattern Classification and Scene Analysis*. John Wiley & Sons, New York, 1973.
75. R. Dunia and S. J. Qin. Joint diagnosis of process and sensor faults using principal component analysis. *Control Engineering Practice*, 6:457–469, 1998.
76. R. Dunia, S. J. Qin, and T. F. Edgar. Multivariable process monitoring using nonlinear approaches. In *Proc. of the American Control Conf.*, pages 756–760, Piscataway, New Jersey, 1995. IEEE Press.
77. R. Dunia, S. J. Qin, T. F. Edgar, and T. J. McAvoy. Identification of faulty sensors using principal component analysis. *AIChE J.*, 42:2797–2812, 1996.
78. R. Engelmore and T. Morgan. *Blackboard Systems*. Addison-Wesley, Menlo Park, CA, 1988.
79. D. F. Enns. Model reduction with balanced realizations: An error bound and a frequency weighted generalization. In *Proc. of the IEEE Conf. on Decision and Control*, pages 127–132, Piscataway, New Jersey, 1984. IEEE Press.
80. P. Fasolo and D. E. Seborg. SQC approach to monitoring and fault detection in HVAC control systems. In *Proc. of the American Control Conf.*, volume 3, pages 3055–3059, Piscataway, New Jersey, 1994. IEEE Press.
81. F. E. Finch, O. O. Oyeleye, and M. A. Kramer. A robust event-oriented methodology for diagnosis of dynamic process systems. *Computers & Chemical Engineering*, 14:1379–1396, 1990.
82. R. A. Fisher. The use of multiple measurements in taxonomic problems. *Ann. Eugenics*, 7:179–188, 1936.
83. I. E. Frank. A nonlinear PLS model. *Chemometrics and Intelligent Laboratory Systems*, 8:109–119, 1990.
84. I. E. Frank, J. Feikema, N. Constantine, and B. R. Kowalski. Prediction of product quality from spectral data using the partial least-squares method. *J. of Chemical Information and Computer Sciences*, 24:20–24, 1983.
85. P. M. Frank. Fault diagnosis in dynamic systems via state estimation - A survey. In S. Tzafestas, M. Singh, and G. Schmidt, editors, *System Fault Diagnostics, Reliability and Related Knowledge-Based Approaches*, volume 1, pages 35–98. Reidel, Dordrecht, Holland, 1987.
86. P. M. Frank. Fault diagnosis in dynamic systems using analytical and knowledge-based redundancy—a survey and some new results. *Automatica*, 26:459–474, 1990.
87. P. M. Frank. Robust model-based fault detection in dynamic systems. In P. S. Dhurjati and G. Stephanopoulos, editors, *On-line Fault Detection and Supervision in the Chemical Process Industries*, pages 1–13. Pergamon Press, Oxford, 1993. IFAC Symposia Series, Number 1.
88. P. M. Frank. Enhancement of robustness in observer-based fault detection. *Int. J. of Control*, 59:955–981, 1994.
89. P. M. Frank. On-line fault detection in uncertain nonlinear systems using diagnostic observer: A survey. *Int. J. of Systems Science*, 25:2129–2154, 1994.
90. P. M. Frank. Analytical and qualitative model-based fault diagnosis - A survey and some new results. *European J. of Control*, 2:6–28, 1996.
91. P. M. Frank and X. Ding. Frequency domain approach to optimally robust residual generation and evaluation for model-based fault diagnosis. *Automatica*, 30:789–804, 1994.
92. P. M. Frank and N. Kiupel. Residual evaluation for fault diagnosis using adaptive fuzzy thresholds and fuzzy inference. In *Proc. of the 13th IFAC*

World Congress, volume N, pages 115–120, Piscataway, New Jersey, 1996. IEEE Press.

93. N. B. Gallagher, B. M. Wise, and C. W. Stewart. Application of multi-way principal components analysis to nuclear waste storage tank monitoring. *Computers & Chemical Engineering*, 20:S739–S744, 1996.
94. E. A. Garcia and P. M. Frank. On the relationship between observer and parameter identification based approaches to fault detection. In *Proc. of the 13th IFAC World Congress*, volume N, pages 25–29, Piscataway, New Jersey, 1996. IEEE Press.
95. H. E. Garcia and R. B. Vilim. Combining physical modeling, neural processing, and likelihood testing for online process monitoring. In *Proc. of the IEEE International Conference on Systems, Man and Cybernetics*, volume 1, pages 806–810, Piscataway, New Jersey, 1998. IEEE Press.
96. G. Geiger. Monitoring of an electrical driven pump using continuous-time parameter estimation methods. In *Prof. of the 6th IFAC Symp. on Identification and Parameter Estimation*, pages 603–608, Oxford, 1982. Pergamon Press.
97. S. Geisser. Discrimination, allocatory and separatory, linear aspects. In *Classification and Clustering*. Academic Press, New York, 1977.
98. P. Geladi and B. R. Kowalski. Partial least-squares regression: A tutorial. *Analytica Chimica Acta*, 185:1–17, 1986.
99. C. Georgakis, B. Steadman, and V. Liotta. Decentralized PCA charts for performance assessment of plant-wide control structures. In *Proc. of the 13th IFAC World Congress*, pages 97–101, Piscataway, New Jersey, 1996. IEEE Press.
100. J. Gertler, W. Li, Y. Huang, and T. McAvoy. Isolation enhanced principal component analysis. *AIChE J.*, 45(2):323–334, 1999.
101. J. J. Gertler. *Fault Detection and Diagnosis in Engineering Systems*. Marcel Dekker, Inc., New York, 1998.
102. M. Gevers and V. Wertz. On the problem of structure selection for the identification of stationary stochastic processes. In *Sixth IFAC Symp. on Identification and System Parameter Estimation*, pages 387–392, Piscataway, New Jersey, 1982. IEEE Press.
103. D. Godbole and R. Sengupta. Tools for the design of fault management systems. In *IEEE Conf. on Intelligent Transportation Systems*, pages 159–164, Piscataway, New Jersey, 1997. IEEE Press.
104. G. H. Golub and C. F. van Loan. *Matrix Computations*. Johns Hopkins University Press, Baltimore, Maryland, 1983.
105. E. Gomez, H. Unbehauen, P. Kortmann, and S. Peters. Fault detection and diagnosis with the help of fuzzy-logic and with application to a laboratory turbogenerator. In *Proc. of the 13th IFAC World Congress*, volume N, pages 235–240, Piscataway, New Jersey, 1996. IEEE Press.
106. J. B. Gomm. On-line learning for fault classification using an adaptive neuro-fuzzy network. In *Proc. of the 13th IFAC World Congress*, volume N, pages 175–180, Piscataway, New Jersey, 1996. IEEE Press.
107. A. J. Gonzalez and D. D. Dankel. *The Engineering of Knowledge-Based Systems*. Prentice-Hall, Englewood Cliffs, New Jersey, 1993.
108. F. Hamelin and D. Sauter. Robust residual generation for F.D.I. in uncertain dynamic systems. In *Proc. of the 13th IFAC World Congress*, volume N, pages 181–186, Piscataway, New Jersey, 1996. IEEE Press.
109. D. Hanselman and B. Littlefield. *The Student Edition of MATLAB: Version 5, User's Guide*. Prentice Hall, New Jersey, 1997.
110. D. Hanselman and B. Littlefield. *Mastering MATLAB 5, A Comprehensive Tutorial and Reference*. Prentice Hall, New Jersey, 1998.

111. T. J. Harris. Assessment of control loop performance. *Can. J. of Chem. Eng.*, 67:856–861, 1989.
112. T. J. Harris and W. H. Ross. Statistical process control procedures for correlated observations. *Can. J. of Chem. Eng.*, 69:48–57, 1991.
113. I. Hashimoto, M. Kano, and K. Nagao. A new method for process monitoring using principal component analysis. In *AIChE Annual Meeting*, 1999. Paper 224a.
114. D. M. Hawkins. Multivariate quality control based on regression-adjusted variables. *Technometrics*, 33:61–67, 1991.
115. F. Hayes-Roth, D. A. Waterman, and D. B. Lenat. *Building Expert Systems.* Addison-Wesley, Reading, Massachusetts, 1983.
116. J. D. Healy. A note of multivariate CUSUM procedures. *Technometrics*, 29:409–412, 1987.
117. D. M. Himes, R. H. Storer, and C. Georgakis. Determination of the number of principal components for disturbance detection and isolation. In *Proc. of the American Control Conf.*, pages 1279–1283, Piscataway, New Jersey, 1994. IEEE Press.
118. D. M. Himmelblau. Use of artificial neural networks to monitor faults and for troubleshooting in the process industries. In *IFAC Symp. On-line Fault Detection and Supervision in the Chemical Process Industry*, Oxford, 1992. Pergamon Press.
119. D. M. Himmelblau, R. W. Braker, and W. Suewatanakul. Fault classification with the aid of artificial neural networks. In *IFAC/IMAC Symp. Safeprocess*, pages 369–373, Oxford, 1991. Pergamon Press.
120. W. W. Hines and D. C. Montgomery. *Probability and Statistics in Engineering and Management Science.* John Wiley & Sons, New York, 3rd edition, 1990.
121. D. Hodouin and S. Makni. Real-time reconciliation of mineral processing plant data using bilinear material balance equations coupled to empirical dynamic models. *Int. J. of Mineral Processing*, 48:245–264, 1996.
122. T. Holcomb and M. Morari. PLS/neural networks. *Computers & Chemical Engineering*, 16:393–411, 1992.
123. Z. Q. Hong and J. Y. Yang. Optimal discriminant plane for a small number of samples and design method of classifier on the plane. *Pattern Recognition*, 24:317–324, 1991.
124. J. C. Hoskins, K. M. Kaliyur, and D. M. Himmelblau. Fault diagnosis in complex chemical plants using artificial neural networks. *AIChE J.*, 37:137–141, 1991.
125. H. Hotelling. Relations between two sets of variables. *Biometrika*, 26:321–377, 1936.
126. H. Hotelling. Multivariate quality control. In Eisenhart, Hastay, and Wallis, editors, *Techniques of Statistical Analysis.* McGraw Hill, New York, 1947.
127. B. N. Huallpa, E. Nobrega, and F. J. V. Zuben. Fault detection in dynamic systems based on fuzzy diagnosis. In *Proc. of the IEEE Int. Conf. on Fuzzy Systems*, volume 2, pages 1482–1487, Piscataway, New Jersey, 1998.
128. Y. C. Huang and X. Z. Wang. Application of fuzzy causal networks to wastewater treatment plants. *Chem. Eng. Sci.*, 54:2731–2738, 1999.
129. R. Hudlet and R. Johnson. Linear discrimination and some further results on best lower dimensional representations. In J. V. Ryzin, editor, *Classification and Clustering*, pages 371–394. Academic Press, New York, 1977.
130. D. R. Hush and B. G. Horne. Progress in supervised neural networks. *IEEE Signal Processing Magazine*, 10:8–39, 1993.
131. J. P. Ignizio. *Introduction to Expert Systems.* McGraw-Hill, Inc., New York, 1991.

132. IMSL. Visual Numerics, Inc., Houston, Texas, 1997. computer software.
133. M. Iri, K. Aoki, E. O'Shima, and H. Matsuyama. An algorithm for diagnosis of system failures in the chemical process. *Computers & Chemical Engineering*, 3:489–493, 1979.
134. ISA. *ANSI/ISA-5.1-1984 Instrumentation Symbols and Identification.* Instrumentation Society of America, Durhan, NC, 1984.
135. R. Isermann. Process fault detection based on modeling and estimation methods: A survey. *Automatica*, 20:387–404, 1984.
136. R. Isermann. Fault diagnosis of machines via parameter estimation and knowledge processing - Tutorial paper. *Automatica*, 29:815–835, 1993.
137. R. Isermann. Integration of fault detection and diagnosis methods. In *Fault Detection, Supervision, and Safety for Technical Processes: IFAC Symposium*, Oxford, 1994. Pergamon Press.
138. R. Isermann. Model based fault detection and diagnosis methods. In *Proc. of the American Control Conf.*, pages 1605–1609, Piscataway, New Jersey, 1995. IEEE Press.
139. R. Isermann. Supervision, fault-detection and fault-diagnosis method - An introduction. *Control Engineering Practice*, 5:639–652, 1997.
140. R. Isermann and P. Ball. Trends in the application of model based fault detection and diagnosis of technical processes. In *Proc. of the 13th IFAC World Congress*, volume N, pages 1–12, Piscataway, New Jersey, 1996. IEEE Press.
141. R. Isermann and B. Freyermuth. Process fault diagnosis based on process model knowledge - Part I: Principles for fault diagnosis with parameter estimation. *J. of Dynamic Systems, Measurement, and Control*, 113:620–626, 1991.
142. J. E. Jackson. Quality control methods for two related variables. *Industrial Quality Control*, 7:2–6, 1956.
143. J. E. Jackson. Quality control methods for several related variables. *Technometrics*, 1:359–377, 1959.
144. J. E. Jackson. *A User's Guide to Principal Components.* John Wiley & Sons, New York, 1991.
145. J. E. Jackson and G. S. Mudholkar. Control procedures for residuals associated with principal component analysis. *Technometrics*, 21:341–349, 1979.
146. R. Jacobs. Increased rates of convergence through learning rate adaptation. *Neural Networks*, 1:295–3075, 1988.
147. E. W. Jacobsen. *Studies on Dynamics and Control of Distillation Columns.* PhD thesis, University of Trondheim, Trondheim, Norway, 1991.
148. L. C. Jain and N. M. M. (Eds.). *Fusion of Neural Networks, Fuzzy Sets, and Genetic Algorithms.* CRC Press, New York, 1999.
149. L. C. Jain, R. P. Johnson, Y. Takefuji, and L. A. Zadeh (Eds.). *Knowledge-Based Intelligent Techniques in Industry.* CRC Press, New York, 1999.
150. R. A. Johnson and D. W. Wichern. *Applied Multivariate Statistical Analysis.* Prentice Hall, New Jersey, 3rd edition, 1992.
151. H. L. Jones. *Failure Detection in Linear Systems.* PhD thesis, Massachusetts Institute of Technology, Cambridge, 1973.
152. P. R. S. Jota, S. M. Islam, T. Wu, and G. Ledwich. Class of hybrid intelligent system for fault diagnosis in electric power systems. *Neurocomputing*, 23:207–224, 1998.
153. T. Kailath. *Linear Systems.* Prentice-Hall, Englewood Cliffs, New Jersey, 1980.
154. C. Kan, A. C. Tamhane, and R. S. Mah. Gross error detection in serially correlated process data. *Ind. Eng. Chem. Res.*, 29:1004–1012, 1990.

155. C. Kan, A. C. Tamhane, and R. S. Mah. Gross error detection in serially correlated process data: 2 dynamic systems. *Ind. Eng. Chem. Res.*, 31:254–262, 1992.
156. N. K. Kasabov. *Foundations of Neural Networks, Fuzzy Systems, and Knowledge Engineering.* MIT Press, Cambridge, Massachusetts, 1996.
157. M. H. Kaspar and W. H. Ray. Chemometric methods for process monitoring and high-performance controller design. *AIChE J.*, 38:1593–1608, 1992.
158. M. H. Kaspar and W. H. Ray. Dynamic PLS modeling for process control. *Chem. Eng. Sci.*, 48:3447–3461, 1993.
159. A. H. Kemna, W. E. Larimore, D. E. Seborg, and D. A. Mellichamp. On-line multivariable identification and control chemical processes using canonical variate analysis. In *Proc. of the American Control Conf.*, pages 1650–1654, Piscataway, New Jersey, 1994. IEEE Press.
160. E. K. Kemsley. Discriminant analysis of high-dimensional data: A comparison of principal components analysis and partial least squares data reduction methods. *Chemometrics and Intelligent Laboratory Systems*, 33:47–61, 1996.
161. E. T. Keravnou and L. Johnson. *Component Expert Systems.* McGraw-Hill, Inc., New York, 1986.
162. P. Kesavan and J. H. Lee. Diagnostic tools for multivariable model-based control systems. *Ind. Eng. Chem. Res.*, 36:2725–2738, 1997.
163. M. Kitamura. Detection of sensor failures in nuclear plant using analytic redundancy. *Trans. Am. Nucl. Soc.*, 34:581–583, 1980.
164. T. Kohonen. Self-organized formation of topologically correct feature maps. *Biological Cybernetics*, 43:59–69, 1982.
165. T. Kohonen. Physiological interpretation of the self-organising map algorithm. *Neural Networks*, 6:895–905, 1993.
166. T. Kohonen. *Self-Organizing Maps.* Springer-Verlag, Berlin, 1993.
167. T. Kohonen. *Self-Organizing Maps.* Springer-Verlag, Berlin, 2nd edition, 1997.
168. B. Koppen-Seliger and P. M. Frank. Fuzzy logic and neural networks in fault detection. In L. C. Jain and N. M. Martin, editors, *Fusion of Neural Networks, Fuzzy Sets, and Genetic Algorithms*, pages 171–209. CRC Press, New York, 1999.
169. K. A. Kosanovich, M. J. Piovoso, K. S. Dahl, J. F. MacGregor, and P. Nomikos. Multi-way PCA applied to an industrial batch process. In *Proc. of the American Control Conf.*, pages 1294–1298, Piscataway, New Jersey, 1994. IEEE Press.
170. T. Kourti and J. F. MacGregor. Process analysis, monitoring and diagnosis using multivariate projection methods. *Chemometrics and Intelligent Laboratory Systems*, 28:3–21, 1995.
171. T. Kourti and J. F. MacGregor. Multivariate SPC methods for process and product monitoring. *J. of Quality Technology*, 28:409–428, 1996.
172. M. A. Kramer. Nonlinear principal component analysis using autoassociative neural networks. *AIChE J.*, 37:233–243, 1991.
173. M. A. Kramer and J. B. L. Palowitch. A rule-based approach to fault diagnosis using the signed directed graph. *AIChE J.*, 33:1067–1078, 1987.
174. M. A. Kramer and F. E. Finch. Development and classification of expert systems for chemical process fault diagnosis. *Robotics and Computer-integrated Manufacturing*, 4:437–446, 1988.
175. M. A. Kramer and F. E. Finch. Fault diagnosis of chemical processes. In S. G. Tzafestas, editor, *Knowledge-based System Diagnosis, Supervision, and Control*, pages 247–263. Plenum Press, New York, 1989.
176. J. Kresta, J. F. MacGregor, and T. Marlin. Multivariate statistical monitoring of process operating performance. *Can. J. Chem. Eng.*, 69:35–47, 1991.

177. J. V. Kresta, T. E. Marlin, and J. F. MacGregor. Multivariable statistical monitoring of process operating performance. *Can. J. of Chem. Eng.*, 69:35–47, 1991.
178. V. Krishnaswami and G. Rizzoni. A survey of observer based residual generation for FDI. In *Fault Detection, Supervision, and Safety for Technical Processes: IFAC Symposium*, pages 35–40, Oxford, 1994. Pergamon Press.
179. W. J. Krzanowski. Between-group comparison of principal components. *J. Amer. Stat. Assn.*, 74:703–706, 1979.
180. A. M. Kshirsagar. *Multivariate Analysis*. Marcel Dekker, New York, 1972.
181. W. Ku, R. H. Storer, and C. Georgakis. Isolation of disturbances in statistical process control by use of approximate models. In *AIChE Annual Meeting*, 1993. Paper 149g.
182. W. Ku, R. H. Storer, and C. Georgakis. Uses of state estimation for statistical process control. *Computers & Chemical Engineering*, 18:S571–S575, 1994.
183. W. Ku, R. H. Storer, and C. Georgakis. Disturbance detection and isolation by dynamic principal component analysis. *Chemometrics and Intelligent Laboratory Systems*, 30:179–196, 1995.
184. W. E. Larimore. System identification, reduced-order filtering and modeling via canonical variate analysis. In *Proc. of the American Control Conf.*, pages 445–451, Piscataway, New Jersey, 1983. IEEE Press.
185. W. E. Larimore. Canonical variate analysis for system identification, filtering, and adaptive control. In *Proc. of the IEEE Conf. on Decision and Control*, pages 635–639, Piscataway, New Jersey, 1990. IEEE Press.
186. W. E. Larimore. Identification and filtering of nonlinear systems using canonical variate analysis. In M. Casdagli and S. Eubank, editors, *Nonlinear Modeling and Forecasting, SFI Studies in the Sciences of Complexity*, pages 283–303. Addison-Wesley, Reading, Massachusetts, 1992.
187. W. E. Larimore. *ADAPTx Automated System Identification Software Users Manual*. Adaptics, Inc., McLean, VA, 1996.
188. W. E. Larimore. Statistical optimality and canonical variate analysis system identification. *Signal Processing*, 52:131–144, 1996.
189. W. E. Larimore. Canonical variate analysis in control and signal processing. In T. Katayama and S. Sugimoto, editors, *Statistical Methods in Control and Signal Processing*, pages 83–120. Marcel Dekker, New York, 1997.
190. W. E. Larimore. Optimal reduced rank modeling, prediction, monitoring, and control using canonical variate analysis. In *Proc. of the IFAC Int. Symp. on Advanced Control of Chemical Processes*, pages 61–66, Alberta, Canada, 1997.
191. W. E. Larimore and D. E. Seborg. Short Course: Process Monitoring and Identification of Dynamic Systems Using Statistical Techniques, Los Angeles, 1997.
192. B. K. Lavine. Chemometrics. *Anal. Chem.*, 70:209R–228R, 1998.
193. L. Lebart, A. Morineau, and K. M. Warwick. *Multivariate Descriptive Statistical Analysis*. John Wiley & Sons, New York, 1984.
194. C. C. Lee. Fuzzy logic in control systems: Fuzzy logic controller - Part I. *IEEE Trans. on Systems, Man, and Cybernetics*, 20:404–418, 1990.
195. C. C. Lee. Fuzzy logic in control systems: Fuzzy logic controller - Part II. *IEEE Trans. on Systems, Man, and Cybernetics*, 20:419–435, 1990.
196. G. Lee, B. Lee, E. S. Yoon, and C. Han. Multiple-fault diagnosis under uncertain conditions by the quantification of qualitative relations. *Ind. Eng. Chem. Res.*, 38:988–998, 1999.
197. S. C. Lee. Sensor value validation based on systematic exploration of the sensor redundancy for fault diagnosis. *IEEE Trans. on Systems, Man, and Cybernetics*, 24:594–605, 1994.

198. T. Li. Expert systems for engineering diagnosis: Styles, requirements for tools, and adaptability. In S. G. Tzafestas, editor, *Knowledge-based System Diagnosis, Supervision, and Control*, pages 27–37. Plenum Press, New York, 1989.
199. L. Ljung. *System Identification: Theory for the User*. Prentice-Hall, Englewood Cliffs, New Jersey, 1987.
200. G. Locher, B. Bakshi, G. Stephanopoulos, and K. Schugerl. A method for an automated rule extraction from raw process data: 1 Process trends, wavelet transformation and decision trees. *Automatisierungstechnik*, 44:61–70, 1996.
201. G. Locher, B. Bakshi, G. Stephanopoulos, and K. Schugerl. A method for an automated rule extraction from raw process data: 2 A case study. *Automatisierungstechnik*, 44:138–145, 1996.
202. C. A. Lowry and W. H. Woodall. A multivariate exponentially weighted moving average control chart. *Technometrics*, 34:46–53, 1992.
203. C. A. Lowry, W. H. Woodall, C. W. Champ, and S. E. Rigdon. A multivariate exponentially weighted moving average control chart. *Technometrics*, 34:46–53, 1992.
204. R. Luo, M. Misra, and D. M. Himmelblau. Sensor fault detection via multiscale analysis and dynamic PCA. *Ind. Eng. Chem. Res.*, 38:1489–1495, 1999.
205. P. R. Lyman. *Plant-wide Control Structures for the Tennessee Eastman process*, M.S. thesis, Lehigh University, 1992.
206. P. R. Lyman and C. Georgakis. Plant-wide control of the Tennessee Eastman problem. *Computers & Chemical Engineering*, 19:321–331, 1995.
207. J. F. MacGregor. Statistical process control of multivariate processes. In *Proc. of the IFAC Int. Symp. on Advanced Control of Chemical Processes*, pages 427–435, New York, 1994. Pergamon Press.
208. J. F. MacGregor, C. Jaeckle, C. Kiparissides, and M. Koutoudi. Process monitoring and diagnosis by multiblock PLS methods. *AIChE J.*, 40:826–838, 1994.
209. J. F. MacGregor and T. Kourti. Statistical process control of multivariate processes. *Control Engineering Practice*, 3:403–414, 1995.
210. R. S. H. Mah and V. Chakravarthy. Pattern recognition using artificial neural networks. *Computers & Chemical Engineering*, 16:371–377, 1992.
211. R. S. H. Mah, K. D. Schnelle, and A. N. Patel. A plant-wide quality expert system for steel mills. *Computers & Chemical Engineering*, 15:445–450, 1991.
212. E. Malinowski. Statistical F-test for abstract factor analysis and target testing. *J. Chemometrics*, 3:46–50, 1989.
213. E. C. Malthouse, A. C. Tamhane, and R. S. H. Mah. Nonlinear partial least squares. *Computers & Chemical Engineering*, 21:875–890, 1997.
214. D. Mandel, A. Abdollahzadeh, D. Maquin, and J. Ragat. Data reconciliation by inequality balance equilibration: A LMI approach. *Int. J. of Mineral Processing*, 53:157–169, 1998.
215. R. Manne. Analysis of two partial-least-squares algorithms for multivariate calibration. *Chemometrics and Intelligent Laboratory Systems*, 2:187–197, 1987.
216. J. Maroldti. A specification method for diagnostic systems in large industrial environments. In S. G. Tzafestas, editor, *Knowledge-based System Diagnosis, Supervision, and Control*, pages 59–79. Plenum Press, New York, 1989.
217. H. Martens and T. Naes. *Multivariate Calibration*. John Wiley & Sons, 1989.
218. T. J. McAvoy. A methodology for screening level control structures in plantwide control systems. *Computers & Chemical Engineering*, 22:1543–1552, 1998.

219. T. J. McAvoy, Y. Nan, and G. Chan. An improved base control for the Tennessee Eastman problem. In *Proc. of the American Control Conf.*, pages 240–244, Piscataway, New Jersey, 1995. IEEE Press.
220. A. Medvedev. State estimation and fault detection by a bank of continuous finite-memory filters. In *Proc. of the 13th IFAC World Congress*, volume N, pages 223–228, Piscataway, New Jersey, 1996. IEEE Press.
221. R. K. Mehra and J. Peschon. An innovations approach to fault detection and diagnosis in dynamic systems. *Automatica*, 7:637–640, 1971.
222. T. Mejdell and S. Skogestad. Estimation of distillation compositions from multiple temperature measurements using partial least squares regression. *Ind. Eng. Chem. Res.*, 30:2543–2555, 1991.
223. T. Mejdell and S. Skogestad. Output estimation using multiple secondary measurements: High-purity distillation. *AIChE J.*, 39:1641–1653, 1993.
224. E. Micheli-Tzanakou, editor. *Supervised and Unsupervised Pattern Recognition*. CRC Press, New York, 1999.
225. P. Miller, R. E. Swanson, and C. E. Heckler. Contribution plots: A missing link in multivariate quality control. *Applied Mathematics & Computer Science*, 8:775–792, 1998.
226. L. A. Mironovski. Functional diagnosis of linear dynamic systems. *Automation and Remote Control*, 40:1198–1205, 1979.
227. L. A. Mironovski. Functional diagnosis of dynamic system – a survey. *Automation and Remote Control*, 41:1122–1142, 1980.
228. K. J. Mo, G. Lee, D. S. Nam, Y. H. Yoon, and E. S. Yoon. Robust fault diagnosis based on clustered symptom trees. *Control Engineering Practice*, 5:199–208, 1997.
229. K. J. Mo, Y. S. Oh, and E. S. Yoon. Development of operation-aided system for chemical processes. *Expert Systems with Applications*, 12:455–464, 1997.
230. D. C. Montgomery. *Introduction to Statistical Quality Control*. John Wiley and Sons, New York, 1985.
231. T. Moody and C. Darken. Fast learning in networks of locally tuned processing units. *Neural Computation*, 1:281–294, 1989.
232. M. Morari and E. Zafiriou. *Robust Process Control*. Prentice-Hall, Englewood Cliffs, New Jersey, 1989.
233. O. Moseler and R. Isermann. Model-based fault detection for a brushless dc motor using parameter estimation. In *Proc. of the 24th Annual Conf. of the IEEE Industrial Electronics Society, IECON*, volume 4, pages 956–1960, Piscataway, New Jersey, 1998.
234. R. J. Muirhead. *Aspects of Multivariate Statistical Theory*. John Wiley & Sons, New York, NY, 1982.
235. D. Mylaraswamy, S. N. Kavuri, and V. Venkatasubramanian. A framework for automated development of causal models for fault diagnosis. In *AIChE Annual Meeting*, San Francisco, California, 1994. Paper 232g.
236. D. Mylaraswamy and V. Venkatasubramanian. A hybrid framework for large scale process fault diagnosis. *Computers & Chemical Engineering*, 21:S935–S940, 1997.
237. Y. Nan, T. J. McAvoy, K. A. Kosanovich, and M. J. Piovoso. Plant-wide control using an inferential approach. In *Proc. of the American Control Conf.*, pages 1900–1904, Piscataway, New Jersey, 1993. IEEE Press.
238. S. Narasimhan and R. S. Mah. Generalized likelihood ratios for gross error identification in dynamic processes. *AIChE J.*, 34:1321–1331, 1988.
239. A. Negiz and A. Cinar. On the detection of multiple sensor abnormalities in multivariate processes. In *Proc. of the American Control Conf.*, pages 2364–2368, Piscataway, New Jersey, 1992. IEEE Press.

240. A. Negiz and A. Cinar. Statistical monitoring of multivariable dynamic processes with state-space models. *AIChE J.*, 43:2002–2020, 1997.
241. A. Negiz and A. Cinar. Monitoring of multivariable dynamic processes and sensor auditing. *J. of Process Control*, 8:375–380, 1998.
242. I. Nikiforov, M. Staroswiecki, and B. Vozel. Duality of analytical redundancy and statistical approach in fault diagnosis. In *Proc. of the 13th IFAC World Congress*, volume N, pages 19–24, Piscataway, New Jersey, 1996. IEEE Press.
243. P. Nomikos and J. F. MacGregor. Monitoring batch processes using multiway principal component analysis. *AIChE J.*, 40:1361–1375, 1994.
244. J. Nouwen, F. Lindgren, W. K. B. Hansen, H. J. M. Verharr, and J. L. M. Hermens. Classification of environmentally occurring chemicals using structural fragments and PLS discriminant analysis. *Environ. Sci. Technol.*, 31:2313–2318, 1997.
245. B. A. Ogunnaike and W. H. Ray. *Process Dynamics, Modeling, and Control.* Oxford University Press, New York, 1994.
246. Y. S. Oh, J. H. Yoon, D. Nam, C. Han, and E. S. Yoon. Intelligent fault diagnosis based on weighted symptom tree model and fault propagation trends. *Computers & Chemical Engineering*, 21:S941–S946, 1997.
247. O. O. Oyeleye, F. E. Finch, and M. A. Kramer. Qualitative modeling and fault diagnosis of dynamic processes by MIDAS. *Chem. Eng. Comm.*, 96:205–228, 1990.
248. R. J. Patton and J. Chen. Optimal unknown input disturbance matrix selection in robust fault diagnosis. *Automatica*, 29:837–841, 1993.
249. R. J. Patton and J. Chen. Observer-based fault detection and isolation: Robustness and applications. *Control Engineering Practice*, 5:671–682, 1997.
250. R. J. Patton, J. Chen, and T. M. Siew. Fault diagnosis in non-linear dynamic systems via neural-networks. In *Proc. of the Int. Conf. on CONTROL*, volume 2, pages 1346–1351, Stevenage, U.K., 1994. IEE Press.
251. R. J. Patton and M. Hou. A matrix pencil approach to fault detection and isolation observers. In *Proc. of the 13th IFAC World Congress*, volume N, pages 229–234, Piscataway, New Jersey, 1996. IEEE Press.
252. R. J. Patton and M. Hou. Design of fault detection and isolation observers: A matrix pencil approach. *Automatica*, 43:1135–1140, 1998.
253. R. J. Patton, S. W. Willcox, and S. J. Winter. A parameter insensitive technique for aircraft sensor fault analysis. *J. of Guidance, Control and Dynamics*, 10:359–367, 1987.
254. L. F. Pau. Survey of expert systems for fault detection, test generation and maintenance. *Expert Systems*, 3:100–111, 1986.
255. R. K. Pearson. Data cleaning for dynamic modeling and control. In *Proc. of the European Control Conf.*, Karlsruhe, Germany, 1999. IFAC. Paper F853.
256. D. W. Peterson and R. L. Mattson. A method of finding linear discriminant functions for a class of performance criteria. *IEEE Trans. Info. Theory*, 12:380–387, 1966.
257. T. F. Petti, J. Klein, and P. S. Dhurjati. Diagnostic model processor. Using deep knowledge for process fault diagnosis. *AIChE J.*, 36:565–575, 1990.
258. J. J. Pignatiello, Jr. and G. C. Runger. Comparisons of multivariate CUSUM charts. *J. of Quality Technology*, 22:173–186, 1990.
259. M. J. Piovoso and K. A. Kosanovich. Applications of multivariate statistical methods to process monitoring and controller design. *Int. J. of Control*, 59:743–765, 1994.
260. M. J. Piovoso, K. A. Kosanovich, and R. K. Pearson. Monitoring process performance in real time. In *Proc. of the American Control Conf.*, pages 2359–2363, Piscataway, New Jersey, 1992. IEEE Press.

261. M. M. Polycarpou and A. T. Vemuri. Learning methodology for failure detection and accommodation. *IEEE Control Systems Intelligence and Learning*, 15:16–24, 1995.
262. M. Pottmann and D. E. Seborg. Nonlinear predictive control strategy based on radial basis function models. *Computers & Chemical Engineering*, 21:965–980, 1997.
263. A. Pouliezos, G. Stavrakakis, and C. Lefas. Fault detection using parameter estimation. *Quality and Reliability Engineering International*, 5:283–290, 1989.
264. B. L. S. Prakasa Rao. *Identifiability in Stochastic Models Characterization of Probability Distributoins.* Academic Press, New York, 1992.
265. P. Purkait, S. Chakravorti, and K. Bhattacharya. TIFDES - An expert system tool for transformer impulse fault diagnosis. *IEE Conference Publication*, 467(5):180–183, 1999.
266. S. J. Qin. Recursive PLS algorithms for adaptive data modeling. *Computers & Chemical Engineering*, 22:503–514, 1998.
267. S. J. Qin and R. Dunia. Determining the number of principal components for best reconstruction. *J. of Process Control*, 10(2):245–250, 2000.
268. S. J. Qin and T. J. McAvoy. Nonlinear PLS modeling using neural networks. *Computers & Chemical Engineering*, 16:379–391, 1992.
269. A. C. Raich and A. Cinar. Statistical process monitoring and disturbance isolation in multivariate continuous processes. In *Proc. of the IFAC Conf. on Advanced Control of Chemical Processes*, pages 427–435, New York, 1994. Pergamon Press.
270. A. C. Raich and A. Cinar. Multivariate statistical methods for monitoring continuous processes: Assessment of discriminatory power disturbance models and diagnosis of multiple disturbances. *Chemometrics and Intelligent Laboratory Systems*, 30:37–48, 1995.
271. A. C. Raich and A. Cinar. Process disturbance diagnosis by statistical distance and angle measures. In *Proc. of the 13th IFAC World Congress*, pages 283–288, Piscataway, New Jersey, 1996. IEEE Press.
272. A. C. Raich and A. Cinar. Statistical process monitoring and disturbance diagnosis in multivariable continuous processes. *AIChE J.*, 42:995–1009, 1996.
273. R. Reiter. A theory of diagnosis from first principles. *Artificial Intelligence*, 32:57–95, 1987.
274. R. R. Rhinehart. A watchdog for controller performance monitoring. In *Proc. of the American Control Conf.*, pages 2239–2240, Piscataway, New Jersey, 1995. IEEE Press.
275. J. A. Romagnoli and G. Stephanopoulos. On the rectification of measurement errors for complex chemical plants. *Chem. Eng. Sci.*, 35:1067–1081, 1980.
276. J. A. Romagnoli and G. Stephanopoulos. Rectification of process measurement data in the presence of gross errors. *Chem. Eng. Sci.*, 36:1849–1863, 1981.
277. E. L. Russell and R. D. Braatz. Fault isolation in industrial processes using Fisher discriminant analysis. In J. F. Pekny and G. E. Blau, editors, *Foundations of Computer-Aided Process Operations*, pages 380–385. AIChE, New York, 1998.
278. E. L. Russell, L. H. Chiang, and R. D. Braatz. *Data-driven Techniques for Fault Detection and Diagnosis in Chemical Processes.* Springer Verlag, London, 2000.
279. E. L. Russell, L. H. Chiang, and R. D. Braatz. Fault detection in industrial processes using canonical variate analysis and dynamic principal component analysis. *Chemometrics & Intelligent Laboratory Systems*, 51:81–93, 2000.

280. M. Sampath, R. Sengupta, S. Lafortune, K. Sinnamohideen, and D. C. Teneketzis. Diagnosability of discrete-event systems. *IEEE Trans. on Automatic Control*, 40:1555–1575, 1995.
281. M. Sampath, R. Sengupta, S. Lafortune, K. Sinnamohideen, and D. C. Teneketzis. Failure diagnosis using discrete-event models. *IEEE Trans. on Control Systems Technology*, 4:105–124, 1996.
282. R. J. Schalkoff. *Pattern Recognition: Statistical, Structural and Neural Approaches*. John Wiley & Sons, New York, 1992.
283. C. D. Schaper, W. E. Larimore, D. E. Seborg, and D. A. Mellichamp. Identification of chemical processes using canonical variate analysis. *Computers & Chemical Engineering*, 18:55–69, 1994.
284. K. D. Schnelle and R. S. H. Mah. A real-time expert system for quality control. *IEEE Intelligent Systems & Their Applications*, 7(5):36–42, 1992.
285. R. J. P. Schrama. Accurate identification for control - The necessity of an iterative scheme. *IEEE Trans. on Automatic Control*, 37:991–994, 1992.
286. D. Shields. Quantitative approaches for fault diagnosis based on bilinear systems. In *Proc. of the 13th IFAC World Congress*, volume N, pages 151–155, Piscataway, New Jersey, 1996. IEEE Press.
287. J. Shiozaki, H. Matsuyama, E. O'Shima, and M. Iri. An improved algorithm for diagnosis of system failures in the chemical process. *Computers & Chemical Engineering*, 9:285–293, 1985.
288. H. T. Simsek, R. Sengupta, S. Yovine, and F. Eskafi. Fault diagnosis for intra-platoon communications. In *Proc. of the 38th IEEE Conf. on Decision and Control*, volume 4, pages 3520–3525, Piscataway, New Jersey, 1999. IEEE Press.
289. O. Simula, E. Alhoniemi, J. Hollmen, and J. Vesanto. Monitoring and modeling of complex processes using hierarchical self-organizing maps. In *Proc. of the IEEE Int. Symp. on Circuits and Systems*, pages 73–76, Piscataway, New Jersey, 1996. IEEE Press.
290. O. Simula and J. Kangas. Process monitoring and visualization using self-organizing maps. In *Neural Networks for Chemical Engineers, Computer-Aided Chemical Engineering*, chapter 14, pages 371–384. Elsevier, Amsterdam, 1995.
291. S. Skogestad and I. Postlethwaite. *Multivariable Feedback Control: Analysis and Design*. John Wiley & Sons, New York, 1996.
292. R. S. Spraks. Quality control with multivariate data. *Australian Journal of Statistics*, 34:375–390, 1992.
293. S. N. Srihari. Application of expert systems in engineering: An introduction. In S. G. Tzafestas, editor, *Knowledge-based System Diagnosis, Supervision, and Control*, pages 1–10. Plenum Press, New York, 1989.
294. R. Srinivasan and V. Venkatasubramanian. Automating HAZOP analysis of batch chemical plants: Part I. The knowledge representation framework. *Computers & Chemical Engineering*, 22:1345–1355, 1998.
295. N. Stanfelj, T. E. Marlin, and J. F. MacGregor. Monitoring and diagnosing process control performance: The single loop case. *Ind. Eng. Chem. Res.*, 32:301–314, 1993.
296. G. Stephanopoulos. *Chemical Process Control - An Introduction to Theory and Practice*. Prentice Hall, Englewood Cliffs, New Jersey, 1990.
297. W. Suewatanakul and D. M. Himmelblau. Fault detection via artificial neural networks. *Engineering Simulation*, 13:967–984, 1996.
298. A. K. Sunol, B. Ozyurt, P. K. Mogili, and L. Hall. A machine learning approach to design and fault diagnosis. In *Int. Conf. on Intelligent Systems in*

Process Engineering, AIChE Symposium Series, volume 92, pages 331–334, 1996.
299. E. Sutanto and K. Warwick. Cluster analysis for multivariate process control. In *Proc. of the American Control Conf.*, pages 749–751, Piscataway, New Jersey, 1995. IEEE Press.
300. B. G. Tabachnick and L. S. Fidell. *Using Multivariate Analysis.* Harper & Row, Cambridge, 1989.
301. E. E. Tarifa and N. J. Scenna. A fault diagnosis prototype for a bioreactor for bioinsecticide production. *Reliability Engineering & System Safety*, 48:27–45, 1995.
302. E. E. Tarifa and N. J. Scenna. Fault diagnosis, direct graphs, and fuzzy logic. *Computers & Chemical Engineering*, 21:S649–S654, 1997.
303. E. E. Tarifa and N. J. Scenna. Methodology for fault diagnosis in large chemical processes and an application to a multistage flash desalination process: Part II. *Reliability Engineering & System Safety*, 60:41–51, 1998.
304. E. E. Tarifa and N. J. Scenna. Methodology for fault diagnosis in large chemical processes and an application to a multistage flash desalination process: Part I. *Reliability Engineering & System Safety*, 60:29–40, 1998.
305. Q. Tian. Comparison of statistical pattern-recognition algorithms for hybrid processing, II: Eigenvector-based algorithm. *J. Opt. Soc. Am. A*, 5:1670–1672, 1988.
306. I. Tjoa and L. Biegler. Simultaneous strategies for data reconciliation and gross error detection of nonlinear systems. *Computers & Chemical Engineering*, 15:679–690, 1991.
307. H. Tong and C. M. Crowe. Detection of gross errors in data reconciliation by principal component analysis. *AIChE J.*, 41:1712–1722, 1995.
308. N. D. Tracy, J. C. Young, and R. L. Mason. Multivariate control charts for individual observations. *J. of Quality Control*, 24:88–95, 1992.
309. L. H. Tsoukalas and R. E. Uhrig. *Fuzzy and Neural Approaches in Engineering.* John Wiley & Sons, New York, 1997.
310. Y. Tsuge, J. Shiozaki, H. Matsuyama, and E. O'Shima. Fault diagnosis algorithms based on the signed directed graph and its modifications. *Ind. Chem. Eng. Symp. Ser.*, 92:133–144, 1985.
311. M. L. Tyler and M. Morari. Optimal and robust design of integrated control and diagnostic modules. In *Proc. of the American Control Conf.*, pages 2060–2064, Piscataway, New Jersey, 1994. IEEE Press.
312. M. L. Tyler and M. Morari. Performance monitoring of control systems using likelihood methods. *Automatica*, 32:1145–1162, 1996.
313. S. G. Tzafestas and P. J. Dalianis. Fault diagnosis in complex systems using artificial neural networks. In *Proc. of The Third IEEE Conf. on Control Applications*, pages 877–882, Piscataway, New Jersey, 1994. IEEE Press.
314. R. Vaidhyanathan and V. Venkatasubramanian. Experience with an expert system for automated HAZOP analysis. *Computers & Chemical Engineering*, 20:S1589–S1594, 1996.
315. S. Valle, W. Li, and S. J. Qin. Selection of the number of principal components: the variance of the reconstruction error criterion with a comparison to other methods. *Ind. Eng. Chem. Res.*, 38:4389–4401, 1999.
316. C. V. Van Loan. Generalizing the singular value decomposition. *SIAM J. Numer. Anal.*, 13:76–83, 1976.
317. P. Van Overschee and B. De Moor. N4SID*: Subspace algorithms for the identification of combined deterministic-stochastic systems. *Automatica*, 30:75–93, 1994.

318. P. Van Overschee and B. De Moor. *Subspace Identification for Linear Systems: Theory - Implementation - Applications.* Kluwer Academic Publishers, Norwell, Massachusetts, 1996.
319. H. Vedam and V. Venkatasubramanian. Signed digraph based multiple fault diagnosis. *Computers & Chemical Engineering*, 21:S655–S660, 1997.
320. H. Vedam and V. Venkatasubramanian. PCA-SDG based process monitoring and fault diagnosis. *Control Enineering Practice*, 7:903–917, 1999.
321. D. R. Vinson, C. Georgakis, and J. Fossy. Studies in plant-wide controllability using the Tennessee Eastman challenge problem, the case for multivariable control. In *Proc. of the American Control Conf.*, pages 250–254, Piscataway, New Jersey, 1995. IEEE Press.
322. N. Viswanadham and K. D. Minto. Robust observer design with application to fault detection. In *Proc. of the American Control Conf.*, pages 1393–1399, Piscataway, New Jersey, 1988. IEEE Press.
323. Z. Vukic, H. Ozbolot, and D. Pavlekovic. Application of expert systems, fuzzy algorithms and neural networks to fault diagnosis in control systems. *Brodogradnja*, 47:41–49, 1999.
324. D. L. Waikar and F. Rahman. Assessment of artificial intelligence techniques for power system protection. *Proc. of the Int. Conf. on Energy Management & Power*, 2:436–441, 1998.
325. E. Walter. *Identifiability of Parametric Models.* Pergamon Press, New York, 1987.
326. L. Wang. *Adaptive Fuzzy Systems and Control - Design and Stability Analysis.* Prentice-Hall, Englewood Cliffs, New Jersey, 1994.
327. X. Z. Wang. *Data Mining and Knowledge Discovery for Process Monitoring and Control.* Springer-Verlag, London, 1999.
328. Y. Wang, D. E. Seborg, and W. E. Larimore. Process monitoring based on canonical variate analysis. In *Proc. of the IFAC Int. Symp. on Advanced Control of Chemical Processes*, pages 523–528, Alberta, Canada, 1997.
329. Z. Wang, Y. Liu, and P. J. Griffin. A combined ANN and expert system tool for transformer fault diagnosis. *IEEE Trans. on Power Delivery*, 13:1224–1229, 1998.
330. Z. Wang, Y. Liu, and P. J. Griffin. Neural net and expert system diagnose transformer faults. *IEEE Computer Applications in Power*, 13:50–55, 2000.
331. K. Wantanabe and D. M. Himmelblau. Instrument fault detection in systems with uncertainties. *Int. J. of Systems Science*, 13:137–158, 1982.
332. K. Wantanabe and D. M. Himmelblau. Fault diagnosis in nonlinear chemical processes. *AIChE J.*, 29:243–261, 1983.
333. K. Watanabe, S. Hirota, L. Hou, and D. M. Himmelblau. Diagnosis of multiple simultaneous fault via hierarchical artificial neural networks. *AIChE J.*, 40:839–848, 1994.
334. K. Watanabe, I. Matsuura, M. Abe, M. Kubota, and D. M. Himmelblau. Incipient fault diagnosis of chemical processes via artificial neural networks. *AIChE J.*, 35:1803–1812, 1989.
335. D. L. Waterman. *A Guide to Expert Systems.* Addison-Wesley, New York, 1986.
336. W. W. S. Wei. *Time Series Analysis.* Addison-Wesley, Reading, Massachusetts, 1994.
337. G. H. Weiss, J. A. Romagnoli, and K. A. Islam. Data reconciliation - An industrial case study. *Computers & Chemical Engineering*, 20:1441–1449, 1996.
338. S. J. Wierda. Multivariate statistical process control, recent results and directions for future research. *Statistica Neerlandica*, 48:147–168, 1994.

339. N. A. Wilcox and D. M. Himmelblau. Possible cause and effect graphs (PCEG) model for fault diagnosis - I Methodology. *Computers & Chemical Engineering*, 18:103–116, 1993.
340. N. A. Wilcox and D. M. Himmelblau. Possible cause and effect graphs (PCEG) model for fault diagnosis - II Applications. *Computers & Chemical Engineering*, 18:117–127, 1994.
341. A. S. Willsky. A survey of design methods for failure detection in dynamic systems. *Automatica*, 12:601–611, 1976.
342. A. S. Willsky and H. L. Jones. A generalized likelihood ratio approach to the detection and estimation of jumps in linear systems. *IEEE Trans. on Automatic Control*, pages 108–112, 1976.
343. B. M. Wise and N. B. Gallagher. The process chemometrics approach to process monitoring and fault detection. *J. of Process Control*, 6:329–348, 1996.
344. B. M. Wise and N. B. Gallagher. *PLS_Toolbox 2.0 for use with MATLAB*. Eigenvector Research, Manson, Washington, 1998.
345. B. M. Wise, N. L. Ricker, D. J. Velkamp, and B. R. Kowalski. A theoretical basis for the use of principal component models for monitoring multivariate processes. *Process Control and Quality*, 1:41–51, 1990.
346. B. M. Wise, N. L. Ricker, and D. F. Veltkamp. Upset and sensor failure detection in multivariate processes. Technical report, Eigenvector Research, Manson, Washington, 1989.
347. S. Wold. Cross-validatory estimation of components in factor and principal components models. *Technometrics*, 20:397–405, 1978.
348. S. Wold, K. Esbensen, and P. Geladi. Principal components analysis. *Chemometrics and Intelligent Laboratory Systems*, 2:37, 1987.
349. S. Wold, N. Kettaneh-Wold, and B. Skagerberg. Nonlinear PLS modeling. *Chemometrics and Intelligent Laboratory Systems*, 7:53–65, 1989.
350. S. Wold, H. Martens, and H. Russwurm. *Food Research and Data Analysis*. Applied Science Publishers, London, 1983.
351. W. H. Woodall and M. M. Ncube. Multivariate CUSUM quality-control procedures. *Technometrics*, 27:285–292, 1985.
352. J. Wuennenberg. *Observer-based Fault Detection in Dynamic Systems*. PhD thesis, University of Duisburg, Germany, 1990.
353. B. J. Wythoff. Backpropagation neural networks. *Chemometrics and Intelligent Laboratory Systems*, 18:115–155, 1993.
354. E. S. Yoon and J. H. Han. Process failure detection and diagnosis using the tree model. In *Proc. of the IFAC World Congress*, pages 126–129, Oxford, 1987. Pergamon Press.
355. P. C. Young. Parameter estimation for continuous-time models - A survey. *Automatica*, 17:23–29, 1981.
356. C. C. Yu and C. Lee. Fault diagnosis based on qualitative/quantitative process knowledge. *AIChE J.*, 37:617–628, 1997.
357. L. Zadeh. Fuzzy sets. *Information and Control*, 8:338–353, 1965.
358. D. A. Zahner and E. Micheli-Tzanakou. Artificial neural networks: Definitions, methods, applications. In E. Micheli-Tzanakou, editor, *Supervised and Unsupervised Pattern Recognition*, pages 61–78. CRC Press, New York, 1999.
359. G. Zhang, M. Xu, and S. Xu. Design of the expert system for the fault diagnosis of 200 MW turbine generator set. *Advances in Modeling & Analysis B*, 40:51–59, 1998.
360. J. Zhang, E. Martin, and A. J. Morris. Fault detection and classification through multivariate statistical techniques. In *Proc. of the American Control Conf.*, pages 751–755, Piscataway, New Jersey, 1995. IEEE Press.

361. J. Zhang, A. J. Morris, and E. B. Martin. Robust process fault detection and diagnosis using neuro-fuzzy networks. In *Proc. of the 13th IFAC World Congress*, volume N, pages 169–174, Piscataway, New Jersey, 1996. IEEE Press.
362. Q. Zhang. A frequency and knowledge tree/causality diagram based expert system approach for fault diagnosis. *Reliability Engineering & System Safety*, 43:17–28, 1994.
363. Q. Zhang, X. An, J. Gu, B. Zhao, D. Xu, and S. Xi. Application of FBOLES - A prototype expert system for fault diagnosis in nuclear power plants. *Reliability Engineering & System Safety*, 44:225–235, 1994.
364. Y. Zhang and A. Kandel. *Compensatory Genetic Fuzzy Neural Networks and Their Applications*. World Scientific, Singapore, 1997.
365. Y. Zhang, X. Li, G. Dai, H. Zhang, and H. Chen. Fault detection and identification of dynamic systems using multiple feedforward neural networks. In *Proc. of the 13th IFAC World Congress*, volume N, pages 241–246, Piscataway, New Jersey, 1996. IEEE Press.
366. J. Zhao, B. Chen, and J. Shen. A hybrid ANN-ES system for dynamic fault diagnosis of hydrocracking process. *Computers & Chemical Engineering*, 21:S929–S933, 1997.
367. A. Zolghadri. Model based fault detection in a multivariable hydraulic process. In *Proc. of the 13th IFAC World Congress*, volume N, pages 253–258, Piscataway, New Jersey, 1996. IEEE Press.

Index